SYSTEMATICS AND BIOGEOGRAPHY
CLADISTICS AND VICARIANCE

SYSTEMATICS
AND
BIOGEOGRAPHY

CLADISTICS AND VICARIANCE

Gareth Nelson and Norman Platnick

NEW YORK   COLUMBIA UNIVERSITY PRESS

Library of Congress Cataloging in Publication Data

Nelson, Gareth
Systematics and biogeography.

Bibliography: p.
Includes index.
1. Biology—Classification. 2. Biogeography.
I. Platnick, Norman, joint author. II. Title.
III. Title: Cladistics and vicariance.
QH83.N4   574'.012   80-20828
ISBN 0-231-04574-3

Columbia University Press
New York   Guildford, Surrey

Copyright © 1981 Columbia University Press
Printed in the United States of America
10 9 8 7 6 5 4 3 2

impunitum numquam beneficium

CONTENTS

PREFACE

The views presented in this volume have their source largely in the work of two biologists, the late Willi Hennig, author of a 1966 book called *Phylogenetic Systematics*, and Leon Croizat, author of a 1964 book called *Space, Time, Form: The Biological Synthesis*, and in the writings of a philosopher of science, Sir Karl Popper. Hennig and Croizat have not found their work particularly compatible (Hennig never cited Croizat, and Croizat [1976] has published negative comments on Hennig), and neither one has indicated any interest in Popper's views or cited them as being compatible with his own.

Yet both Hennig and Croizat have made substantial, and substantially similar, contributions in (1) pointing out major inadequacies in some conventional methods in systematics and biogeography, respectively, and (2) suggesting significantly improved methods for those fields. We believe that the contributions of both Hennig and Croizat can be readily (and fruitfully) understood within the context of Popper's view of the nature and growth of scientific knowledge, and that the ideas of all three men are largely compatible. At the same time, were it possible for all three to read this book, each might disagree with large parts of it. The reader can judge to what extent we have been successful in synthesizing and extending their contributions, and what value the resulting perspective may have.

Systematics and biogeography are sciences rich in past accomplishments and future prospects—so rich, in fact, that novice and professional alike may fail to grasp any unity in the myriads of facts and interpretations already in hand or potentially available. In this volume we experiment with a perspective that allows past work to be related to questions still open. Our results are to some extent traditional: Aristotle, Linnaeus, and the elder Candolle again emerge as prime movers; and much modern commentary pales into insignificance. Our results are to some extent novel: cladograms and their agreement in the geographical dimension emerge as the keys to the future of our sciences. The keys have precedents in the older notions of natural groups (taxa) and natural areas (regions). In combination they amount simply to the maxim that earth and life evolve together, both in general and in particular.

Although our experiment might be judged successful in the sense that it gives results and offers prospects, the reader may find the text difficult to follow sequentially. The introduction (chapter 1) aside, the historical accounts (chapters 2 and 6) pose the least difficulty and are the easiest places to begin. The results (chapters 4 and 8) are commentaries on the state of the art and will be appreciated best by readers with some knowledge of the current literature. The analytical chapters (3 and 7) are apt to prove obstructive to readers other than those committed to mastering the logic underlying our argument, such as we have been able to formulate it. We intended chapter 5 as a synthesis of the material preceding it, with reference to the temporal dimension, but it seems to us now something of a digressionary interlude.

We are deeply indebted to James S. Farris, Charles W. Harper, Jr., and David L. Hull for commenting in detail on large parts of the text. We also thank the following colleagues for their assistance with various aspects of this project: J. W. Atz, R. Batten, R. Birdsong, A. Boucot, R. Brady, D. Brothers, L. Brundin, H. D. Cameron, S. Coats, L. Constance, J. Cracraft, L. Croizat, R. A. Crowson, N. Eldredge, W. K. Emerson, W. Fink, E. S. Gaffney, S. J. Gould, M. Hecht, W. Hennig, L. H. Herman, R. F. Johnston, D. B. Kitts, A. Kluge, J. Larson, S. Løvtrup, J. G. Lundberg, N. Macbeth, L. Marcus, M. C. McKenna, C. D. Michener, M. F. Mickevich, E. C. Olson, L. Parenti, C. Patterson, K. R. Popper, F. H. Rindge, D. E. Rosen, H. H. Ross, M. Ruse, B. Schaeffer, D. Schlee, R. Schmid, R. T. Schuh, H.-P. Schultze, M. D. F. Udvardy, F. Vuilleumier, H.-E. Wanntorp, E. O. Wiley, E. O. Wilson, P. Wygodzinsky, and R. Zangerl. Finally, we thank J. Barbaris and V. Morales of the American Museum, who drafted most of the illustrations, and Joe Ingram, Joan McQuary, Vicki P. Raeburn, and their colleagues at Columbia University Press for their interest, cheerfulness, and forbearance.

Gareth Nelson and Norman Platnick
American Museum of Natural History
New York City, August 1980

SYSTEMATICS AND BIOGEOGRAPHY
CLADISTICS AND VICARIANCE

INTRODUCTION

1

COMPARATIVE BIOLOGY: SPACE, TIME, AND FORM

THE SCIENCE OF COMPARATIVE BIOLOGY

THE TWO BIOLOGIES

Biology today has become a vast conglomeration of subdisciplines, each of which has its own language and literature and poses its own set of barriers to understanding or interpretation by the nonspecialist. It is not surprising, therefore, that many thoughtful biologists have attempted to gain a perspective on their science by grouping these numerous subdisciplines together, in various ways, into two or more main branches of biology. To be successful, these attempts must demonstrate the existence of unique features unifying each branch, features that go beyond the mere truism that all the separate fields are concerned, in some way, with the study of life.

Probably the most common route taken toward this end involves the concept of levels of organization. The living world can certainly be viewed as a series of increasingly encompassing and complex levels, proceeding from the molecule and cell through the individual organism on up to the ecosystem. For some reason, perhaps because we ourselves are organisms, it seems natural to divide biology into those subdisciplines operating at the level of organisms and above (organismic biology) and those concerned only with parts of organisms. That this distinction actually can serve to separate working biologists into two camps is attested by the fact that many university biology departments are so divided. It could be questioned, however, whether this division has proved healthy, either for the departments or for biology. The groupings do not seem to offer much in the way of unifying features; does a biochemist studying a pathway of protein synthesis in bacteria have anything more in common with a functional morphologist investigating the mechanics of a turtle jaw than with a behaviorist exploring the courtship rituals of a pair of spiders?

Another possible approach centers not on levels of organization but on the theory of evolution. To Mayr (1961:1501; also Rosa 1933), for example, the term "biology" is a label for "two largely separate fields," evolutionary biology and functional biology. Mayr contrasted these fields with regard to the types of questions asked and the kinds of causes investigated. Evolutionary biology asks "why?" questions and studies ultimate causes, factors that govern changes in the genetic programs of organisms, whereas functional biology asks "how?" questions and studies proximate causes, factors that govern the responses of organisms and their parts to the environment. These groupings of disciplines do have unifying features and seem, therefore, to provide a useful perspective. Unfortunately, at least one important area of biology is entirely omitted from this classification; as Mayr indicated, "descriptive structural" studies fall into neither field. This omission could be remedied by adopting an enlarged "structural and functional biology" or simply a "nonevolutionary biology," but the unifying features of that branch disappear in either case.

Perhaps, then, biologists can be heuristically clustered not so much by the level of organization of the phenomena that they study, the causes of those phenomena, or the types of questions asked about them, as just by the underlying intentions of various biological studies. Investigations can seemingly inquire about either the uniformity of life or the diversity of life, aspects that can be referred to as general biology and comparative biology, respectively. The general biologist usually works on a single species and regards it as an experimental tool, hoping to discover in it properties that may prove to be general. To such biologists, diversity is only a hindrance; often laboratory strains of a species, specially bred to show the smallest possible amount of variation among individuals, will be preferred.

Commendable as such studies may be, they are most often thwarted (at least in an ultimate sense); in the long run, the properties discovered by a general biologist in one species are usually discovered to hold true for some, but not all, other species. Relatively few properties prove to be true for all living organisms. As a result, the outcome of continued studies in general biology has been the accumulation of data on both the uniformity and the diversity of life, on the similarities and the differences among living organisms. To the comparative biologist falls the task of attempting to understand this accumulation of data.

It is with comparative biology, the science of diversity, that this book is concerned, and the book has a twofold purpose. It is an inquiry into (1) the theoretical structure of comparative biology, and (2) the nature of our knowledge of organismic diversity.

COMPARATIVE BIOLOGY AND ITS ELEMENTS

Historically, comparative biology grew through the steady accumulation of data on diversity in the attributes of organisms, and its early practitioners had their hands full differentiating, describing, and naming enormous numbers of newly discovered biological entities and processes. These activities have been of tremendous importance, but they are not ends in themselves. The mere compilation of data about the world in which we live, no matter in how highly ordered a fashion, is not sufficient for understanding the world. Data so compiled and ordered are still only data in search of interpretation.

With the advent of evolutionary theory around the beginning of the last century, a means of interpreting the data was provided. If life has evolved, we can conclude that biological phenomena (entities and processes) are diverse because they have become diverse. In other words, a study of the diversity of life, when viewed through the lens of a theory of evolution, becomes equivalent to a study of the history of life. That history, of course, has two dimensions: there is a history of life in time and a history in space. Comparative biology, if it is to allow us to understand the living world, must thus deal with three distinguishable elements: (1) similarities and differences in the attributes of organisms, (2) the history of organisms in time, and (3) the history of organisms in space. The role that comparative biology plays in dealing with the interfaces between these three elements or factors is concisely expressed in the title of a book by Croizat (1964), "Space, Time, Form: The Biological Synthesis." Form, in this context, refers not only to the structure of organisms, but to all their attributes, be they structural, functional, or behavioral.

There are few, if any, subdisciplines of biology that do not or cannot contribute information on the attributes of organisms relevant to comparative biology, but there are four conventionally delineated subdisciplines that basically belong to comparative biology. Two of them are (1) systematics, which is concerned primarily with form (in the broad sense) and secondarily with time, and (2) biogeography, which is concerned primarily with space and, like systematics, secondarily with time. Systematics and biogeography together form the core of comparative biology. Two other subdisciplines, embryology (broadly construed to include within its scope all developmental processes) and paleontology, differ from systematics and biogeography in being primarily concerned with time and secondarily with form (and, in the case of paleontology, to some extent with space as well).

The order in which the elements of space, time, and form (and the respective disciplines primarily associated with them) are treated in this book is the reverse of the order just given and that used in the title of Croizat's work. There are two reasons for this reversal. The first is that systematics, in providing classifications that summarize existing knowledge about the attributes of organisms, is a necessary practical prerequisite to the other fields. If our systematics is inadequate, it will scarcely be possible to do adequate work in biogeography, paleontology, or embryology. The second reason for the reversal, and one of the themes of this book, is that hypotheses about the history of organisms in time are tested by statements about their attributes, and that hypotheses about the history of organisms in space are tested by statements about their history in time.

METHODS IN COMPARATIVE BIOLOGY

Much of this book is devoted to questions of method, of how we might go about analyzing the massive amounts of data on life's diversity that are available to us. As a result, much of the book is concerned with areas that are (and no doubt remain) intrinsically problematical. The reason for this is relatively simple: we have no way to evaluate our methods scientifically. It may seem paradoxical to suggest that comparative biology (or any science) must adopt methods without itself being able to attest to their efficacy. But the fact is that we use our methods in an attempt to solve problems. If we already knew the correct solutions to those problems, we could easily evaluate and choose among various competing methodologies: those methods which consistently provide the correct solutions would obviously be preferred. But of course, if we already knew the correct solutions, we would have no need of the methods.

Considerations of scientific methodology, then, typically involve questions that are philosophical rather than scientific. From this we can conclude that one's general philosophy of science may greatly influence methodological discussions and decisions, and that it is therefore incumbent on scientists engaging in such discussions or making such decisions to present, as explicitly as possible, the philosophical point of view from which they argue.

Some comparative biologists, for example, seem to adopt a philosophical position of extreme empiricism, viewing science as nothing more than ordered observation. To such workers, much of what is here considered comparative biology appears to be simply impossible to do, and they are quick to criticize such work on the grounds that it is speculative. If our knowledge is limited to those things we can (supposedly) observe directly, the task of studying the history of life is indeed fraught with insurmountable difficulties, since we can hardly observe directly the past history of present-day organisms. But such a philosophical view characterizes (if anything) technology, not science; it denies to science precisely those processes that are most characteristic of it: the proposal and testing of hypotheses.

This alternate view, proposed and explored by Popper (1959), emphasizes that all scientific knowledge is hypothetical, conjectural, and speculative, consisting either of general statements (theories) that can never be confirmed or verified but only falsified, or of particular statements that use class names and therefore imply other conjectural and unverifiable theories. The task of the scientist, in Popper's view, is to propose solutions to problems and to test those proposed solutions as rigorously as possible, in the hope of showing them to be wrong if they are. Those theories that could possibly be shown to be wrong, but which we have not yet succeeded in refuting, are accepted (on a tentative basis), not as being necessarily true but as being possibly closer to the truth than those competing theories already refuted.

In general, then, the methods we prefer are those that allow the generation of testable (as opposed to irrefutable and therefore unscientific) hypotheses, and those that allow the generation of more severely (rather than less severely) testable hypotheses. As Popper has shown, those conjectures that are most easily testable are also those with the greatest information content and are therefore apt to be the most useful.

The remainder of this chapter consists of an introduction to the problems of comparative biology, the methods used in attempting to solve those problems, and some of the basic concepts involved with those attempts. The remainder of the book will consider these matters in detail.

PROBLEMS OF FORM: SYSTEMATICS

THE PURPOSES OF CLASSIFICATION

If a biologist discovers a property of an organism, one question immediately raised is "How general is it?" Often the answer is that it has some, but limited, generality: it is true of some, but not all, organisms. Should a second such property be discovered, we can ask not only "How general is it?" but "Is it more or less general than the first property?" If these questions can be answered, sets of organisms can be recognized about which it is possible to make general statements, statements like "All these organisms (1, 2, 3, 4, 5), and no others, have property A" and "All the organisms that have property B form a subset (1, 2, 3) of those that have property A." By this means, sets and subsets of organisms can be defined. When these sets and subsets are given names, a classification results.

Often it is found that two or more properties have precisely the same degree of generality, that is, that all those organisms that have property A, and no others, also have properties B, C, and D. When this happens, the name of that set of organisms acquires utility as a means of information storage. In referring to the set by its name, we can indicate that there is a set of organisms about which general statements can be made, without actually having to list all the properties known to occur in the organisms belonging to the set. If we then organize our data on these properties by associating them with the name of the set, we also have a means of retrieving the data, should we ever need to do so.

Thus classifications obviously perform an essential function in information storage and retrieval. They allow us to deal with tremendous amounts of data by subsuming a great deal of information into single words (the names of sets of organisms), so that we are not forced to deal with all the data we have simultaneously. But is this task of data organization *the* essential function of classification? If it were, would we not, as scientists, leave the task of constructing classifications, of organizing our huge pile of data, to technicians, just as we leave the task of organizing our huge pile of publications to librarians?

Why, then, do scientists concern themselves with constructing classifications? Perhaps because classifications serve not only to summarize information we already have, but also to predict information that we do not yet have. For example, if there is a set of organisms, all of which share properties (A, B, C, D, E) that no other organisms have, and we find another organism about which we know only that it has properties A and B, can we not predict that it will have properties C, D, and E as well? Similarly, if we already know that we can make general statements about some properties of a set of organisms (1, 2, 3, 4, 5) and about a subset of that set (1, 2, 3), can we not predict that for other properties that we do not yet know, some may be true of the set (1, 2, 3, 4, 5), others may be true of the subset (1, 2, 3), but none should be true for other incongruent subsets like (3, 4, 5)?

To the extent that there is order in nature, to the extent that existing classification is an accurate hypothesis about that order, and to the extent that our hypotheses about properties and their distribution among organisms are correct, such predictions will be successful. If they are not successful, we have discovered an interesting problem, either in our perception of properties and their distribution, in our classification, or in nature. Classifications, then, are useful not merely as data summaries but also as hypotheses about order in nature. These hypotheses, once tested and corroborated, can be used in studies of aspects of nature other than the attributes of organisms. It is as hypotheses that classifications are ultimately useful.

THE ELEMENTS OF CLASSIFICATION

Biological classification is a very old enterprise, arising from human-kind's ability to generalize about the natural world. The type of generalizations of concern may be termed kinds. For example, one may observe a particular organism and give it a name: "cat." Early in life, humans learn that names such as "cat" apply not to one particular organism, but rather to a kind of organism. There is not only one cat, there are many. And there are even different varieties of them.

Biological classification is an attempt to specify all of the different kinds of organisms. In its modern form, it is basic to understanding the natural world. But the enterprise of classification is problematical. Nature is rich in different kinds of organisms, and persons disagree on exactly what kinds there are. When disagreement occurs, a person is apt to reflect on the nature of that disagreement and may ask: what is a "kind"? Most biologists have been too busy to give much thought to this general question. For most practical problems of humanity and science, it does not matter. With reference to a particular kind, an appropriate answer is usually available. Asked "what is a mammal?", a biologist is apt to respond with a definition: "a tetrapodous vertebrate with homothermy, hair, internal fertilization, milk glands, and a dentary-squamosal articulation." Or the response may consist of examples: "a monotreme, marsupial, or placental." Further inquiry is apt to proceed away from the general question "what is a kind?" to more particular questions, such as "what is hair?" or "what is a monotreme?" To such particular questions, answers can usually be given that satisfy most day-to-day curiosity. But if there is to be a general theory of classification, the general question must be answered. Most biologists might answer it in a preliminary way by stating that there are two kinds of "kinds": species and groups of species. As a result, the initial question (what is a kind?) may be restated in two forms: (1) "what is a species?" and (2) "what is a group?"

The second question will be explored at length in this book, but the first question will be considered only here, and only briefly. Unlike the general question (what is a kind?), the question "what is a species?" has been extensively debated by biologists. Numerous different species concepts have been proposed: there are biological species, evolutionary species, morphological species, polytypic species, phenetic species, ecological species, paleontological species, essentialistic species, nominalistic species, and doubtless many others. Most considerations of this topic have been attempts to define the word "species," but as Popper has pointed out, controversies over the possible definitions of terms are not themselves productive, because all definitions utilize other words that are themselves in need of definition, resulting in a never-ending process (an infinite regress). For example, almost all definitions of the word "species" that have been proposed utilize the word "population"— species are populations, or groups of populations, that meet one or more criteria. But the word "population" is itself in need of definition, and is fully as difficult to define as is the word "species."

All of these varied species concepts do share some elements in common. All of them admit that we can never study species as wholes, but only samples of them, and all of them provide criteria by which, in some cases, we may be able to say that a given sample of organisms represents not one but two (or more) species. None of them, however, can ever guarantee the integrity of a sample; none can ever guarantee that a sample contains only one species. The most we can say is that we have not yet been able to differentiate species within the sample. So no matter what species concept a biologist claims to use, there is an upper limit to the number of species, set by the number of samples that can be differentiated in some way. To a very large extent, this is the species concept actually used in practice: those samples that a biologist can distinguish, and tell others how to distinguish (diagnose), are called species.

This, however, is not in itself a sufficient concept, because there are samples that can be distinguished but which do not appear to exist independently in nature. In many groups of organisms, for example, we can distinguish samples representing males and females; or eggs, larvae, pupae, and adults. We find, however, that males by themselves do not produce other males, or larvae other larvae, so that these samples, by themselves, have no independent existence in nature. Thus the concept of species must include a criterion of self-perpetuation: males and females together; eggs, larvae, pupae, and adults together; form self-perpetuating species.

If we call any attribute of an organism by which we can distinguish samples a character, we can say that to be diagnosable, a sample of specimens must have a unique set of characters. It need not have even a single character that is unique to it, but the total set of its known characters must be different from that of all other known samples, or we will not be able to distinguish it. In this book, then, species are simply the smallest detected samples of self-perpetuating organisms that have unique sets of characters. As such, they include as species the "subspecies" of those biologists who use that term.

THE STRUCTURE OF CLASSIFICATIONS

Classification, in the sense used here, is the biological or scientific system of named groups and subgroups of the various kinds of organisms. Named groups have their own scientific names, e.g., Eukaryota (nucleate organisms), Angiospermae (flowering plants), Metazoa (multicellular animals). In the biological system, the groups are ranked hierarchically: the Eukaryota may be considered a kingdom, the Metazoa a phylum, and the Angiospermae a class. What groups should be recognized, and how they should be subdivided, are the primary concerns of a theory of classification (systematics). What names and ranks the groups and subgroups should have are important but secondary concerns. These affect not only the biologists who create classifications, but all other persons who have to learn and use them.

Since the time of Linnaeus, the biological system has been conceived as a hierarchy with specified levels, or ranks. The ranks commonly in use today include:

> Kingdom
> Phylum
> Class
> Order
> Family
> Tribe
> Genus
> Species.

The ranks provide a system of categories to accommodate groups of organisms. Groups fitted into this system, so that each group has a unique name and a specified rank, are called taxa (singular: taxon). An example is provided by Simpson's (1945) classification of mammals, which are ranked as a class. His Class Mammalia is a taxon, which includes all mammals, and which is subdivided into two subtaxa:

Class Mammalia
  Subclass Prototheria (monotremes)
  Subclass Theria (marsupials and placentals).

The ranks, or categories, of genus and species have a particular significance, for they are the basis of binomial nomenclature. The purpose of binomial nomenclature is to give a different name to each individual species of organism, and to permit different species to be grouped: several species may be grouped into one genus (plural: genera). In this system, the name of each individual kind of organism is a binomial, consisting of two parts, generic and specific. *Homo sapiens* (humankind) is an example of a binomial (the generic part of the name is *Homo*, and the specific part is *sapiens*). Together, both parts form one name. Binomials, once created, do not necessarily remain constant. Although *Homo sapiens* is the name of the only living species presently placed in the genus *Homo*, future usage might change. The chimpanzee (*Pan troglodytes*) and gorilla (*Gorilla gorilla*) could conceivably join humankind in the genus *Homo*, in which case they would be known as *Homo troglodytes* and *Homo gorilla*, respectively. Rules of binomial nomenclature have been elaborated for plants (ICN, 1978), animals (ICN, 1964), and bacteria (ICN, 1975) in the hope of ultimately stabilizing the names of the individual kinds of organisms, but the rules place no restriction on the limits of genera and other higher taxa (tribes, families, etc.). The use of higher ranks, or categories, is a matter of interpretation and tradition.

Aside from the rules of binomial nomenclature and certain conventions for forming names of higher taxa, there is no single theory of classification that a biologist is obliged to accept and use, and classificatory practice varies. Consider groups such as the conventional families Hominidae (containing one species, *Homo sapiens*) and Curculionidae (containing over fifty-thousand species of beetles). Do these families share some property, other than their specified membership and scientific name, that makes them groups of the biological system? Do these families share some property that makes them families rather than, for example, orders or genera? To either question, no answer is possible that is generally accepted within the scientific world, for there is no generally accepted theory of classification.

One feature, however, is common to all hierarchical classifications: they can be represented by branching diagrams. For example, the following hierarchy:

> Class Mammalia
> Subclass Prototheria
> Subclass Theria
> Infraclass Metatheria
> Infraclass Eutheria

can be represented by a branching diagram, or dendrogram:

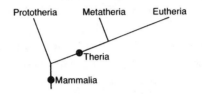

If we call the information in the classification about what groups there are, and how they are subdivided, the primary content of the classification; and the information about what the names and ranks of the groups and subgroups are, the secondary content; we can see that all of the primary content and half the secondary content (the names) of the classification are also contained in the branching diagram. The dendrogram specifies that the group Theria must be higher in rank than the groups Metatheria and Eutheria, and that the Mammalia must be higher in rank than the Prototheria and Theria, but it does not specify what particular ranks those groups must have.

Much of this book is about branching diagrams and their meaning in the field of comparative biology. Their most obvious use is to depict evolutionary genealogies, and in that role they are variously known as evolutionary, or phyletic, trees. They are also used in a more general sense to depict the structure of knowledge, particularly knowledge about the similarities and differences of organisms. A consistent terminology is adopted in this book in order to differentiate between these usages:

1. *Phyletic trees* depict aspects of evolutionary genealogies.
2. *Cladograms* depict structural elements of knowledge.

One theme of this book is that the sense of a dendrogram is best understood not in a genealogical sense (as a phyletic tree), at least not at first, but rather in a general sense relating to the structure of knowledge (as a cladogram). The theme may be appreciated by consideration of an example involving comparisons of the axial skeleton of three different kinds of organisms, a lamprey, a shark, and a perch:

Lamprey: a notochord
Shark: a vertebral column of articulated cartilages
Perch: a vertebral column of articulated bones.

Comparison of the characteristics of these three species shows that there are two general characters:

1. All three species have an axial skeleton.
2. The shark and perch have an axial skeleton in the form of a vertebral column.

The general characters may each be represented by a branching diagram:

And the two branching diagrams may be combined:

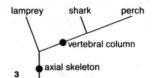

The general characters may also be summarized in the form of a hierarchy:

I. Organisms with an axial skeleton (lamprey, shark, perch)
   IA. Organisms with a notochord (lamprey)
   IB. Organisms with a vertebral column (shark and perch).

And the various terms of the hierarchy may be given proper scientific names and rendered in a formal classification:

Phylum Chordata (organisms with an axial skeleton)

Subphylum Cyclostomata (notochord)
Subphylum Gnathostomata (vertebral column).

The branching diagram (3), the hierarchy, and the classification may all be considered as identical summaries of the same two general characters. The summaries are useful because they communicate knowledge that there are general characters. This knowledge is useful because without it there would exist only specific items of information, such as:

Lamprey: a notochord
Shark: a vertebral column of articulated cartilages
Perch: a vertebral column of articulated bones.

Knowledge of general characters is useful for another reason: it is the basis for expectations about order in nature. The two general characters specify a certain pattern of order that may be summarized in yet another way, for general statements can be made for two groups:

Group 1: Lamprey + Shark + Perch
Group 2: Shark + Perch.

The expectation that may be associated with these statements is that the pattern contained in them is true not only for two general characters of the axial skeleton, but will prove true also for all other general characters that might subsequently be discovered. This *expectation of generality*—that nature is ordered in a certain specifiable pattern— is the hypothesis embodied in a cladogram. It is the cladogram that specifies the pattern:

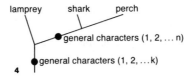

Since the theory of organic evolution became generally accepted in the latter half of the nineteenth century, branching diagrams have been used as representations of evolution, such that the lines of the diagram represent lineages extending through time. The simple branching diagram of the lamprey, shark, and perch may be considered from an evolutionary perspective. From this perspective, the diagram specifies that the shark and perch have evolved from a common ancestral species (A2), and that this ancestral species and the lamprey evolved from an older ancestral species (A1):

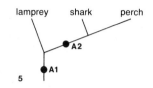

From this perspective, ancestral species A1 eventually evolved into two species: the lamprey and ancestral species A2. And species A2 itself evolved into two species: the shark and the perch. Therefore, the diagram specifies the evolutionary relationships between the species. The diagram is a phyletic tree. Comparison between the cladogram (4) and the tree (5) reveals only one difference: in the cladogram the lines represent general characters, whereas in the tree the lines represent common ancestral species. The difference may seem slight or stupendous, depending upon one's point of view and system of values.

The difference will be discussed at length in this book, of which one major theme is that a phyletic tree is a concept derived from, and subsidiary to, a cladogram. A phyletic tree is a cladogram interpreted in an evolutionary context. The remainder of this introductory discussion, however, will be concerned only with cladograms—with the way they summarize evidence, with possible problems in the evidence, and with their success as hypotheses about order in nature.

THE EVIDENCE FOR CLASSIFICATIONS
Branching diagrams, thoughts about them, and interpretations of them, have a complex history. We have seen that branching diagrams have the same primary information content as hierarchical classifications. Yet unless branching diagrams are related to the real world of organisms, they may not appear very useful. The present section explores branching diagrams in their role as summaries of real information.

Consider an example of three species (A, B, C) and five characters (v, w, x, y, z) distributed among the species as follows (plus signs indicate that a species has the attribute):

Characters

Species	v	w	x	y	z
A	+	–	–	+	+
B	–	+	–	+	+
C	–	–	+	–	+

In order to explore branching diagrams in their role of summaries, it is not important what species A, B, and C might be, or what characters v, w, x, y, and z might be. Given the information that these species exist, and that they have the characters indicated, certain true statements can be made, the statements can be represented by elements of branching diagrams, and the statements can be added together, summarized, and represented in more complex diagrams. Two kinds of summaries will be considered: an explicit summary, which specifies all of the information; and an implicit summary, which does not specify all of the information, but which is consistent with the information. The explicit summaries are here termed *character cladograms*; the implicit summaries are merely cladograms in the ordinary sense. The information provided by each of the five characters can be summarized as follows:

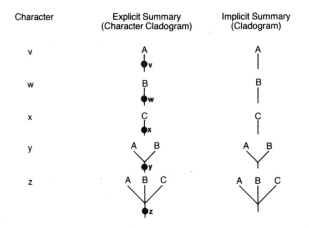

| Character | Explicit Summary (Character Cladogram) | Implicit Summary (Cladogram) |

Character v indicates that there is a taxon A, characterized (defined) by v; character y indicates that there is a group AB, characterized (defined) by y. These character cladograms can be added together in various ways, to produce complex character cladograms. There are ten possibilities for character pairs:

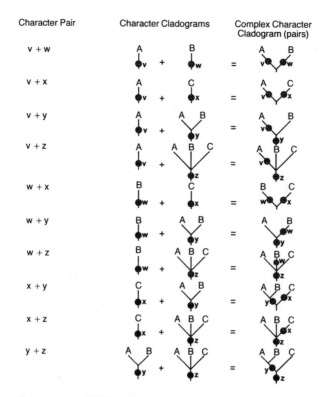

Character Pair	Character Cladograms	Complex Character Cladogram (pairs)

There are also ten possibilities for character triplets:

Character Triplet	Complex Character Cladogram (Triplets)
v + w + x	
v + w + y	
v + w + z	
v + x + y	
v + x + z	
v + y + z	

| w + x + y |
| w + x + z |
| w + y + z |
| x + y + z |

Similarly, there are five possibilities for character quadruplets:

Character Quadruplet	Complex Character Cladogram (quadruplets)
v + w + x + y	
v + w + x + z	
v + w + y + z	
v + x + y + z	
w + x + y + z	

There is, however, only one possibility for all five characters together. This possibility will be illustrated by adding the characters, and the character cladograms, one at a time:

Characters	Character Cladograms
v	
v + w	
v + w + x	
v + w + x + y	
v + w + x + y + z	

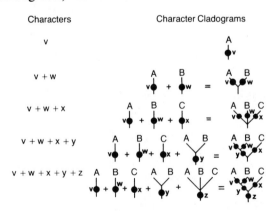

Finally, the same information will be added, in the same stepwise sequence, but this time with implicit elements and cladograms:

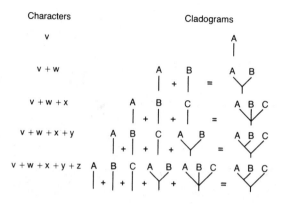

There are thus many true character cladograms possible even for a rather small amount of data. All of the different character cladograms are true, so far as they go in representing the data. But there is only one cladogram involved. The cladogram is defined by two general characters (y and z), each of which defines a group (AB and ABC, respectively). The cladogram is a true summary of the data. This can be seen by adding two more general characters to the data set:

Species	t	u	v	w	x	y	z
A	+	+	+	−	−	+	+
B	+	+	−	+	−	+	+
C	+	−	−	−	+	−	+

Characters

Many additional character cladograms can now be constructed, but there is still only one cladogram involved, the groupings of which (AB and ABC) are defined by the general characters (t, u, y, and z). In fact, an infinite number of additional characters can be added to the data set, and so long as they have the same distributions as characters v, w, x, y, or z, the cladogram will still be a true summary of the data.

More complex cladograms result in the same way from data sets on larger numbers of taxa. For example, given the following data:

| | | | Characters | | |
Species	v	w	x	y	z
A	+	+	−	−	−
B	+	+	−	−	−
C	+	−	−	−	+
D	+	−	−	+	+
E	+	−	+	+	+
F	+	−	+	+	+

the information can be implicitly summarized:

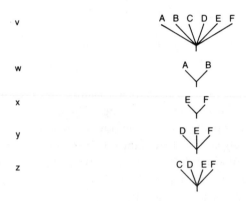

and added together to produce a cladogram:

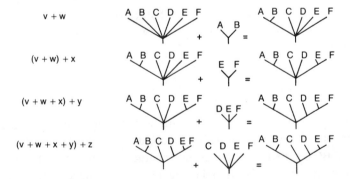

Again, the addition of an infinite number of characters distributed like characters v, w, x, y, or z still produces only a single cladogram, which remains a true summary of the existing data, and a prediction about the structure of any new data on species A–F that we might acquire.

THE PROBLEMS OF CLASSIFICATION

Problems can arise in classifications if additional data fail to conform to the pattern shown by previously known information, that is, if instead of finding repetitions of the expected order, we find incongruent general characters like p or q in the following example:

Species	Characters						
	v	w	x	y	z	p	q
A	+	−	−	+	+	+	−
B	−	+	−	+	+	−	+
C	−	−	+	−	+	+	+

If nature is orderly and our hypotheses about general characters and their distribution are correct, the incongruence should not exist. How might we decide wherein the problem lies? It is possible that nature, in this case, is not orderly, and that characters like y, p, and q all reflect real (but random) aspects of nature. There is no direct way for us to actually determine whether the disorder that we perceive exists in nature or only in our own hypotheses. The purpose of science, however, is to discover and explain regularities in nature, and if we give up the search for regularities, we also give up the game of science. So we may assume that the order really does exist and that we, and not nature, are responsible for the apparent disorder.

This means that either our original hypotheses about general characters distributed like y were wrong, and that our grouping (AB) is therefore wrong, or that our hypotheses about general characters distributed like p or q are wrong. We might, therefore, try to reexamine the general characters and attempt to find out which ones are the culprits. How, after all, might we have made the error(s)? There seem to be three possibilities, which will be examined in detail with the aid of actual examples.

Suppose, for instance, that we take four species (a robin, a woodpecker, a bat, and a mouse) and observe the following general characters: (a) the robin and the woodpecker both have feathers; (b) the robin and the woodpecker both lay eggs; (c) the robin, the woodpecker, and the bat all have wings; (d) the bat and the mouse both have hair; and (e) the bat and the mouse both have milk glands. The data can be summarized as follows:

Species	Characters a	b	c	d	e
Robin	+	+	+	–	–
Woodpecker	+	+	+	–	–
Bat	–	–	+	+	+
Mouse	–	–	–	+	+

We can see that general characters a, b, d, and e together define a single cladogram, but that character c is incongruent with d and e. Perhaps something is wrong with one or more of those characters. If we investigate the wings of robins, woodpeckers, and bats more thoroughly, we may find that they are very different: the bones in the two bird wings are hollow and have very different arrangements and shapes from those in the bat wing. Thus there might be two characters involved, rather than one: the wings of birds (with hollow bones) and the wings of bats (without hollow bones and with a different structure). If so, the data are really:

Species	Characters a	b	c	c′	d	e
Robin	+	+	+	–	–	–
Woodpecker	+	+	+	–	–	–
Bat	–	–	–	+	+	+
Mouse	–	–	–	–	+	+

and the cladogram is:

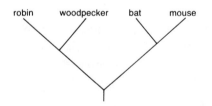

Suppose, instead, that we take three species (a lamprey, a perch, and a human) and observe the following general characters: (a) the lamprey and the perch both have gills; (b) the perch and the human both have a vertebral column of articulated bones; (c) the perch and the human both have jaws; and (d) the perch and the human both have paired appendages. The data can be summarized as follows:

Characters

Species	a	b	c	d
Lamprey	+	−	−	−
Perch	+	+	+	+
Human	−	+	+	+

We can see that general characters b, c, and d have the same generality but that character a is incongruent with the others. Perhaps something is wrong with our general character "gills." Unlike our previous general character "wings," there does seem to be a single character "gills," involving gill slits and aortic arches. Perhaps our mistake was of a different sort: are we sure that the general statement about the lamprey and the perch is not also true for the human? Do humans not also have gills? If we investigate the embryology of humans, we find that there is a stage in development in which human embryos do indeed have gill slits and aortic arches. We do not perceive gills in adult humans, because they have been transformed, during development, into other structures, but humans do, nonetheless, have gills. They merely have them in a modified form. So the data are really:

Characters

Species	a	b	c	d
Lamprey	+	−	−	−
Perch	+	+	+	+
Human	+	+	+	+

and the cladogram is:

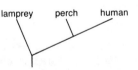

Here again we could say that there is a second character (a′) involved, namely the "modified gills" of adult humans, but this would not be a general character in a cladogram of these three species.

In the first kind of mistake, with wings, we initially thought of different characters (bird wings and bat wings) as being basically the same; in the second kind of mistake, with gills, we initially thought of the same characters (perch gills and human gills) as being basically different. In both cases, it is the incongruence between two or more characters that tells us we have made a mistake, either by investigating the character in insufficient detail (wings) or by ignoring the possibility of character transformation (gills). The second type of error is much more common than the first in actual classifications. In some cases, as with gills, we may be lucky enough to find ontogenetic evidence of the character transformation; in other cases, we may have to detect character transformations in a different way.

For example, two kinds of jaws are found among spiders: some spiders have jaws that work with an up-and-down motion, whereas others have jaws that work with a side-to-side motion. As a result, spiders have frequently been divided into two groups: the tarantulalike spiders (with vertically mobile jaws) and the true spiders (with horizontally mobile jaws). When other characters are examined, however, incongruences are found: some tarantulalike spiders, for example, have characters of the silk-spinning organs and the nervous system that they share with the true spiders rather than the other tarantulas. Thus we have three groups of spiders (two groups of tarantulalike forms and the true spiders) that share the following general characters of (a) the jaws, (b) the silk-spinning organs, and (c) the nervous system:

	Characters		
Groups	a	b	c
Tarantulas I	+	–	–
Tarantulas II	+	+	+
True Spiders	–	+	+

We can see that general character a is incongruent with the others. We might question first whether the vertically mobile jaws of the two groups of tarantulas are actually the same character, but we find no evidence of differences in their structure or function. We might then question whether perhaps young spiders of one or more of the groups have a different type of jaw from adults of those groups, which have a type that is thus the result of an ontogenetic character transformation, but no evidence of such a transformation has been found by spider embryologists.

Is there another possible source of evidence about character transformations? We might ask, for example, what the closest relatives of spiders are, and what kind of jaws they have. There are numerous characters (call them d, e, and f) which indicate that there is a group including all spiders, and other characters (call them g, h, and i) which indicate that there is a larger group including both spiders and amblypygids (the tailless whipspiders of tropical regions). We can thus draw a character cladogram for amblypygids and the three groups of spiders:

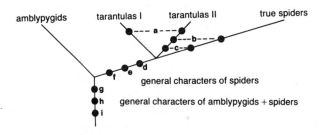

The character cladogram shows the incongruence between general character a and characters b and c. If we investigate amblypygids, we find that they do not have character b or c, but that they do have character a! Thus we cannot use character a to define a group "tarantulas I + tarantulas II," because it actually defines a much larger group, including at least "amblypygids + tarantulas I + tarantulas II." If general characters d through i are correct, we can hypothesize that there has been a character transformation in an evolutionary sense: that true spiders do have vertically mobile jaws, but that they have them in a modified form (in which they have rotated ninety degrees so that they now move horizontally), which we can call character a':

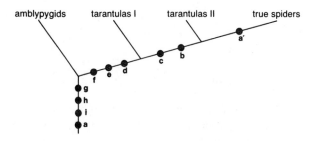

This technique of hypothesizing character transformations is called "outgroup comparison." Its use requires that, for the purpose of analysis at one level of the cladogram (within spiders, for example), we assume that another level of the cladogram (amblypygids + spiders, in this case)

is correct. This method is legitimate only because the assumptions made for analysis at one level of the cladogram can be questioned and tested independently. If, for example, it is subsequently discovered that general characters d, e, and f are incorrect, and that other general characters unite tarantulas with amblypygids rather than with true spiders, then our conclusions about spider jaws can also be shown to be incorrect. Moreover, outgroup comparison is just a shortcut. We could, after all, examine *all* other known organisms to see if any of them has character a and arrive eventually at the same result.

If we examine the original incongruence between character a and characters b and c, we can see that the problem was not that character a is invalid, but that we were using it at the wrong level of the cladogram (that is, just as with gills, we had originally underestimated its degree of generality). From this, we can draw a general conclusion: any character can be used at only one level of the cladogram, and when we find that we are using a character twice (for example, using general character a to define a group "tarantulas I + tarantulas II" when it also defines a larger group), we have found a mistake.

So mistakes are possible if we misperceive the identity of a character (bird and bat wings) or the generality of a character (perch and human gills, amblypygid and spider jaws). There is one other kind of mistake, less commonly made in practice but certainly not unknown there. Suppose that we again take a lamprey, perch, and human and add to our previously resolved data set a fifth character (e), the lamprey and the perch both have no written language:

	Characters				
Species	a	b	c	d	e
Lamprey	+	−	−	−	+
Perch	+	+	+	+	+
Human	+	+	+	+	−

We can see that character e is incongruent with b, c, and d. We can also analyze that character in detail and discover that although it does appear to be the same character in lampreys and perches, in a sense it is true also of humans, since human embryos and infants also have no written language. We can consider that during human development the character "no written language" is transformed into "written language present." We could also discover the character transformation by observing that the character "no written language" does not really define a group "lamprey + perch" but a much larger group including at least all nonhuman organisms. We could then see that all living organisms have either no written language or what must be a modified form of that character (written language present), and that the character "no written language" is therefore of no use in constructing classifications. But the character can also be questioned on a more fundamental level: is the absence of an attribute in two or more organisms a general character? It is certainly a character, in the sense that it can be used to distinguish samples of organisms, and it might prove very useful in the practical task of identifying samples, but is it of any use to us in trying to construct a cladogram (and thereby a classification)? The answer is no; the absence of a written language in lampreys and perches is no more a general character than is the absence of Cadillac engines in their stomachs or totem poles on their skulls. There is an infinite number of attributes that are lacking in any organism; general characters must refer to the presence of attributes, not to their absence.

The examples given here may seem silly, but such characters have been used to define groups by comparative biologists in the past. Early workers, for example, divided animals into Vertebrata and Invertebrata, but the absence of vertebrae (like the absence of Cadillac engines) is not a general character. For many years, insects were divided into two groups, the Pterygota (insects with wings) and Apterygota (wingless insects), but the absence of wings is not a general character. It is not surprising that groups such as Invertebrata and Apterygota have been abandoned by systematists: since they are not based on patterns of general characters, they have had no success in predicting the structure of newly accumulated data on the organisms they include.

The example of the Apterygota does raise an interesting problem, however. The group Apterygota, as used by more modern entomologists, included such wingless insects as springtails and silverfish. But there are other groups of insects, such as fleas, which have no wings but were nonetheless placed by those workers in the Pterygota, not the Apterygota! Such systematists realized that there is a difference between the absence of a character and the loss of a character, and argued that whereas springtails and silverfish merely lack wings, fleas have lost them. In other words, some characters which might appear to be simple absences are the result of character transformations. Some comparative biologists have argued that loss characters have no information content for classification, but that is true only for absent characters, not "lost" ones. Characters that we call "losses"—characters that are the result of transformations—can always be rephrased as modifications; they have the same information content as other general characters. Those systematists who placed the fleas in the Pterygota were able to argue that fleas have wings that have been modified (because of their ectoparasitic mode of life) in such a way that, to the superficial observer, they appear to have no wings. If we examine in detail the thorax of a flea, for example, we find that it does indeed have features indicative of wings, features that are lacking in springtails and silverfish. Thus fleas share general characters of the Pterygota (wings present, resulting in thoracic modifications) and have, in addition, a general character of their own (modified wings).

We can summarize all these arguments by saying simply that if the order we seek in nature exists, incongruent characters indicate only that we have erred by considering as a general character features that are really (1) two or more general characters, (2) single general characters used at the wrong level, or (3) not general characters at all.

THE SUCCESS OF CLASSIFICATIONS
The material on classification presented above can be summarized with the aid of some elementary concepts relevant to a larger aspect of the problem. Let us start with a species A:

A

We might observe certain things to be true of it:

A
•

And we might file them away, appropriately, in an accumulation of observations about species A. Some of the things would, however, be true of other species besides A. So our observations would be incomplete, without a statement specifying that our observations pertain to A uniquely or, alternatively, to some other species, too. The other species could be few or numerous. Let's consider two:

A   B

Whichever two we choose, we would observe some things to be true of A uniquely, some true for B uniquely, and some true for both:

A   B
●   ●
● - - ●

And what do we do with this information? File it away in three accumulations, one for A, one for B, and one for A + B? Well, we might try, but it will not work, for we are no better off than before in being able to specify the generality of our observations. Besides, there probably is no accumulation set aside just for species A and B.

Let's consider three species:

A   B   C

Whatever species we choose, we might observe some things to be true of each, other things to be true of each possible pair, and still other things to be true of all three:

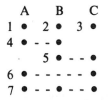

We will not ask what to do with all of these observations, because by now it should be obvious that an impossibly large number of accumulations would be required to handle all the possible information of this kind.

Rather, let's focus on the pattern displayed by the information. The pattern consists of all possible elements. The pattern is random, in the sense that it can be duplicated by tossing three coins (A, B, C). Every so often they will all come up the same, giving us element 7. If coin A comes up heads, and the others tails, we would have elements 1 and 5, and so on. For what it's worth, each of the seven elements has an equal probability of coming up, and about eight tosses will produce all seven elements, on the average.

To some persons, this kind of pattern is the real world: for them, ultimate reality is chaotic. These persons tend not to become scientists, but if they do, they tend to become frustrated. To other persons, this kind of pattern is unsatisfactory as a representation of the real world; these persons tend to assume that the randomness stems from faulty observations and incorrect hypotheses. The problem is to see order in apparent chaos through critical observation and hypothesis testing. Some of these persons become scientists, and systematists.

The basic postulate of systematics is that what can be learned about any three species exemplifies a pattern of the form: given three species, two of them are more closely related to each other than either is to the third. In the case of species A, B, and C, for example, we might have reason to believe that A and B are more closely related to each other than either is to C. Never mind what, precisely, is meant by "related." We would still have things that are true, each for A, B, and C, and things true for all three (a residue, as it were, of randomness):

```
        A    B    C
    1 •  2 •  3 •
    7 • - - • - - •
```

But in between we would have simplified the picture:

```
        A    B    C
    1 •  2 •  3 •
    4 • - - •
    7 • - - • - - •
```

Interestingly, we would still not have learned that, for example, item 1 really is unique to A, or item 2 to B, or item 3 to C—there still being the possibility that these observations are true for species yet unexamined by us. We would, however, have discovered an element of pattern—item 4—an element of a different kind than the observations that we accumulate. The element of pattern can be restated in a taxonomic, or systematic, form: there is a group, including A and B, but excluding C. Now, there is a place—an accumulation, if you like—for information of that kind, and that is systematics in its traditional guise.

So far, so good, you might say. But what happened, you might ask, to the other elements, to elements 5 and 6?

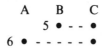

```
        A    B    C
       5 • - - •
   6 • - - - - - •
```

Suffice it to say that, in one of the three senses discussed in the last section, they are not true statements, and let us look at another aspect of the problem.

We have three species A, B, C. We observe 99 things to be true for species B and C, and only one thing to be true for species A and B:

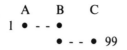

The information exemplifies two possible patterns:

Given this information, is a choice between these two patterns problematical? This question, believe it or not, causes systematists to part company. One group regards the choice as nonproblematical. The views of this group will be discussed below under the heading of phenetic taxonomy. The other group regards the choice as problematical, inasmuch as there are two patterns exemplified by the information. A representative of this second group might state: "These things are all very well so far as they go, but I need to know more about them."

Let's supply this representative with some additional information by mentioning some things true for various vertebrate animals. The lone thing turns out to be lungs, present in both species A and B, absent in C. Of the 99, the list begins as follows:

1. Fins
2. Gills
3. A slimy skin
4. Aquatic habits
5. Lays eggs
6. Cold-blooded
7. No spoken language,

and so on. About this time, our representative interrupts: "I think I understand the nature of the problem. But first let me ask you: are you sure that these things that you say are true for B and C are not, in some sense, also true for A?" Well, let's reconsider the list:

1. Fins? Species A has arms and legs. Could they not be construed as fins of a sort?
2. Gills? Species A, when an embryo, has gill slits and aortic arches. Not gills exactly, but pretty close.
3. A slimy skin? Species A develops internally, in its maternal parent. And while in utero it is a bit slimy.
4. Aquatic habits? Species A in utero is immersed in a miniature sea.
5. Lays eggs? Yes, but internally (and sometimes externally, when writing books).
6. Cold-blooded? Yes, in utero (and all too frequently in later life as well).
7. No spoken language? Yes, when young, but that isn't a general character anyway.

Suppose, then, that with a little scrutiny, all 99 turn out to be true for all three species. This leaves the following picture:

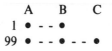

Not very interesting, but not problematical. And what does it add up to? Just a simple illustration of two points: that science is a way of viewing things as problematical; and that the evidence for classification is problematical and deserves to be viewed as such.

With regard to our problem and its outcome, what do we have? The observations are interesting in themselves, but what is more impressive is the pattern—the AB component of it—as a general summary of our observations. If we have done our job well, and have discovered the correct pattern, it will be a general summary of whatever other observations we might make, now or in the future. The pattern, therefore, has a truth of its own. Discovering that kind of truth is one task of systematics, and indeed of biology. Why so? Because it is a truth of the real world, or so we may infer. The alternative is to accept the idea that the real world is randomly organized—that the real world, in this regard, contains no truth at all.

By patterns, we mean branching diagrams or, alternatively, classifications. In this case, our AB observation—lungs—is a defining character of a group, or taxon, of vertebrates—the Teleostomi (the group that includes lungfishes, humans, birds, "reptiles," mammals, and amphibians). Groups or taxa are the kind of items that systematics accumulates, and we found one in our comparisons of species A, B, and C. The species? Yes, indeed:

A	B	C
Human	Lungfish	Shark

Patterns such as this one are interesting for other reasons. One is the fashion these days to discuss processes of evolution. Some biologists expound on processes as if all worthwhile general knowledge is contained therein. Now what, one might ask, are processes of evolution? Do they not all presuppose the existence of a nonrandom pattern such as the one we have considered? No patterns—in general, no processes. No patterns, nothing to explain by invoking one or another concept of process. In short, a process is that which is the cause of a pattern. No more, no less. Pattern analysis is, in its own right, both primary and independent of theories of process, and is a necessary prerequisite to any analysis of process.

Interestingly, although systematists might disagree with each other with regard to the problem presented above, and choose either the pattern supported by the lone character or by the 99 characters, they would nonetheless agree on the ultimate goal of their enterprise: a stable classification. Stability has two aspects, operating largely at the level of binomial nomenclature or at the level of higher classification. It is obvious that if the names of individual species change very frequently, they will lose much of their usefulness as means of information storage and retrieval. But as we have seen, information storage is only one of the functions of classification, prediction being another. Predictions are embodied not so much in the naming of individual species as in the grouping of species—in higher taxa. It might seem that here stability is not a proper goal at all. After all, if classifications are hypotheses about order in nature, and systematics is a science, the appropriate goal is presumably the growth of our knowledge, not its stagnation; and we should always be eager to discard our current classifications in favor of improved ones that have increased predictive value.

The solution to this seeming paradox is a simple one. In reality, stability and predictive value are not in conflict with each other, they are one and the same thing. The most stable classification is the one that most successfully predicts the structure of newly acquired data, and those classifications that are not successful in predicting new data will have no stability. They will be discarded in favor of more successful classifications—classifications that have higher predictive value and, hence, in the long run, greater stability.

In the above discussion, several matters that may be considered problematical have been glossed over. We will only list them here, for they will be discussed later in this book:

1. What are general characters, and how is knowledge of them achieved?
2. What are the contexts in which branching diagrams can be interpreted?
3. What is the relation between classification and the different contexts in which branching diagrams can be interpreted?
4. Are general characters summarized equally well by branching diagrams and classifications interpreted in different contexts?

All of these questions belong to the field of science called biological systematics, or, more simply, systematics or taxonomy. This field of science has undergone an unusual ferment in recent years, with much argument over basic questions such as those posed above (see Dupuis [1979] for a survey and bibliography of this recent literature). Within the field of systematics, there is no general agreement about how these questions should be answered. To one degree or another, the first part of this book, devoted to form, is an attempt to answer them.

PROBLEMS OF TIME: ONTOGENY AND PALEONTOLOGY

ONTOGENY AND GENERAL CHARACTERS

We have seen that systematics, even if considered in a nonphyletic sense (at the level of cladograms rather than phyletic trees), involves not only the element of form but also the element of time: the concept of character transformation implies transformation through time. If the concept of character transformation is derived in any particular case from the study of ontogeny, it can be considered a direct technique of classification; if the concept is derived from outgroup comparison, it can be considered an indirect technique of classification, since it depends not only on the observation of character distributions or changes, but also on a hypothesized higher-level classification. The spans of time involved in these two techniques are, of course, very different: in the first case, only the lifetime of a single organism is involved; in the second, large segments of the history of life.

Outgroup comparison resolves the different levels at which one general character (such as the vertically mobile jaws of amblypygids and spiders) and a second general character, representing a modification of the first (such as the horizontally mobile jaws of true spiders), may be used to define groups. The results can be viewed as indicating that the first character is primitive and the second advanced, or derived, relative to each other. Similar considerations apply to ontogenetic character transformations: features that appear early in development can be considered primitive relative to the modifications of those features that appear later in development. This symmetry of argument about primitive and derived characters led Haeckel, in the last century, to the so-called biogenetic law: that the order of development reflects the order of evolution. Indeed, the word "evolution" originally referred to ontogenetic development (Bowler 1975). But the use of ontogenetic data, or of outgroup comparison, in classification need not be justified by reference to any theory of recapitulation; both techniques can be seen as applications of a single principle, parsimony, involving only the preference for simpler hypotheses over more complex ones. Suppose, for example, that we have two species, one with gill slits in the pharynx of the adult (species A) and one without gill slits in the adult pharynx (species B). Obviously, both species share a general character (the pharynx). In a phyletic context, we can assume that they share a common ancestor, which also had a pharynx. But we cannot, with this information alone, reject hypotheses that the common ancestor of A and B either did or did not have gill slits in the pharynx of adults (i.e., that either the presence, or the absence, of gill slits is primitive).

Suppose, however, that we study the ontogeny of species A and B, and discover that early in development both species have pharyngeal gill slits, which subsequently either remain as slits (species A) or close (species B). The young stages of these species obviously share a second general character (gill slits); in a phyletic context, the presence of gill slits in young stages can be considered primitive for the two species. But what about the adults? There are two possibilities: (1) the presence of gill slits in adults, as in young stages, is primitive, or (2) the presence of gill slits in adults is derived. With reference to species A and B, possibility (1) requires that species B has gained an attribute (closed gill slits in adults) that was lacking in the common ancestor, whereas possibility (2) requires that species A has lost an attribute (closed gill slits in adults) that was present in the common ancestor. But of course, for species A to have lost the attribute, it must first have been acquired by the common ancestor, and that required prior gain of the attribute is equivalent to the entire change implied by possibility (1). Thus these possibilities can be diagrammed:

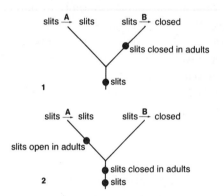

and it can be seen that possibility (2) involves both character transformations required by possibility (1): the acquisition of gill slits, and the acquisition of closed gill slits in adults, plus a third character transformation, the acquisition of open gill slits in adults. Possibility (1) is the more parsimonious, and can be preferred on that basis.

Outgroup comparison is even more obviously an application of the parsimony criterion. To use the example of spider jaws once more, it is possible that the vertically mobile jaws of amblypygids and the two groups of tarantulalike spiders represent a modified form of the horizontally mobile jaws found in true spiders (possibility 1) rather than the reverse transformation presented above (possibility 2):

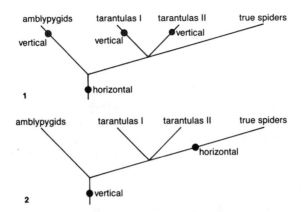

But possibility (1) is less parsimonious, and possibility (2) is thus to be preferred.

One might ask why a parsimony criterion should be used—after all, how do we know that evolution has actually been parsimonious? The answer, of course, is that we don't: we don't know whether evolution was always, sometimes, or even never parsimonious. We cannot observe the path of evolution directly, but only its results, and we can only attempt to reject some hypotheses about that path in favor of others. But there are a tremendous number of possible hypotheses; take, for example, the presence of hair in mammals. It is possible that hair has been acquired independently in every species of mammal (that is, that the most recent common ancestor of any two species of mammals had no hair); it is also possible that two particular mammalian species acquired their hair together, from their own most recent common ancestor, but that all other mammals acquired their hair independently of those two and of each other. The number of such hypotheses possible for even a single character is immense, and when entire sets of characters are considered, approaches close enough to infinity to approximate it for all practical purposes. None of these hypotheses can be rejected on grounds that they are impossible, but almost all of them can be rejected on grounds that more parsimonious alternatives are available. In short, if we do not prefer the most parsimonious hypothesis, we have no basis for preferring any one of these numerous alternatives over the others.

The parsimony argument, as applied to these direct and indirect techniques of classification, may seem straightforward enough. But it is not unproblematical, and the problems stem from neoteny. Some biologists (for example, Gould 1977) have attempted to distinguish between phenomena caused by a retardation of somatic development, resulting in an adult animal that has retained juvenile features (neoteny), and phenomena caused by an acceleration of gonadal development, resulting in a precociously reproductive juvenile animal (paedogenesis or progenesis), and have used the term paedomorphosis to include both kinds of phenomena. But we concur with De Beer's judgment (1951:52) that "as there are no hard and fast distinctions between them, they are all included here under neoteny." If, for instance, we know that in our example with gill slits, species A is neotenic, and that the presence of open gill slits in the "adults" of A is the result of a neotenic retention, then possibility (2) would seem to be the correct one, despite the parsimony argument. An actual case would be the classic example of neoteny, the axolotl, reproducing while still having functional gills (a larval character of salamanders). But what has been falsified here: the ontogenetic technique of classification, or the assumption that reproducing axolotls with gills are adults? After all, the neoteny was detected in this case by the discovery that under certain conditions, axolotls do transform (from "adults" to adults). These problems will be investigated in the section of this book devoted to the element of time.

PALEONTOLOGY AND GENERAL CHARACTERS
When one begins to investigate the attributes of organisms in relation to their history, and to interpret cladograms in a phyletic context, questions come into play about the fossil remains of organisms, the information they provide, and the role their study (paleontology) plays in comparative biology. Paleontologists seem to fall into three camps with regard to these questions. One group contends that paleontology is the primary historical subdivision of comparative biology, in that only the fossil record provides us with direct evidence of the path of evolution. To these workers, the taxa that appear earlier in the fossil record are the ancestors of the taxa that appear later, and the truth of evolution is there, in the rocks, waiting patiently to be revealed. If we crack enough rocks, the true pattern of evolution will emerge. A second group contends that this is not the case at all, that the fossil record only provides data on organismic diversity, just as does the study of extant

organisms, and that these data must be analyzed and interpreted in precisely the same way as neontological data. Still a third group of paleontologists agrees largely with the views of the second group but contends that paleontology provides a second indirect technique of classification (the "paleontological method"). This indirect technique is founded on the assumption that those features (not taxa) that appear earlier in the fossil record are "ancestral" to (more primitive than) those features that appear only later. To these workers there exists what Agassiz, in the last century, called the "threefold parallelism," involving the ontogenetic, comparative, and paleontological methods.

Given any two species, extinct or extant, they might share one of two possible kinds of relationships: (1) a relationship of common ancestry, and (2) an ancestor-descendant relationship. These two kinds of relationship are not mutually exclusive; the second is a subset of the first. If life on earth had a common origin, then any two species that we might choose will be related by common ancestry at one level or another (A and B have a common ancestor), even if the level is that of all organisms. Some of these pairs of species might also have an ancestor-descendant relationship (A and B have a common ancestor, and that ancestor is A). Some paleontologists (belonging to the first group mentioned above) have argued that an ultimate aim of comparative biology is to resolve the ancestor-descendant relationships that might exist among all organisms. But this aim can be questioned at a fundamental level, because only the relationship of common ancestry, and not the ancestor-descendant relationship, can be applied to all possible groups of organisms. So two problems are raised by the views of this group of paleontologists: (1) Are studies of ancestor-descendant relationships primary to comparative biology? and (2) Even if they are not primary, can they nonetheless be successfully carried out? These problems will be investigated in detail in chapter 3.

The notion of a "paleontological method" applicable only at the level of features, rather than taxa, can also be viewed as problematical. Suppose, for example, that the earliest known fossil of an amblypygid or spider has horizontally mobile jaws, and that vertically mobile jaws appear only later in the fossil record. Does this mean that our conclusions from the comparative method are wrong? Or merely that the earliest amblypygids and spiders happen not to have been fossilized, or happen not yet to have been found? And what would happen if, in some case, the ontogenetic and comparative methods agreed, but the paleontological method differed from them? Frequently, reference is made to the quality of the fossil record: if the record is "good," the paleontological method is supposed to be dependable; if it is not so good, it is not so dependable. But how can we tell when the record is "good"? When it agrees with the ontogenetic or comparative techniques? The problems of possible contradictions between the direct (ontogenetic) and indirect (comparative and "paleontological") techniques, and the possible effects of neoteny on each of them, will be explored in chapter 5.

PROBLEMS OF SPACE: BIOGEOGRAPHY

DISTRIBUTION AND EARTH HISTORY

Of the three elements of comparative biology, space has undoubtedly proven to be the most elusive in the history of the field thus far. Exactly why this should be the case is not easy to say. One factor might be that at least a tentative classification must be available before biogeographic investigation can begin: one must first know that there is a certain taxon before one can investigate either its distribution or the causes of that distribution. To some extent, then, biogeography must lag behind systematics. But perhaps a more important factor has been the absence of a clear-cut task, a major question that all biogeographers could agree is the focus of their diverse studies. The task of systematics, in comparison, is certainly clear: to produce an adequate classification of living organisms, a task involving the analysis of patterns in the distribution of attributes among taxa. But what is the task of biogeography? One could answer similarly—the analysis of patterns in the distribution of taxa among the various regions of the world. But what exactly is a biogeographic pattern?

It might be fair to say that two very different kinds of answers have been given, and are still being given, to this question. One approach (ecological biogeography) is to examine the distribution of a taxon in an attempt to find out why it may have the limits it actually has. Essentially, this involves asking the question in a negative way: why is it that the taxon does not occur more widely than it actually does? Many answers are possible: organisms have temperature requirements, humidity requirements, food requirements, and so on down a long list. If we look hard enough, and long enough, we will always find some "explanation" for the absence of a taxon from the regions neighboring those areas where it actually occurs. But what do we learn, in a positive sense, from this process, aside from the truism that the requirements of the organisms are met in the areas where they actually do occur?

Early in the history of comparative biology it was recognized that such answers are inadequate, in that one can find areas on different continents, for example, which appear to have the same temperature, rainfall, elevation, and the like, but that nonetheless are inhabited by different species. If present-day features of the environment are not sufficient to explain distributional patterns, what could the explanation be? The only alternative seems to be history: we can conclude that present-day distributions are the result of the history of taxa and of the areas in which they have lived. Historical biogeography, then, has two components: the history of life on earth, and the history of the earth itself. It is with these two components, and the relationships between them, that the last part of this book will deal.

To deal with the history of life we must start with the attributes of organisms, as summarized in cladograms, and interpret this information in a historical context. The question thus arises: must we deal with the history of life at the level of ancestor-descendant relationships, or only at the level of relationships of common ancestry? Since only the latter can be applied to all possible sets of organisms, we can for biogeographic purposes restrict ourselves, at least initially, to phylograms that indicate only the recency of common ancestry, and not the details of that ancestry as portrayed in phyletic trees. Phylograms are derived from, and subservient to, cladograms, and studies of historical biogeography are based on, and subservient to, phylograms.

To deal with the history of the earth is more difficult. This topic, after all, is the subject of an independent science, historical geology. Yet much of biogeography must deal with earth history, and biogeographers of the past have often become ensnared in the dogma of historical geology, as conceived at various times. This can be seen in a simple problem illustrating the relationships between phylograms and earth history.

Suppose, for example, that we have a group of six species, for which the following cladogram has been constructed:

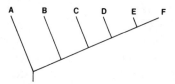

and suppose further that species A, B, C, D, and E occur only in South America but that species F occurs only in Africa. What can we conclude about the history of this group? Many biogeographers of the past (and some of the present) would view this situation as un-problematical: here we clearly have a group that is South American in origin, one member of which has dispersed to Africa. In a phyletic context, we can assign distributions to the ancestors of these taxa (whatever they were), just as we assign characters to them:

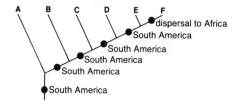

From this point of view, all one has to do to estimate the history of a group in space is to take an estimate of the history of that group in time (a phylogram) and lay it on a map. True, one can imagine other possible scenarios of dispersal: the immediate ancestor of F might have dispersed first to North America, then across the North Atlantic to Europe, and only then to Africa, but surely we can dismiss such explanations as being unparsimonious, at least until such time as specimens of the group are found in North America or Europe.

Wherein, then, lies the problem, and why has this kind of argument been used so frequently? For most of the past century, historical geologists argued that the basic geography of our planet, the continents and their relative positions, has remained stable throughout the long history of life. Of course, Wegener and a few other thoughtful souls did argue on behalf of a peculiar theory called continental drift, and cited the distribution of organisms as important evidence thereof, but this theory was carefully tucked away by their colleagues in the file labeled "historical curiosities." And there were a few foolhardy biologists who, despite widespread skepticism, maintained strenuously that the distributions of the groups they studied cannot be explained without reference to land connections that no longer exist. But in general, given a distribution like the one shown above, and the "knowledge" that South America and Africa are sitting just where they've always been, biogeographers lost no time in constructing dispersal explanations. For the more popular groups, particularly of higher vertebrates, different workers developed different dispersal routes, and argued vociferously for their own preferences.

Scientific theories have a way, sometimes, of returning from oblivion, even after they have been carefully laid to rest. And with the coming of plate tectonics, we all "know" now that South America and Africa are not today just where they have always been. During the interval, some biogeographers realized that dispersal is not the only possibility for explaining disjunct distributions. Conventional wisdom about process, after all, tells us that speciation, at least in animals, usually involves a process of geographic isolation (i.e., is usually allopatric), and occurs when a formerly continuous population is divided by the appearance of a barrier (a process called vicariance). If we find related species in different areas, we do not need to assume that there has been dispersal between the areas, only that a barrier has appeared between them:

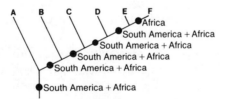

What was seen in the first model as a dispersal event (from South America to Africa) is seen in the second model as the appearance of a barrier dividing South America (and the ancestors of E) from Africa (and the ancestors of F). The implication is that at least some older ancestors occurred in both South America and Africa (i.e., that those two areas shared taxa in common).

DISPERSAL AND VICARIANCE

Dispersal hypotheses are of two general types, one of which is vicariance in disguise. In this first type of hypothesis, an ancestral species enlarges its range through time and is then fragmented into two disjunct ranges, the populations of which differentiate through time, ultimately to form two allopatric (geographically separated) descendant species. Implied is some causal factor, the appearance of a barrier, responsible for the fragmentation of the range of the ancestral species. The reason why this example of "dispersal" is really vicariance is that the postulated dispersal takes place prior to the appearance of the barrier and prior to the fragmentation of the range of the ancestral species. The effect of the postulated dispersal is only the creation of primitive cosmopolitanism (a requirement of the vicariance model).

The second, classic dispersalist model postulates dispersal over a preexisting barrier. In this case, an ancestral species, by means of "accidental crossing" of a barrier, expands its range and, in the process, simultaneously fragments its range. The effect of the postulated dispersal is immediate isolation and disjunction. The populations in the disjunct areas subsequently differentiate into two allopatric species.

In both cases, the existence of a barrier is implied. In the vicariance model, dispersal, if it takes place at all, occurs in the absence of a barrier; in the dispersalist model, dispersal occurs across a barrier. The explanations offered by both models amount to a correlation between a particular disjunction and a particular barrier: according to the vicariance model, the disjunction and barrier are the same age; according to the dispersalist model, the disjunction is younger than the barrier.

Both models allow the possibility that primitive cosmopolitanism may be achieved by an ancestral species that enlarges its range through the means of dispersal characteristic of the species. The models differ with respect to the causal factors invoked to explain disjunctions and, ultimately, allopatric differentiation. In the case of vicariance, disjunction is caused by the appearance of a barrier that fragments the range of an ancestral species; in the case of dispersal, disjunction is caused by dispersal of an ancestral species across a preexisting barrier. The causal factors may thus be isolated: (1) vicariance: the appearance of a barrier; (2) dispersal: dispersal across a preexisting barrier.

The vicariance model predicts that if we could find a single group of organisms that (1) had a primitive cosmopolitan distribution (i.e., whose ancestors were worldwide in distribution), (2) had responded (by speciating) to every geological or ecological vicariance event that occurred (i.e., to every barrier that appeared) after the origin of its ancestors, (3) had undergone no extinction, and (4) had undergone no dispersal, we could, by reconstructing the interrelationships of its members, arrive at a detailed description of the history in space of the ancestral biota of which the ancestors of the group were a part. Since we would also have arrived at a detailed description of the history of the world from the time of the first speciation event within the group to the present, we could, by correlating the sequence of branching points thus reconstructed with the sequence of events indicated by studies in historical geology, arrive at a chronology of the biogeographic events.

That extant distribution patterns are diverse, and do not all obviously correspond to each other in every detail, is evidence that at least criteria (2)—(4) do not always prevail in nature. Since under the vicariance model sympatry (the occurrence of taxa in the same area) is evidence of dispersal, the fact that we find numerous sympatric taxa at any given locality is evidence that much dispersal has occurred. The fossil record provides abundant evidence that extinction has occurred. Finally, the fact that within any biota some taxa are very widespread, and others very localized, is evidence that not all members of a biota need respond (by speciating) to every vicariance event.

Clearly, then, neither dispersal nor vicariance explanations can be discounted a priori as irrelevant for any particular group of organisms, and it might seem that the ideal method of biogeographic analysis would be one that allows us to choose objectively between these two types of explanations for particular groups.

TESTING BIOGEOGRAPHIC HYPOTHESES
Popper (1959) has presented the view that scientific explanations differ from unscientific ones only by virtue of their falsifiability (i.e., that we must be able to test and potentially reject any explanation that is to be considered scientific). Because the vicariance and dispersal models differ with regard to the age of disjunctions and barriers, it might appear that a critical test between them could be made by investigating these ages.

Both models appear to be open to falsification through statements made about the age of a particular disjunction and the age of a particular barrier. But the age of either one, in any particular case, is problematical. The minimum age of a taxon is that of the oldest known fossil attributed to it or to its most closely related taxon (an age that can always be augmented by the discovery of an older fossil). The age of a barrier can be estimated through studies of historical geology. Both types of studies (paleontology and geology) are subject to wide margins of error. Both types of studies can, however, falsify a correlation between a particular disjunction and a particular barrier. The vicariance type of correlation is falsified if one of the disjunct taxa is shown to be older than the barrier; the dispersal type of correlation is falsified if one of the disjunct taxa is shown to be the same age as, or older than, the barrier.

Both types of correlation are potentially falsified, therefore, through discovery of older fossils. But in each case, it is a particular correlation between a disjunction and a barrier, not the model itself, that is falsified. Both models include ad hoc protection from falsification through discovery of older fossils. The ad hoc principle is the rejection of a particular barrier (apparently of the wrong age to explain a disjunction) in favor of another (an older barrier, if one can be found; or if not found, postulated).

It is apparent that, in any particular case, neither model is exposed to a critical test by the dating of barriers and disjunctions, and that both models incorporate approximately the same ad hoc protection against falsification. In the absence of a critical test by the dating of barriers and disjunctions, we can proceed by adopting one model and attempting to refute other implications it may have that are not shared by the second model.

TESTING DISPERSAL HYPOTHESES

Assume that we have three allopatric taxa (A, B, and C) distributed in three corresponding areas (a, b, and c), and that we have tested and corroborated hypotheses that the three taxa form a group and that taxa B and C are more closely related to each other than either is to A. We could construct a dispersal explanation to the effect that the common ancestor of the group was originally found only in area a and is represented there today by taxon A, that some members of that ancestral taxon dispersed to area b and subsequently speciated there, and that some members of this second taxon subsequently dispersed to area c and eventually speciated there (example 1):

How might we test this hypothesis? To test any explanation, we must be able to deduce from it some prediction with which additional data can either agree or disagree. What can we deduce from the above explanation (1)? Because the dispersal capabilities of these organisms may or may not be similar to those of other groups, we can make no prediction about what patterns other groups that occupy these areas might show. Because the postulated dispersal events involve only movements of ancestors of the three taxa, we can make no prediction about what the dispersal capabilities of the three extant taxa might be. However, it seems that we might be able to make some predictions about the distributions of fossil specimens of the group that we might find.

Since we have hypothesized that taxon C evolved within area c, we should presumably be able to find fossil specimens attributable to taxon C within area c if suitable deposits exist there. Suppose, however, that we find a fossil attributable to taxon C in area a; does this falsify our dispersal hypothesis? The presence of C in area a could be accounted for by yet another postulated dispersal that occurred after the speciation of C, and we could accommodate the additional data without abandoning the hypothesis in favor of a vicariance explanation. Suppose instead that we had found a fossil taxon in areas b and c that cannot be attributed to one of the extant taxa but that, because it both shares the general characters uniting B and C and lacks the unique features distinguishing them, has to be added to the cladogram as in example 2:

Our hypothesis requires that B and C had a common ancestor that for
at least some period of time occurred in both areas (since the "founders"
of taxon C did not speciate in transit), so the fossils are consistent with
our explanation. Similarly, if the fossil taxon found in areas b and c
shared only some of the general characters uniting B and C, it has
therefore to be added to the cladogram as in example 3:

Our hypothesis is still tenable as it does not specify the length of time
for which the common ancestor of B and C may have occurred in area c
before it speciated there. A similar fossil found only in areas a and c
could also be accommodated in a dispersal hypothesis by postulating an
initial dispersal from a to c and a second dispersal from c to b.

Suppose, however, that we found a single fossil taxon in areas a, b,
and c that has to be added the cladogram as in examples 4 or 5:

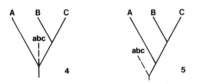

Our hypothesis is that area c was populated by a taxon that had
originated in area b (i.e., that had been isolated from a taxon in area a);
thus, there should never have been a single taxon found in all three areas.
Must we now abandon the dispersal hypothesis? It is possible that the
presence of the fossil taxon in areas b and c represents an independent
set of invasions into those areas from area a, and that the center of origin
of the common ancestor of all four taxa was indeed only area a. If we
accept this possibility, it appears that any distribution pattern whatsoever can be explained by dispersal if we are willing to postulate a
sufficient number of separate dispersal events; this would mean that
dispersal explanations can never be rejected and are therefore unscientific under Popper's criterion.

To prevent this untestability, we might adopt a methodological rule
that requires us to minimize the number of parallel dispersals (dispersals

from one given area to a second given area), in the same way that we seek to minimize the number of parallel acquisitions of general characters in cladograms by adopting a methodological rule that requires us to choose the most parsimonious hypothesis of relationships that will account for any given set of character distributions. Given this methodological rule, we could recognize that examples 4 and 5 require parallel dispersals from area a to area b, and possibly to area c as well, and that we must therefore abandon the dispersal explanation (1) in favor of a vicariance hypothesis (example 6):

In other words, given such data, we would have to abandon the hypothesis that area a was the center of origin of the group in favor of a hypothesis of primitive cosmopolitanism (i.e., that the common ancestor of the group occurred in all the areas in which its descendants occur today). It should be noted that the methodological rule used to render dispersal explanations testable is not one of parsimony in regard to the number of postulated dispersals (which would always lead to a vicariance explanation) but only in regard to the number of postulated parallel dispersals.

Clearly, then, a system in which we always adopt a dispersal explanation of allopatric patterns and use the discovery of cosmopolitan fossil taxa to reject those explanations is possible. But is it sufficient for our desired purpose, to distinguish all cases of vicariance from cases of dispersal? Since there may be many groups whose distributions are due to vicariance alone, but which may not be thus resolved because the relevant fossils are unavailable, the initial adoption of dispersal explanations may greatly overestimate the number of groups whose distributions are the result of dispersal.

TESTING VICARIANCE HYPOTHESES
Perhaps we should therefore choose a vicariance hypothesis as our initial explanation of allopatric distribution patterns. How might we test such a hypothesis for the same taxa and distributions considered above?

From explanation 6, we can deduce that two vicariance events occurred, one of which divided area abc into two smaller areas (a and bc) and one of which subsequently subdivided area bc into two still smaller areas (b and c). If those geological or ecological events did occur during earth history, they should have affected other organisms living in area abc. Thus, if a vicariance event did divide area abc, we should be able to find other taxa living in area a that have their closest relatives in area bc, and other taxa living in area b that have their closest relatives in c. In other words, we can test explanation 6 by converting the cladogram of taxa A, B, and C (reflecting, when interpreted phyletically, the relative recency of their common ancestry) into a cladogram of *areas* a, b, and c (reflecting the relative recency of their common ancestral biotas and the relative recency of the geological or ecological events involved). The converted cladogram thus states that areas b and c share a more recent common ancestral biota with each other than either of them do with area a, and that area bc was fragmented only after it was isolated from area a.

If we examine the interrelationships among taxa of other groups extant, or known to be extinct, in area abc and find that all of the groups (or more of them than could be expected by chance alone) correspond to the area cladogram, explanation 6 is corroborated. Since there are only three possible dichotomous cladograms for three areas, one-third of the groups examined might be expected to conform to explanation 6 by chance alone. As larger problems (involving four, five, or more areas) are examined, however, the proportion of groups that could be expected to share a given pattern of area interrelationships by chance alone decreases rapidly.

There is another aspect, however, to such larger problems. Suppose that upon subsequent investigation, we find that groups that do correspond to some general allopatric pattern do not also correspond in their higher-level relationships. Assume, for example, that we have detected a general pattern among the taxa distributed in area abc corresponding to explanation 6, but that the cladograms of some of the groups sharing the pattern relate taxa in area abc only to taxa in areas to the north (say to areas d through j, below), and that the cladograms of other groups sharing that pattern relate taxa in area abc only to taxa in areas to the south (say to areas k through r):

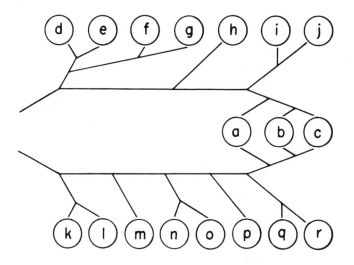

In other words, we find that two larger and mutually exclusive general patterns share some elements in a limited area (abc), where they are sympatric. Here vicariance alone is not a sufficient explanation of the pattern, and some dispersal is implied.

How we might resolve the nature of the dispersal involved will be considered below; suffice it to say here (1) that the larger pattern requires not only a single case of dispersal but at least one case of biotic dispersal, i.e., of the dispersal of several elements of a biota into the same new area, promoting cosmopolitanism of taxa, (2) that the episode of biotic dispersal can be dated as having occurred before the division of area abc but after the division of either area hij or area pqr, or both, (3) that there are events in earth history (such as the fusion of India with mainland Asia or the appearance of the Panamanian isthmus) that provide opportunities for episodes of biotic dispersal, and (4) that the vicariance model requires episodes of cosmopolitanism to have occurred in order to account for present-day sympatry among members of many different groups.

Clearly, then, a system in which we always adopt a vicariance explanation of allopatric patterns and use the absence of a general correspondence (or a more general correspondence than could be expected by chance alone) among other groups with similar distributions, or the discovery of sympatry among general patterns, to reject those explanations is also possible. But is it sufficient for our desired purpose, to distinguish all cases of vicariance from cases of dispersal? Since there may be many general allopatric patterns shared by groups whose distributions are due only to unidirectional and sequential ("stepping-stone") dispersal (like that shown in explanation 1), but which may not be thus resolved because the sequence and destinations of the dispersal events have been the same for each component group of the pattern, the initial adoption of vicariance explanations may greatly overestimate the number of groups whose distributions are the result of vicariance.

PATTERNS OF DISTRIBUTION

If it is true that neither critical testing by the dating of barriers and disjunctions, nor the initial adoption of either dispersal or vicariance explanations, provides a method sufficient in theory (much less in practice) to distinguish unambiguously instances of vicariance from instances of dispersal, what are the implications? Perhaps the question is impermeable to direct analysis because the results of both dispersal and vicariance can be expressed only in one and the same way: a pattern of taxon distributions among different areas, and because the empirical data available to us allow us to retrieve directly only the pattern and not its cause. If so, and if we do not wish to abandon the problem entirely, we must seek a way to answer the question indirectly.

If, as indicated above, a general allopatric pattern could be produced by vicariance or by sequential biotic dispersal, there is only one kind of information that can be obtained from distributional data—information about the relative recency of connections (common ancestral biotas) among different areas. If there is a general allopatric pattern corresponding to explanation 1 above, areas b and c share a more recent connection than either of them do with area a. That connection might have involved, for example, an actual land connection (and vicariance) or merely changes in the relative interdistances of areas at various times in the past (and the resulting possibilities for biotic dispersal). In either case, of course, the general pattern of area interconnections tells us something about the history of those areas. Thus, distributional data seem sufficient to resolve a pattern of interconnections among areas that reflects their history, but not to specify the nature of those interconnections.

Questions regarding earth history and the interconnections of areas are open to several tests, however, of which biotic distribution is only one. Stratigraphy, paleomagnetism, geochemistry, and other similar sources of data from historical geology contribute independent historical hypotheses. The possibility exists, therefore, of using distributional and geological data as reciprocally illuminating sources of evidence. Thus, having used biotic distribution to specify a pattern of interconnections among areas, we might, in at least some cases, be able to use data from historical geology to specify the nature of the interconnections themselves.

Take, for example, the case of the sympatric general patterns (areas a-r) shown above. In what way can the history of area abc account for the pattern of interconnections specified? There seem to be four possibilities. It is possible that dispersal occurred only within area abc; this could happen if area abc is actually a composite of two smaller areas (each belonging to one of the larger patterns) that have been joined together, and that are no longer discernible as separate areas because of an episode of biotic dispersal between the two after their merger. It is also possible that area abc is anciently a part of the southern area, and that a vicariance event resulted in a shift of area abc or a piece of northern land toward each other, permitting dispersal of parts of the northern biota into area abc (and possibly vice versa). Similarly, area abc could be anciently a part of the northern area with biotic dispersal from the south being facilitated by a vicariance event. Finally, it is possible that area abc represents new land that emerged between the northern and southern areas and was populated by biotic dispersal from both areas.

Correlations of the sequence of connections indicated by the two general biotic patterns with the sequence of connections indicated by geological data can allow us to date the episode of biotic dispersal within fairly precise limits. Hence it seems likely that geological data would permit us to resolve the question of which of the four possible geographical or ecological events was actually involved, and thereby allow us to specify the nature of the interconnections between area abc and the areas united by each general pattern.

Clearly, then, if resolution of the nature of interconnections among areas can only be accomplished through the use of independent data from historical geology, the first question we should ask when confronted with disjunct distribution patterns is not, "Is this pattern the result of vicariance or dispersal?" but "Does this pattern correspond to a general pattern of area interconnections (and thus reflect the history of those areas) or not?" What is needed is a method of analysis that will allow us to determine whether two given distribution patterns correspond to each other or not, so that we can test a hypothesis that the pattern of relationships of areas indicated by one group is a general one. After a hypothesized pattern is corroborated as general (or as more general than could be expected by chance alone), we may be able to ascribe it to vicariance or dispersal by the use of independent evidence of earth history.

DISTRIBUTION AND EARTH GEOGRAPHY

The most elementary questions of historical biogeography thus concern areas of endemism and their interrelationships. We may ask first whether there are areas of endemism, that is, whether the areas to which taxa are restricted are geographically nonrandom. This question is relatively unproblematical; no one would deny that there are areas (be they islands such as Hawaii, New Zealand, and the Galapagos, or isolated mountain ranges, such as the Himalayas or Andes, or deserts, such as the Sahara or Namib) that are each inhabited by taxa of many groups of organisms which occur nowhere else.

But then the question arises: given a list of certain areas of endemism, are the interrelationships of the endemic taxa geographically nonrandom, and, if so, what does the pattern (or patterns) of area relationships tell us about the history of the areas and the taxa which inhabit them? If the world has had only one history, after all, we might expect to find similar patterns, caused by the same events, in group after group; there should be a set of patterns of great generality. This generality can be explored with reference to a hypothetical problem involving three areas:

A   B   C

In each area we find 100 species. The 100 species of area A occur nowhere else. Of the 100 species in area B, 99 also occur in area C, leaving only one each unique to areas B and C:

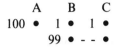

The question is: how are the three areas interrelated? Does the matter seem problematical? We hope so, and we hope you ask: "Well, what are the interrelationships of the species that live in the areas?"—a crucial question. Here is one possible answer. There are three species interrelated thus:

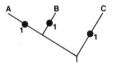

And there are 198 species interrelated thus:

Now, then: is the matter still problematical? Does the problem seem familiar? What, then, are the interrelationships of the areas? There are two possibilities to consider:

It is no disgrace to choose possibility (2). Those who do will find themselves in the distinguished company of most of the biogeographers who have lived and worked during the past 200 years. We hope, however, to demonstrate in this book that to do so is a mistake. But to what kind of larger problem is the decision between these two possibilities relevant?

Current studies indicate that the oldest rocks surviving in existing ocean basins date back only to the Jurassic age, about 150 million years ago. According to the concepts of plate tectonics already popularized, since that time the Atlantic Ocean has opened, separating the New World from Europe and Africa. During the process, India drifted north from a position alongside Africa and collided with Asia. Australia separated from Antarctica and drifted north to its present location. If we start with the present configuration, and put the time machine in reverse, we would observe the Atlantic to become smaller and eventually disappear; India, and perhaps other lands bordering the Indian Ocean to the east, to come together, in the western Indian Ocean, with Madagascar; Australia to move southward, eventually attaching to Antarctica.

This concept, therefore, leaves to the continents their identities pretty much as we know them today, India and its environs being the notable exception. And what of the Pacific Ocean? According to this concept, the Pacific has always been there. Indeed, it was larger, and has grown smaller through time. But where is the old Pacific seafloor—the pre-Jurassic seafloor? According to the usual idea, the old seafloor has been subducted. Gone, forever and ever.

Why is all this relevant, if plate tectonics leaves the continents with their modern identities? For about 100 years, biogeographers have had a certain concept of pattern, very stable—at least among zoologists. The pattern is that specified by the concepts of biogeographical regions: the Nearctic, Palaearctic, Oriental, Neotropical, Ethiopian, and Australian. These six classic regions were developed by Sclater and Wallace in the last century on the basis of bird and mammal distributions, respectively. As formulated by them, the regions conform, in the main, to modern continental geography. The Nearctic Region, for example, is basically North America; the Neotropical, basically South America, and so on. The basis for the concept of these regions, as developed by Sclater and Wallace, may conform to possibility (2) of our problem: the regions may have been based on the distributions of widespread taxa. With few exceptions over the years, the arguments of Sclater and Wallace were convincing to other biogeographers—Croizat being the notable exception in recent times. He viewed matters differently, arguing that biogeographical regions for terrestrial organisms do not correspond to present continental geography, but rather to present ocean basins. What does this mean? It means, for example, that organisms in North America are of two sorts: those with relatives in Europe and those with relatives in Asia. It means that North America is not an integral thing, biogeographically. It means that North America is an amalgam of parts of two worlds: an Atlantic and a Pacific. It means that the Nearctic Region is not otherwise a part of the real world.

Now, this is not a book on geophysics. We don't have the answers for the Pacific and its historical development. But we do have a question: if the New World and the Old have separated and drifted apart in the Atlantic sector; India and Madagascar in the Indian; and Australia and Antarctica down south, what mass of continental material has been separated in the Pacific? The tentative answer is: quite a bit. Much of western North America, South America, and other bits and pieces of continental structure bordering the entire Pacific basin. The tentative answer is that there was one or more Pacific landmass that fragmented, and that the fragments spread apart and ultimately attached themselves to the continental masses around the present Pacific basin.

What, then, would this mean for the classical concept of regions? That they all, probably, are hybrids; that they are all mixtures. If so, the big question is: to what extent can the mixtures be sorted out? This is a problem of pattern analysis par excellence. It is a problem to be solved by analysis along the lines of possibility (1) above, involving the interrelationships of narrowly, rather than widely, distributed taxa. It is, admittedly, a problem for the future. But all indications are that it will work out pretty well.

So the sections below on biogeography, like the rest of the book, pose problems of pattern analysis. Our purpose is to suggest that such problems are worthy of consideration, in themselves and in relation to process, too. In themselves, the problems are not difficult to understand. What is difficult is to form a judgment about them. We present our own judgments, but we know enough of history, science, and human nature in general not to expect ready assent, or even to particularly value it. But we do not pose these problems of pattern analysis lightly, for they are fundamental to much of modern biology.

FORM

2

SYSTEMATIC HISTORY:
KINDS OF BRANCHING DIAGRAMS

NATURAL KINDS

The historical origin of systematics lies in the beginnings, if not earlier, of language, knowledge, and thought; in short, in the beginnings of humanity. Being an old endeavor, its history is not capable of being known and understood in its entirety or with exactitude. All one can hope for, even after a lifetime of specialized study, is to have some idea of its history and to have reached some judgment of its relevance.

Theophrastus, born in Greece in 382 B.C., was a student of both Plato and Aristotle. He was also an avid student of plants, and he attempted to summarize what was known about plants at that time. His writings reveal a concern with systematic problems and classification. His consideration of ivy, what he called *kittos*, is an example (Hort 1916):

The ivy also has many forms; one kind grows on the ground [*epigeios*], another grows tall [*hyphos airomenos*], and of the tall-growing ivies there are several kinds. However the three most important seem to be the white [*leukos*], the black [*melas*], and the helical [*helix*]. And of each of these there are several forms. Of the white, one is white only in its fruit [*leukokarpos*], another in its leaves also [*leukophyllos*]. Again to take only white-fruited sorts, one of these has its fruit well formed close and compact like a ball; and this kind some call *korymbia*, but the Athenians call it the "Acharnian" ivy. Another kind is smaller and loose in growth like the black ivy. There are also variations in the black kind, but they are not so well marked.

The helical present the greatest differences; the principal difference is in the leaves, which are small, angular and of more graceful proportions, while those of the ivy proper are rounder and simple; there is also difference in the length of the twigs, and further in the fact that this tree is barren. For, as to the view that the helical by natural development turn into the ivy, some insist that this is not so, the only true ivy according to these being that which was ivy from the first; (whereas if, as some say, the helical invariably turns into ivy, the difference would be merely one of age and condition, and not of kind, like the difference between the cultivated and the wild pear). However the leaf even of the full-grown helical is very different from that of the ivy, and it happens but rarely and in a few specimens that in this plant a change in the leaf occurs as it grows older, as it does in the abele and the castor-oil plant. There are several forms of the helical, of which the three most conspicuous and important are the green "herbaceous" kind [*poodes*] (which is the commonest), the white [*leuke*], and the variegated [*poikile*], which some call the "Thracian" *helix*. (pp. 273–75, slightly altered)

Theophrastus' understanding of the variability of ivy was portrayed in a diagram (figure 2.1) by a recent commentator. His purpose was to show the structure of Theophrastus' classification. So far as is known, Theophrastus himself never drew diagrams of the sort; but he probably would have recognized the diagram as a true representation of his classification of ivy and its variability.

Theophrastus' account includes several problematical items of interest. A modern student might wonder, for example, whether all of the various

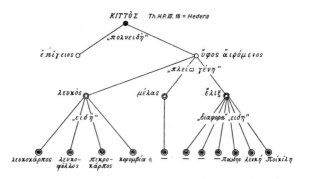

Figure 2.1. A branching diagram representing Theophrastus' concept of the kinds and interrelationships of ivy (his *kittos*), of which there are two basic sorts, procumbent and erect (*epigeios* and *hyphos airomenos*); of the latter there are three sorts, white, black, and helical (*leukos, melas,* and *helix*); of the white there are four sorts, white-fruited, white-leaved, pointed-fruited, and that called *korymbia* (*leukokarpos, leukophyllos, pikrokarpos,* and *korymbia*); of the helical, there are six sorts, of which three are named, herbaceous, white, and variegated (*poodes, leuke,* and *poikile*). After C. Váczy (1971), Les origines et les principes du développement de la nomenclature binaire en botanique. Taxon 20:573–90; figure 2, p. 583.

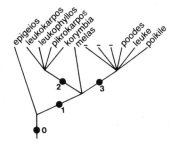

Figure 2.2. The structural components of Theophrastus' classification of ivy (cf. figure 2.1). 0, *kittos*; 1, *hyphos airomenos*; 2, *leukos*; 3, *helix*.

kinds of ivy are distinct "species," according to one or another definition of that term; or one might be curious about the structure of the classification. Indeed, one may ask whether the structure is merely a logical one or, alternatively, whether it has some real existence that is independent of the thoughts of Theophrastus.

The structural components of the classification are four (numbered 0–3 in figure 2.2). Each component may be conceived as representing a group of kinds (or species) of ivy. Component 0 represents the entire spectrum of kinds; component 1, only the erect (*hyphos airomenos*); component 2, the white (*leukos*); component 3, the helical (*helix*).

If the structure of the classification is specified by its components (0–3), then the question about the nature of its existence may be rephrased: do the components have some real existence that is independent of the thoughts of Theophrastus? Or, in other words, did Theophrastus acquire some real knowledge of the kinds and inter-relationships of ivy *in addition* to his observations, or information, that some ivy is procumbent; some other is upright, of which some is white, and some other black, and so on?

It is difficult to imagine how Theophrastus might have answered questions either about "species" or about his classification of ivy. There is little evidence that such questions occurred to him in connection with his investigations of ivy or, for that matter, any other plants. He did assert, however, that "since our study becomes more illuminating if we distinguish different kinds, it is well to follow this plan where it is possible" (Hort 1916:23). But he did not comment in detail as to what could be illuminated by distinguishing different kinds, or what makes such classification possible or impossible. These matters had already been discussed, to some degree, by his teachers, Plato and Aristotle, and one may assume that Theophrastus had little or nothing of his own to add to their teachings.

Plato did not directly consider organisms and their classification, but he did consider the problem of "natural kinds" that have real existence. His chief concern was to explain their existence by means of his theory of ideas, or forms, of which knowledge is innate. Plato's contribution is, thus, on a philosophical, not methodological, level. He explained neither a method by which natural kinds might be discovered; nor a method by which kinds once thought natural might be discovered, in truth, to be artificial. His thoughts, however, are among the earliest on the subject of natural kinds—a subject of interest persisting into modern times among philosophers (Quine 1969).

As analyzed by a modern commentator (Moravcsik 1973), Plato's scheme of grouping allows for four classes of groups:

Class 1: groups that are wholes, have parts, but are not themselves parts.
Class 2: groups that are wholes, have parts, and are themselves parts.
Class 3: groups that are wholes, have no parts, but are themselves parts.
Class 4: groups that are not wholes and, therefore, have no parts.*

With Theophrastus' conception of the different kinds of ivy as an example (figures 2.1, 2.2), the members of the various classes may be specified:

Class 1: component 0 (*kittos*).
Class 2: components 1, 2, and 3 (*hyphos airomenos, leukos, helix*).
Class 3: *epigeios, leukokarpos,* etc.
Class 4: none.

If *kittos* is a whole, then it has two parts (*epigeios* and *hyphos*

---

*Class 3 might be better defined as: groups that are not wholes, have no parts, but are themselves (whole) parts of wholes. And class 4 as: groups that are not wholes, have no parts, and are not themselves (whole) parts of wholes. Thus, *kittos* is a whole with parts; components 1, 2, and 3 are both wholes (with parts) and (whole) parts of wholes; and *epigeios, leukokarpos,* etc., are only parts (but they, too, are whole parts) of wholes. A group composed only of *epigeios* and *leukokarpos* is neither a whole (with parts) nor a (whole) part of a whole; *epigeios* and *leukokarpos* are not parts of a whole other than *kittos.*

airomenos); and if *hyphos airomenos* is a whole, then it has three parts (*leukos, melas,* and *helix*); also, if *leukos* is a whole, then it has four parts (*leukokarpos, leukophyllos, pikrokarpos,* and *korymbia*); similarly, if *helix* is a whole, then it has six parts (three unnamed and *poodes, leuke,* and *poikile*); finally, if *epigeios, leukokarpos,* etc., are wholes, then they have no parts, or at most no parts that are named. Given the initial concept, the part-whole relationships follow directly and obviously (figures 2.1, 2.2).

Consider a group (or component) composed only of *epigeios* and *leukokarpos.* Is it a whole? Does it have parts? Relative to Theophrastus' concept (figures 2.1, 2.2), *epigeios* may be seen to be a part of component 0; *leukokarpos,* a part of component 2, which is a part of component 1, which itself is a part of component 0. The only whole that could be said to have *epigeios* and *leukokarpos* as parts is component 0. Put another way, *epigeios* and *leukokarpos* by themselves do not form a group that is a whole with parts. Together they would form a group that is a member of class 4. For Plato, groups of classes 1–3 are natural and real, and are therefore worth naming (because, in theory, they correspond to innate knowledge). Groups of class 4, in contrast, are unnatural and unreal, and are therefore not worth naming (because, in theory, they do not correspond to innate knowledge).

Indeed, Plato argued, to name a group of class 4 is simply to make a mistake:

the kind of mistake a man would make who, seeking to divide the class of human beings into two, divided them into Greeks and barbarians. This is a division most people in this part of the world make. They separate the Greeks from all other nations making them a class apart; thus they group all other nations together as a class, ignoring the fact that it is an indeterminate class made up of peoples who have no intercourse with each other and speak different languages. Lumping all this non-Greek residue together, they think it must constitute one real class because they have a common name "barbarian" to attach to it. Take another example. A man might think that he was dividing number into its true classes if he cut off the number ten thousand from all others and set it apart as one class. He might go on to invent a single name for the whole of the rest of number, and then claim that because it possessed the invented common name it was in fact the other true class of number—"number other than ten thousand." Surely it would be better and closer to the real structure of the forms to make a central division of number into odd and even or of humankind into male and female. A division setting Lydians or Phrygians or any other peoples in contradistinction to all the rest can only be made when a man fails to arrive at a true division into two groups each of which after separation is not only a portion of the whole class to be divided but also a real subdivision of it. (Skemp 1952; *Statesman* 262d)

How might one decide that a group, such as Theophrastus' *helix* is a true division or real group—a whole? Plato did not say:

Plato gives no mechanical procedure for finding natural kinds. Plato does not think that there are any such procedures. He is not giving a discovery procedure, he is explicating the . . . configurations that obtain once we have discovered natural kinds. He does not tell us *how* to arrive at them; he tells us what things look like *when* we have arrived at them. (Moravcsik 1973:344)

Aristotle was more empirically oriented than Plato, more interested in the problem of defining natural groups, and more active in investigating the world of plants and, especially, animals. He, too, was concerned with causal explanation of natural kinds, but for him the explanation lies not in the notion of form or idea (in Plato's sense) but in the notions of essence and, ultimately, of final cause—that purpose that an essence might be said to serve.

For Aristotle, in short, a group is a real one if (and because) it has an essence and final cause unique to itself. That possibility can be evaluated through empirical investigation of the organisms. Through investigation, essences may be discovered and defined, at least to some degree, and final causes reasoned out.

Despite his extensive studies of animals, Aristotle never made a formal classification of them. Also, he seems never to have broken entirely from the notion of innate knowledge of natural kinds. He wrote, for example (Ogle 1912; Book I, 643b): "The method then that we must adopt is to attempt to recognize the natural groups, following the indications afforded by the instincts of mankind, which led them for instance to form the class of Birds and the class of Fishes. . . ." Aristotle recognized several major groups of organisms, and later commentators arranged them in a series in accordance with Aristotle's notions of their honor, excellence, or nobility (Ogle 1882):

Man
Viviparous quadrupeds (mammals)
Cetacea (whales)
Birds

Scaly quadrupeds (reptiles and amphibians)
Fishes
Malacia (cephalopod molluscs)
Malacostraca (crustaceans)
Insects
Ostracoderma (molluscs excepting cephalopods)
Zoophytes
Plants
Inanimate nature

Aristotle's series may be portrayed also as a branching diagram that includes 12 structural components (figure 2.3:0–11). So far as is known, Aristotle never drew diagrams of the sort (nor even set out a formal classification), but he probably would have recognized the diagram, and the series upon which it is based, as a true representation of his views.

Aristotle was not altogether clear that the groups designated as components are to be understood as having essences; and if they are, what the essences might be. But he did offer some characters that might be considered defining characters. Living nature (component 1), for example, has at least a nutritive faculty. Animals (component 2) have in addition sensitive and appetitive faculties. Higher animals (component 3) have in addition a locomotory faculty; and so on. Man, of course, has all of these faculties as well as an intellectual faculty unique to himself (Ogle 1882).

Aristotle held that each true species has an essence, and that its

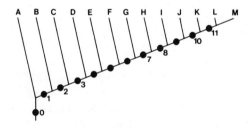

Figure 2.3. The structural components implied by Aristotle's series of natural kinds. A, inanimate nature; B, plants; C, zoophytes; D, Ostracoderma; E, insects; F, Malacostraca; G, Malacia; H, fishes; I, oviparous quadrupeds; J, birds; K, Cetacea; L, viviparous quadrupeds; M, Man. 1, living organisms; 2, animals; 3, higher animals; 7, animals with blood; 8, animals with lungs; 10, viviparous animals.

essence is not to be defined by a single character: "No single differentia [character], I repeat, either by itself or with its antecedents, can possibly express the essence of a species" (Ogle 1912; Book I, 644a). Thus, if the groups (components) of the classification are real, and have essences in his sense, they could not adequately be defined by single characters. Nor could a judgment be made that a particular group is real unless its essence was already known with reasonable completeness and certainty. Such would require additional defining characters and some understanding of final causes. Presumably, then, any group could be judged real if it was defined by characters numerous enough to suggest a (whole) essence consonant with an intelligible final cause, or purpose, in the order, working, and economy of nature.

Because Aristotle referred to groups such as "bloodless animals" (Anaima), some commentators have suggested that a different sort of classification of animals could be attributed to Aristotle:

Enaima (animals with blood)
  Viviparous: Man
                  Cetacea
                  Quadrupeds
  Oviparous:
    Perfect eggs: Birds
                      Scaly Quadrupeds
    Imperfect eggs: Fishes
Anaima (bloodless animals)
  Perfect eggs: Malacia
                 Malacostraca
  Special eggs: Insects
  Generative slime: Ostracoderma
  Spontaneous generation: Zoophytes

This classification can also be represented by a branching diagram (figure 2.4). Comparison of figures 2.3 and 2.4 reveals some differences in the structural components. One may ask whether, if components 2–11 are real, components 12–15 are also real in Aristotle's sense, each with an essence unique to itself? Aristotle's commentators are not clear on this point.

Aristotle was clear, however, that a group like the Anaima (equivalent to the modern "Invertebrata") is not adequately defined by a single

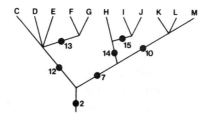

Figure 2.4. The structural components implied by the classification given by various commentators on Aristotle. 2, animals; 7, animals with blood (*Enaima*); 10, viviparous animals with blood; 12, animals without blood (*Anaima*); 13, bloodless animals with perfect eggs; 14, oviparous animals with blood; 15, oviparous animals with perfect eggs (cf. figure 2.3).

character (see above), and that a character such as "bloodless" is not even an adequate character. He considered negative characters, such as "bloodless," in a discussion of what he called "dichotomous division" (Ogle 1912):

> Privative [negative] terms inevitably form one branch of dichotomous division. . . . But privative terms in their character of privatives admit of no subdivision. For there can be no specific forms of a negation, of Featherless for instance or of Footless, as there are of Feathered and of Footed. . . . From this it follows that a privative term, being insusceptible of differentiation, cannot be a generic differentia [group character]. . . . (Book I, 642b)

In justification of the division of animals into bloodless (Anaima) and blooded (Enaima), one commentator wrote (Ogle 1882):

> These form two great groups, those that have blood, and those whose nutritive fluid is not true blood but something analogous to it; a division which coincides with the modern one, introduced by Lamarck, into Vertebrata and Invertebrata. To this division of animals, into those with blood and those without, it is objected that the one group has but a negative character. The objection is drawn from Aristotle's own quiver, and is equally fatal to Lamarck's Invertebrata. Aristotle's division may, however, be so expressed as to avoid this criticism. Animals whose nutritive fluid is red, and animals whose nutritive fluid is white or colourless. (p. xxvi)

Of the classifications that might be attributed to Aristotle (figures 2.3, 2.4), certain structural components conflict (e.g., components 3 and 12). How Aristotle himself might have resolved the conflict is, fortunately or not, a matter of interpretation, which varies among his many commentators (Balme 1962).

Two possibilities for resolution may be considered here. Of two conflicting components, one or both might be false (or unreal) or, alternatively, both might be true (real). The first possibility implies that there might exist a single system of natural kinds; the second possibility implies two or more systems of natural kinds. Resolution, therefore, amounts to a judgment that nature, with respect to its natural kinds, is either simple or complex.

Aristotle's series of natural kinds, from inanimate matter to man, reflects a progression, from the simple to the complex, with respect to Aristotle's notion of life-force (*psyche* or soul). In this progression, some commentators see the notion of evolution, and they see Aristotle as an evolutionist (e.g., Nordenskiöld 1928). Others see the progression as an idea that became incorporated into human thought of later ages. Lovejoy (1936) termed this idea of progression, with its many implications, "the great chain of being," and considered it an important organizing force of subsequent intellectual history. Quite possibly the majority of present-day humanity still views the natural world as some kind of "great chain."

One reason for the "success" of the idea of progression was its adaptation to the spiritual and religious world, particularly as worked out by Plotinus (A.D. 205–270). According to him, existence exemplifies certain "natural kinds" of a sort more abstract than the plants and animals of living nature (Parker 1967):

The One
Reason
Soul
Body
Matter

Compared with Aristotle's, Plotinus' series is abbreviated, and an additional item (The One) has been added to the top. For Plotinus, the idea of progression through the series allows the possibility that humans, through diligent effort, may progress upward toward The One (God). It is possible to view Plotinus' series as a branching diagram with four structural components (figure 2.5), and to inquire whether the components specify natural kinds with real existence. These and related matters were discussed at great length during the Middle Ages. In the meantime, study of organisms languished.

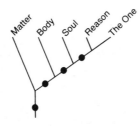

Figure 2.5. The four structural components implied by Plotinus' series (see text).

One of the favorite diagrams of the Middle Ages is the Tree of Porphyry, named after Plotinus' principal disciple and popularizer (figure 2.6). The Tree is said to have been present "in all the old logics" (Baldwin 1911:714). The sense of it is, perhaps, most easily grasped when it is rendered as a branching diagram with its six structural components (figure 2.7). Some of the components are equivalent to Aristotle's (figure 2.3; Corporea = 0, Animatum = 1, Sensibile = 2). There is a new component (Rationale) that includes the immortals. As for the mortals, only two are mentioned ("Sortes" is said to be a corruption of "Socrates"). The diagram is interesting, for it conforms in its main features with Aristotle's notion of "dichotomous division," based on single characters, present or absent.

The Middle Ages waned and passed away with the discovery of the classical Greek and Roman cultures, and with the renewed interest in the natural world, which was soon found to be more complex than the divisions of the Tree of Porphyry. One result was a new generation of branching diagrams, which were called "tables."

John Wilkins (1668) published an early (but not the first) series of such tables (an earlier series was published by Fredericus Caesius in 1651). Wilkins attempted to show the nature and relationships of all things with names, which he attributed to 40 "genuses" (numbered I-XL in figure 2.8) and classified in a scheme with 23 structural components (figure 2.9). Of particular interest is his treatment of organisms (genera VIII-XVIII, figure 2.10). Except for the novel inclusion of certain stones and metals with plants, as "vegetative" (component 9) because all exhibit a vegetative faculty (growth), most of Wilkins' genera and components are those of Theophrastus for plants (herbs [component

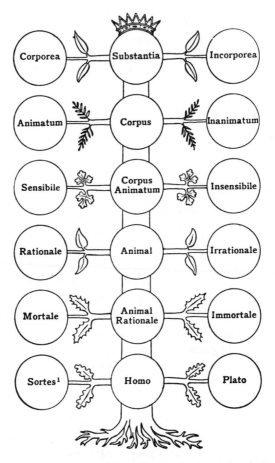

Figure 2.6. The Tree of Porphyry (*Baum des Porphyrius*; *Arbre de Porphyre*; *Scala Ternaria di Porferio*). After Baldwin (1911), figure on p. 714.

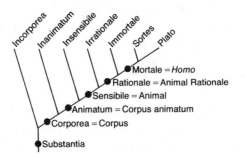

Figure 2.7. The six structural components of the Tree of Porphyry (cf. figure 2.6).

The General Scheme.

All kinds of things and notions, to which names are to be assigned, may be distributed into such as are either more

```
┌ General ; namely those Universal notions, whether belonging more properly to
│                                                    ( GENERAL.  I
│   ( Things ; called TRANSCENDENTAL ⟨ RELATION MIXED.  II
│   {                                                ( RELATION OF ACTION.  III
│   ( Words ; DISCOURSE.  IV
· Special ; denoting either
  ( CREATOR.  V
  ⟨ Creature ; namely such things as were either created or concreated by God, not
    excluding several of those notions, which are framed by the minds of men,
    confidered either
    ( Collectively ; WORLD.  VI
    ⟨ Diftributively ; according to the several kinds of Beings, whether such as do
       ┌ Substance ;                                                ( belong to
       │ ┌ Inanimate ; ELEMENT.  VII
       │ ⟨ Animate ; confidered according to their several
       │ └ ┌ Species ; whether
       │     │ ┌ Vegetative
       │     │ │                              ( STONE.  VIII
       │     │ │ ( Imperfect ; as Minerals, ⟨ METAL.  IX      ( LEAF.  X
       │     │ │ {                  ( HERB confid. accord. to the ⟨ FLOWER.  XI
       │     │ │ ( Perfect ;as Plant, ⟨ SHRUB.  XIII           ( SEED-VESSEL. XII
       │     │ │                    ( TREE.  XIV
       │     │ │             ( EXANGUIOUS.  XV
       │     │ ⟨ Senfitive ;⟨           ( FISH.  XVI
       │     │ │   ( Sanguineous ; ⟨ BIRD.  XVII
       │     │ │   Parts ;( PECULIAR.  XIX ( BEAST.  XVIII
       │     └     ( GENERAL.  XX
       · Accident ;
         ┌                        ( MAGNITUDE.  XXI
         │       Quantity ;⟨ SPACE.  XXII
         │                        ( MEASURE.  XXIII
         │                        ( NATURAL POWER.  XXIV
         │                        │ HABIT.  XXV
         │       Quality ; whether⟨ MANNERS.  XXVI
         │                        │ SENSIBLE QUALITY.  XXVII
         │                        ( SICKNESS.  XXVIII
         │            ( SPIRITUAL.  XXIX
         ⟨       Action⟨ CORPOREAL.  XXX
         │            │ MOTION.  XXXI
         │            ( OPERATION.  XXXII
         │                                        ( OECONOMICAL.  XXXIII
         │                          ( Private.    ⟨ POSSESSIONS.  XXXIV
         │ Relation ; whether more ⟨              ( PROVISIONS.  XXXV
         │                          {              ( CIVIL.  XXXVI.
         │                          (              │ JUDICIAL.  XXXVII
         ·                          ( Publick.    ⟨ MILITARY.  XXXVIII
                                                   │ NAVAL.  XXXIX
                                                   ( ECCLESIASTICAL.  XL.
```

Figure 2.8. A classification of "all kinds of things and notions." After Wilkins (1668), table on p. 23.

12], shrubs [genus XIII], and trees [genus XIV]) and those of Aristotle for animals (Anaima [genus XV] and Enaima [component 14]).

Wilkins engaged two naturalists, John Ray and Francis Willughby, to prepare his tables of plants and animals, respectively. Wilkins stipulated that the tables be organized in groups of three, each with three subdivisions, and, where possible, in pairs of contrasting elements, leading to groups of nine. Willughby's table of the major groups of fishes (figure 2.11) illustrates some of these points. He divides the fishes into

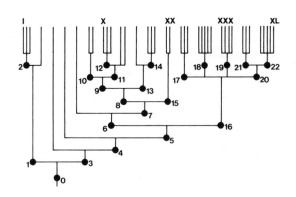

Figure 2.9. The structural components of Wilkins' classification. I–XL, Wilkins' 40 genera. Components: 0, All things and notions; 1, General; 2, Things; 3, Special; 4, Creature; 5, Distributively; 6, Substance; 7, Animate; 8, Species; 9, Vegetative; 10, Imperfect; 11, Perfect; 12, Herb; 13, Sensitive; 14, Sanguineous; 15, Parts; 16, Accident; 17, Quantity; 18, Quality; 19, Action; 20, Relation; 21, Private; 22, Public.

viviparous and oviparous, oblong and flat, saltwater and freshwater; and his divisions result in nine major groups (I–IX).

Ray, who was Willughby's teacher as well as friend and co-worker, commented that this organization results in an artificial rather than natural classification (Raven 1942):

> I was constrained in arranging the Tables not to follow the lead of nature, but to accommodate the plants to the author's prescribed system. This demanded that I should divide herbs into three squadrons or kinds as nearly equal as possible; then that I should split up each squadron into nine "differences" as he called them, that is subordinate kinds, in such wise that the plants ordered under

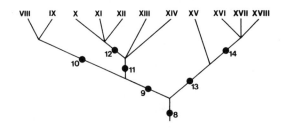

Figure 2.10. Wilkins' classification of Species (component 8). Genera: VIII, Stone; IX, Metal; X, Leaf; XI, Flower; XII, Seed-vessel; XIII, Shrub; XIV, Tree; XV, Exanguious; XVI, Fish; XVII, Bird; XVIII, Beast. Components: as in figure 2.9.

Of Fiſh.

FISH may be diſtributed into ſuch as are
Viviparous; and ſkinned; whoſe figure is either
　{OBLONG and roundiſh. I.
　{FLAT or thick. II.
Oviparous; whether ſuch as do generally belong to
　Salt water; to be further diſtinguiſhed by their
　　Finns on the back; whether ſuch, the *rays* of whoſe *finns* are
　　　{*Wholly ſoft* and flexile. III.
　　　{*Partly ſoft*, and partly *ſpinous*; having
　　　　{TWO FINNS on the back. IV.
　　　　{But ONE FINN. V.
　　Figure; whether
　　　{OBLONG. VI.
　　　{FLAT. VII.
　　CRUSTACEOUS COVERING. VIII.
Freſh water; being ſcaly. IX.

Figure 2.11. Francis Willughby's table of fishes. After Wilkins (1668), table on p. 132.

each "difference" should not exceed a fixed number; finally that I should join pairs of plants together or arrange them in couples.

What possible hope was there that a method of that sort would be satisfactory, and not manifestly imperfect and ridiculous? I frankly and openly admit that it was; for I care for truth more than for my own reputation. (letter to Lister, 7 May 1669; p. 182)

Ray's concern with natural, as opposed to artificial, groups has, of course, an old historical background extending back at least to classical Greece. In his history of botany, Sachs, for example, commented (1890):

Systematic botany, as it began to develope in the 17th century, contained within itself from the first two opposing elements; on the one hand the fact of a natural affinity, . . . and on the other the desire . . . of arriving . . . at a classification of the vegetable kingdom which should satisfy the understanding. These two elements of systematic investigation were entirely incommensurable. . . . This incommensurability between natural affinity and *a priori* grounds of classification is everywhere expressed in the systems embracing the whole vegetable kingdom. . . . (p. 7)

It is not surprising, therefore, that Ray's tables of plants were soon criticized by a fellow botanist, Robert Morison, who later published various tables and diagrams of his own (1672). In a plant genus that he termed "Umbellae Semine rostrato," Morison included nine species (figure 2.12). The names that he applied to the species are interesting, for

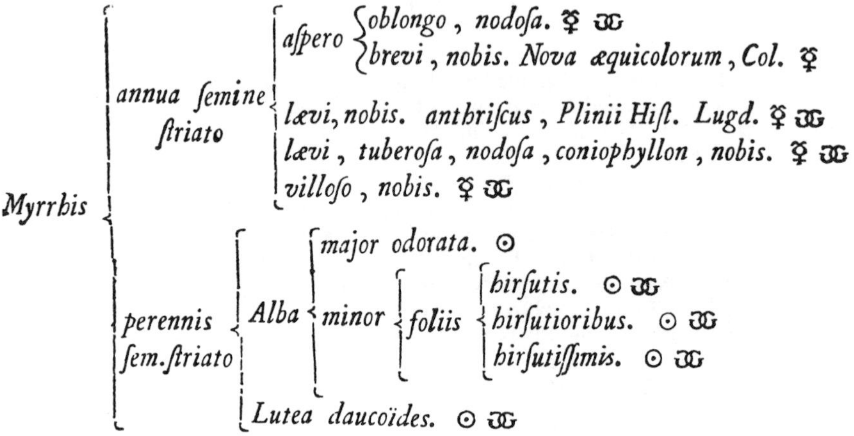

Myrrhis
- annua femine ſtriato
  - aſpero
    - oblongo, nodoſa. ☿ ♎
    - brevi, nobis. Nova æquicolorum, Col. ☿
  - lævi, nobis. anthriſcus, Plinii Hiſt. Lugd. ☿ ♎
  - lævi, tuberoſa, nodoſa, coniophyllon, nobis. ☿ ♎
  - villoſo, nobis. ☿ ♎
- perennis ſem.ſtriato
  - Alba
    - major odorata. ⊙
    - minor
      - foliis
        - hirſutis. ⊙ ♎
        - hirſutioribus. ⊙ ♎
        - hirſutiſſimis. ⊙ ♎
  - Lutea daucoïdes. ⊙ ♎

Figure 2.12. Robert Morison's table of species of some umbelliferous plants. After Morison (1672), table on p. 44.

they exemplify the practice, common at the time, of including numerous descriptive terms—all of the terms, in fact, that appear as items in his tables:

Species 1: Myrrhis perennis semine striato, alba, major, odorata.
    A [English]. *Sweet Chervill, or Cicely.* G. [French]. Myrrhis ou cicutaire de *Matthiol.*
Species 2: Myrrhis perennis sem. striato, alba, minor, foliis hirsutis.
    A. *Lesser montaine Chervill, or Cicely with hairie leaves.*
    G. Petit Myrrhis des montaignes aux fueilles veluës.
Species 3: Etc.

Morison's diagrams (figure 2.13) are of a unique style similar to the Tree of Porphyry, but differing in omitting purely negative characters, and in having a complex branching pattern that accommodates more than two terminal items. In addition to the nine species listed in his table (figure 2.12), Morison includes three others in his genus VIII, under the heading "Umbellae Semine rostrato." He includes also two species of a related "Genus IX"—"Umbellae Semine rotundo, seu testiculato." That Morison's figures should resemble the Tree of Porphyry is not remarkable, for similar figures have been published even in recent times (figures 2.14–2.15).

The tabular style of diagram was used extensively by Ray in his later

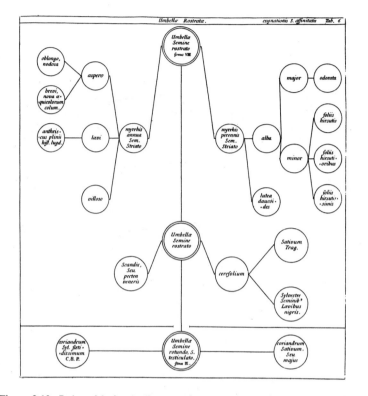

Figure 2.13. Robert Morison's diagram of genera and species of some umbelliferous plants. After Morison (1672), table [plate] 6.

publications (e.g., 1682) and by other naturalists, who were probably influenced in this respect by him. His notion of artificiality in classification clearly stemmed not from the tabular arrangement itself, but from the notions of Wilkins who "was fascinated by the harmonies of nature and a mystic sense of the significance of numbers; and thought that his trinitarian arrangement, three squadrons and three times three differences, corresponded to the symmetry inherent in the nature of things" (Raven 1942: 183).

Of the many tabular arrangements of plants of these early times, one of the more interesting is that of Carl Linnaeus, published in 1735 in the first edition of his *Systema Naturae* (figure 2.16). Some years later (1783) it was translated from Latin into English (figure 2.17). Linnaeus

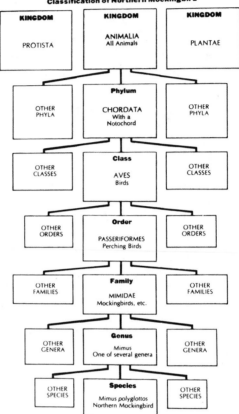

Figure 2.14. A "classification of Northern Mockingbird." After L. L. Short (1975), *Birds of the World* (New York: Bantam Books), figure on p. 16.

entitled his table "Key of the Sexual System" and designed it for the purpose of enabling a person to identify specimens of plant species with which he is unfamiliar, provided that the plant is in blossom. The system features 24 classes (I–XXIV), each divided into one or more orders. In all, the system includes 98 orders, grouped by means of 32 structural components, of which 24 are designated classes (figure 2.18). Although Linnaeus' *Systema* also included accounts of stones (3 orders, 11 genera), minerals (3 orders, 18 genera), fossils (3 orders, 21 genera), and animals (figure 2.19; 6 classes, 25 orders, and 200 genera), his emphasis was on plants, particularly their flowers, which he used as a means for identification of his 24 classes, 98 orders, and some 800 genera of plants. Later editions of his *Systema Naturae* became notable for their use of binomial nomenclature and their extensive lists of species, but the first edition is not binomial in its nomenclature, and for plants it lists only genera.

Linnaeus' sexual system of plants excited interest partly because its language presented plants from a new perspective which reflected his belief, inspired by the recent discovery of sexuality in plants, that the reproductive parts of the flower correspond to human genitalia. In rendering the various aphorisms of Linnaeus' *Fundamenta Botanica* (1736) and *Philosophia Botanica* (1751), Stafleu notes:

The flower precedes the fruit as generation precedes delivery ([aphorism] 141). The subsequent aphorisms contain a detailed description of the structure of the anthers and the stigma and a discussion of the actual proof that pollen fertilizes

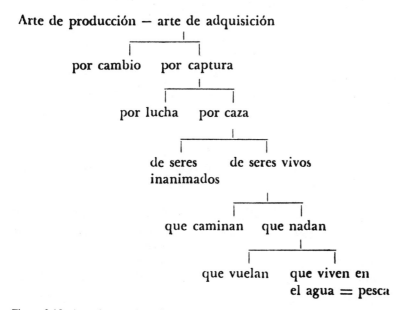

Arte del pescador de caña

Arte de producción — arte de adquisición

por cambio   por captura

por lucha   por caza

de seres   de seres vivos
inanimados

que caminan   que nadan

que vuelan   que viven en
el agua = pesca

Figure 2.15. A modern analog of the Tree of Porphyry. After O. A. Ghirardi (1972), Tiempo y Evolucion (Cordoba: Universidad Nacional de Cordoba), figure on p. 15.

REGNUM VEGETABILE
CLAVIS SYSTEMATIS SEXUALIS
NUPTIAE PLANTARUM.
Actus generationis incolarum Regni vegetabilis.
Florescentia.
PUBLICÆ.
Nuptiae, omnibus manifestae, aperte celebrantur.
Flores unicuique visibiles.
MONOCLINIA.
Mariti & uxores uno eodemque thalamo gaudent.
Flores omnes hermaphroditi: stamina cum pistillis in eodem flore.
DIFFINITAS.
Mariti inter se non cognati.
Stamina nulla sua parte connata inter se sunt.
INDIFFERENTISMUS.
Mariti absque subordinatione.
Stamina longitudine indeterminata.

1. MONANDRIA.	7. HEPTANDRIA.
2. DIANDRIA.	8. OCTANDRIA.
3. TRIANDRIA.	9. ENNEANDRIA.
4. TETRANDRIA.	10. DECANDRIA.
5. PENTANDRIA.	11. DODECANDRIA.
6. HEXANDRIA.	12. ICOSANDRIA.
	13. POLYANDRIA.

SUBORDINATIO.
Mariti certi reliquis praeferuntur.
Stamina duo semper reliquis breviora sunt.

14. DIDYNAMIA.	15. TETRADYNAMIA.

AFFINITAS.
Mariti propinqui & cognati sunt.
Stamina coherent vel inter se, vel cum pistillo.

16. MONADELPHIA.	19. SYNGENESIA.
17. DIADELPHIA.	20. GYNANDRIA.
18. POLYADELPHIA.	

DICLINIA (a δίς bis & κλίνη thalamus) duplex thalamus.
Mariti & Feminae distinctis thalamis gaudent.
Flores masculi & feminei in eadem specie.

21. MONOECIA.	23. POLYGAMIA.
22. DIOECIA.	

CLANDESTINAE.
Nuptiae clam instituuntur.
Flores oculis nostris nudis vix conspiciuntur.
24. CRYPTOGAMIA.

Figure 2.16. Linnaeus' *Clavis Systematis Sexualis*, first published in 1735 in the first edition of his *Systema Naturae*, but reproduced here in a modern format. After Stearn (1957), table on p. 26.

the pistil. The analogy between the floral parts and the animal genitalia is given, and some of the more subsidiary parts are sometimes presented in the light of striking analogies: the calyx is the nuptial bed (*thalamus*), the corolla the curtains (*aulaeum*) . . . "the calyx might be regarded as the *labia majora* or the foreskin; one could regard the corolla as the *labia minora*" (146). (1971:56–57)

In listing the definitions of classes and orders Linnaeus adopted a metaphorical scheme based on love, marriage, and adultery. The nonrepetitious parts of his scheme are given below in their entirety, based on the English translation of 1783:

Class I. Monandria. One male. One husband in marriage (one stamen in a hermaphrodite flower).

Class II. Diandria. Two males. Two husbands in the same marriage (two stamens in a hermaphrodite flower).

Class XI. Dodecandria. Twelve males. Twelve to nineteen husbands in the same marriage (twelve stamens to nineteen in a hermaphrodite flower).

Class XII. Icosandria. Twenty males. Generally twenty husbands, often more (stamens inserted on the calyx [not on the receptacle] in a hermaphrodite flower).

Class XIII. Polyandria. Many males. Twenty husbands or more in the same marriage (stamens inserted on the receptacle, from 20 to 1000 in the same flower with the pistil).

VEGETABLE KINGDOM

KEY OF THE SEXUAL SYSTEM

MARRIAGES OF PLANTS.

Florescence.

⌈ *PUBLIC MARRIAGES.*

Flowers visible to every one.

⌈ IN ONE BED.

Husband and wife have the same bed.

All the flowers hermaphrodite: stamens and pistils in the same flower.

⌈ WITHOUT AFFINITY.

Husbands not related to each other.

Stamens not joined together in any part.

⌈ WITH EQUALITY.

All the males of equal rank.

Stamens have no determinate proportion of length.

1. ONE MALE.	7. SEVEN MALES.
2. TWO MALES.	8. EIGHT MALES.
3. THREE MALES.	9. NINE MALES
4. FOUR MALES.	10. TEN MALES.
5. FIVE MALES.	11. TWELVE MALES.
6. SIX MALES.	12. TWENTY MALES.
	13. MANY MALES.

⌊ WITH SUBORDINATION

Some males above others.

Two stamens are always lower than the others.

14. TWO POWERS.	15. FOUR POWERS.

WITH AFFINITY

Husbands related to each other.

Stamens cohere with each other, or with the pistil.

16. ONE BROTHERHOOD.	19. CONFEDERATE
17. TWO BROTHERHOODS.	MALES.
18. MANY BROTHERHOODS.	20. FEMININE MALES.

⌊ IN TWO BEDS.

Husband and wife have separate beds.

Male flowers and female flowers in the same species.

21. ONE HOUSE.	23. POLYGAMIES.
21. TWO HOUSES.	

⌊ CLANDESTINE MARRIAGES.

Flowers scarce visible to the naked eye.

24. CLANDESTINE MARRIAGES.

Figure 2.17. The English translation of Linnaeus' *Key of the Sexual System.* After Stearn (1957), table on p. 27.

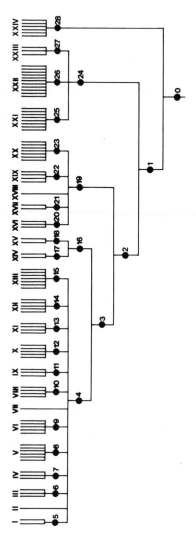

Figure 2.18. The orders, classes, and structural components of Linnaeus' sexual system of plants as published in the first edition of his *Systema Naturae* (1735). Components (with names and numbers of classes and orders in parentheses): 0, Regnum Vegetabile; 1, Publicae; 2, Monoclinia; 3, Diffinitas; 4, Indifferentismus; 5, Class (Monandria, I; Mono-, Digynia); 6, Class (Triandria, III; Mono-, Di-, Trigynia); 7, Class (Tetrandria, IV; Mono-, Di-, Tetragynia); 8, Class (Pentandria, V; Mono-, Di-, Tri-, Tetra-, Penta-, Polygynia); 9, Class (Hexandria, VI; Mono-, Di-, Tri-, Hexa-, Polygynia); 10, Class (Octandria, VIII; Mono-, Di-, Tri-, Tetragynia); 11, Class (Enneandria, IX; Mono-, Tri-, Hexagynia); 12, Class (Decandria, X; Mono-, Di-, Tri-, Penta-, Decagynia); 13, Class (Dodecandria, XI; Mono-, Di-, Tri-, Dodecagynia); 14, Class (Icosandria, XII; Mono-, Di-, Tri-, Penta-, Polygynia); 15, Class (Polyandria, XIII; Mono-, Di-, Tri-, Tetra-, Penta-, Hexa-, Polygynia); 16, Subordinatio; 17, Class (Didynamia, XIV; Gymno-, Angiospermia); 18, Class (Tetradynamia, XV; Fructu Siliculoso, Fructu Siliquoso); 19, Affinitas; 20, Class (Monadelphia, XVI; Pent-, Dec-, Polyandria); 21, Class (Diadelphia, XVII; Hex-, Decandria); 22, Class (Syngenesia, XIX; Monogamia, Polygamia Superflua, Polygamia Frustranea, Polygamia Necessaria); 23, Class (Gynandria, XX; Di-, Tri-, Tetr-, Pent-, Hex-, Dec-, Polyandria); 24, Diclinia; 25, Class (Monoecia, XXI; Mon-, Tri-, Tetr, Pent-, Polyandria, Mono-, Polyadelphia, Syngenesia); 26, Class (Dioecia, XXII; Di-, Tri-, Tetr-, Pent-, Hex-, Oct-, Ene-, Dec-, Polyandria, Monodelphia, Syngenesia); 27, Class (Polygamia, XXIII; Mon-, Di-, Trioecia); 28, Class (Cryptogamia, XXIV; Arbores, Filices, Musci, Algae, Fungi, Lithophyta). Classes not represented by components: II (Diandria; Monogynia), VII (Heptandria; Monogynia); XVIII (Polyadelphia; Polyandria).

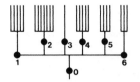

Figure 2.19. The "orders" and structural components of Linnaeus' system of animals, as published in the first edition of his *Systema Naturae* (1735). Components (all are classes except component 0, with orders in parentheses): 0, Regnum Animale; 1, Quadrupedia (Anthropomorpha, Ferae, Glires, Jumenta, Pecora); 2, Aves (Accipitres, Picae, Macrorhynchae, Anseres, Scolopaces, Gallinae, Passeres); 3, Amphibia; 4, Pisces (Plagiuri, Chondropterygii, Branchiostegi, Acanthopterygii, Malacopterygii); 5, Insecta (Coleoptera, Angioptera, Hemiptera, Aptera); 6, Vermes (Reptilia, Testacea, Zoophyta).

Class XIV. Didynamia. Two powers. Four husbands, two taller than the other two (four stamens: of which the two nearest are longer).

Class XV. Tetradynamia. Four powers. Six husbands, of which four are taller (six stamens: of which four are longer, and the two opposite ones shorter).

Class XVI. Monadelphia. One brotherhood. Husbands, like brothers, arise from one base, as if from one mother (stamens are united by their filaments into one body).

Class XIX. Syngenesia. Confederate males. Husbands joined together at the top (stamens are connected by the anthers forming a cylinder [seldom by the filaments]).

Class XX. Gynandria. Feminine males. Husbands and wives growing together (stamens are inserted on the pistils [not on the receptacle]).

Class XXI. Monoecia. One house. Husbands live with their wives in the same house but have different beds (male flowers and female flowers on the same plant).

Class XXII. Dioecia. Two houses. Husbands and wives have different houses (male flowers and female flowers are on different plants).

Class XXIII. Polygamia. Polygamies. Husbands live with wives and concubines (hermaphrodite flowers, and male ones, or female ones in the same species).

Class XXIV. Cryptogamia. Clandestine marriages. Nuptials are celebrated privately (flowers concealed within the fruit, or in some irregular manner).

For most classes (I–XIII) the orders are based on the number of female elements:

Monogynia, Digynia, Trigynia, etc. One female, two females, three females, etc. (according to the number of pistils).

For class XIV, Didynamia, there are two orders:

Gymnospermia and Angiospermia (seeds naked or enclosed in a pericarp).

For class XV, Tetradynamia, there are two orders:

Fructu Siliculoso and Fructu Siliquoso (seeds in small or large pods).

For six classes (XVI–XVIII, XX–XXII), the orders are based on the number of male elements:

Monandria, Diandria, Triandria, etc. (according to the number of stamens).

For class XIX, Syngenesia, there are four orders:

1. Monogamia. Monogamy.* Many marriages with promiscuous intercourse (many florets furnished with stamens and pistils; the flowers of these are vulgarly called composites).
2. Polygamia Superflua. Superfluous polygamy. The married females are fertile, and thence the concubines superfluous (the hermaphrodite flowers of the disk are furnished with stigmas, and produce seeds; and the female flowers also, which constitute the circumference, produce seeds likewise).
3. Polygamia Frustranea. Frustraneous polygamy. The married females are fertile, and the concubines barren (the hermaphrodite flowers of the disk are furnished with a stigma, and produce seeds; but the florets which constitute the circumference, having no stigmas, produce no seeds).
4. Polygamia Necessaria. Necessary polygamy. The married females are barren, and the concubines fertile (the hermaphrodite flowers, from defect of the stigma of the pistil, produce no seed; but the female flowers in the circumference produce perfect seeds).

For class XXIII, Polygamia, the three orders are based on the disposition of hermaphrodite and male or female flowers on the same or different individual plants of a species:

Monoecia, Dioecia, and Trioecia (according to the number of types of plants).

For class XXIV, Cryptogamia, there are six orders:

1, Arbores. 2, Filices. 3, Musci. 4, Algae. 5, Fungi. 6, Lithophyta.

There is little doubt that Linnaeus' sexual system of plants was in itself a significant stimulus to the study of plants and their systematics: "By a bold stroke of the pen the nebulous world of plants was made to act like husbands and wives in unconcerned freedom and everybody prepared to grasp the meaning of Monoecia and Dioecia, Syngenesia and Poly-

---

*Linnaeus later split the order Monogamia into two orders, using a new term ("Polygamia Aequalis"—equal polygamy) with the old definition, and the old term (Monogamia) with a new definition: "Flosculus non composita, sed simplex" (Linnaeus 1754:345).

gamia without effort" (Croizat 1945:55). The system also had its detractors, among whom the most notable was Johann Siegesbeck who

attacked it harshly on the ground that "such loathsome harlotry" ("scortationes quasi detestabiles") as several males to one female would never have been permitted in the vegetable kingdom by the Creator and asked how anyone could teach without offence "so licentious a method" ("methodus talem lascivam") to studious youth. He is remembered today only through the unpleasant small-flowered weed which Linnaeus named *Sigesbeckia.* (Stearn 1957:25)

In presenting the sexual system of plants, Linnaeus declared that it is artificial, designed for the purpose of identification of specimens. He asserted also that a natural system is desirable, that it had been sought after by other botanists and himself, but had yet to be achieved. He later (1738) considered the various systems of plants that had been previously proposed, rendering each of them in a tabular form (few of the original authors had done so). He even analyzed their authors in a like manner, which served as his table of contents (figure 2.20). He then gave his own "Fragments of a Natural Method"—a list of 65 unnamed and undefined orders containing various numbers of genera—which he later (1751) revised and supplied with names derived from plant habit (Larson 1967).

Linnaeus' first natural order (1738; Ordo I) contains six genera (*Arum, Calla, Dracontium, Piper, Acorus, Saururus*). The first three are members of class XX, *Piper* is a member of class II, and *Acorus* and *Saururus* are members of class VI, of the sexual system (figure 2.21). The second natural order (Ordo II) contains six genera unrepresented in his sexual system of 1735. The third natural order (Ordo III) contains nine genera (*Alpinia, Amomum, Canna, Costus, Curcuma, Kaempferia, Maranta, Thalia, Musa*). The first eight are members of class I, and *Musa* is a member of class XXIII, of the sexual system.

For the 15 genera listed above, the conflict in the groupings of the sexual and natural systems may be appreciated with reference to figure 2.21. In this comparison, the natural system (to the right) contradicts components 2 and 4 of the sexual system (to the left), but does not contradict any of the named classes and orders of the sexual system. Addition of other genera as arranged in Linnaeus' natural orders (1738) would, however, probably cause conflict with the classes and orders of the sexual system.

Form

Ordo secundum quem METHODI exhibentur.

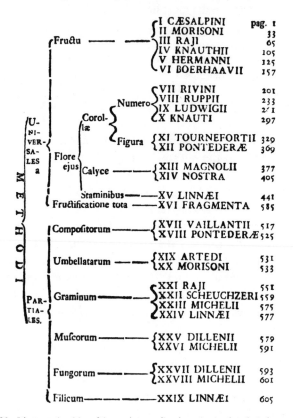

Figue 2.20. Linnaeus' table of botanists, reflecting the basis of their general systems (universales) and their areas of botanical specialization (partiales). After Linnaeus (1738), table of contents.

In reviewing and summarizing the early literature, and in proposing novel artificial and natural systems of his own, Linnaeus may be viewed as having placed the study of systematic botany on a new and modern basis. The central problem, for him and for later workers as well, is the meaning and significance of artificial and natural systems. In Linnaeus' words:

When we are constructing artificial Systems, we assume some principle at will, in accordance with which we make our classification, and so in making this the characters are easily discerned, and accordingly names giving the essential character are easier to make. But in a Natural System it is more difficult to dig

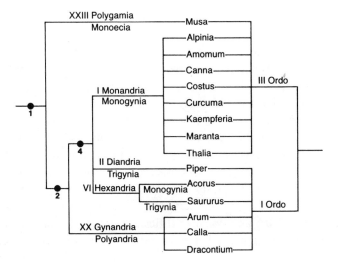

Figure 2.21. Comparison between Linnaeus' sexual and natural systems of classification of 15 plant genera. Left, sexual system; right, natural system. The arrangement of the genera in the sexual system follows Linnaeus (1754); and in the natural system, Linnaeus (1738).

out both the essential characters and essential names to express them. (Hort and Green 1938:112)

In the words of a botanist-historian:

It is the custom to describe these systems, of which those of Cesalpino, Morison, Ray, Bachmann (Rivinus), and Tournefort are the most important, by the one word "artificial"; but it was by no means the intention of those men to propose classifications of the vegetable kingdom which should be merely artificial, and do no more than offer an arrangement adapted for ready reference. . . . it is plain from the objections which every succeeding systematist makes to his predecessors, that the exhibition of natural affinities was more or less clearly in the minds of all as the main object of the system. . . . But a new departure dates from Linnaeus himself, since he was the first who clearly perceived the existence of this discord [between artificial and natural systems]. (Sachs 1890: 7–8)

No one ever gave more attention to artificial systems than Michel Adanson, who over a period of years constructed 65 such systems, each embracing all plants. At the same time he worked out what he thought were 58 natural families of plants. Within his 65 artificial systems, there were nearly 600 subdivisions (classes), each defined by a single character. In comparing the natural families with the classes, he

Form

observed that no natural family was uniquely defined by any one of the 600 characters (Adanson 1763–64).

The nature of Adanson's comparisons is illustrated in figure 2.22, which portrays his system number 33, a simple one with only three classes (represented by the vertical columns in the figure). The distribution of the 58 families is shown by their inclusion in one or more of the three columns. Some families are included in only one column (the 35 families numbered to the right of the three columns); the other families are included in two or three columns (the distribution of families 19, 30, and 31 was unspecified by Adanson). The three classes leave 35 families intact, and split the other 20 families. But no one family is the sole member of any one class. Adanson observed such to be true for all of the 600 classes of his artificial systems. He concluded that searching for defining characters of natural groups is futile, because none seems to exist.

Linnaeus, Adanson, and other early botanists agreed that there are natural groups—that is, that certain groups of species really exist in nature. Within the Aristotelian tradition of the times, their existence could have but one implication: namely, that their existences (essences) could become known through study and could be captured in definitions. Adanson disagreed flatly with this implication, and sought through his comparisons and argument to prove it false.

Certainly his comparisons suggest that erecting artificial systems might be an inefficient way to search for defining characters of natural groups—a suggestion that had been already made by Linnaeus, although he did not pursue its investigation to the elaborate degree found in Adanson's studies. Linnaeus had a different view of the problem, for he believed that defining characters could be found only after the natural groups, and most if not all of the species that constitute them, had been discovered (Larson 1967). This belief was the basis for one of Linnaeus' often-quoted aphorisms (Smith 1814):*

A genus should furnish a character, not a character form a genus; or, in other words, . . . a certain coincidence of structure, habit, and perhaps qualities, among a number of plants, should strike the judgement of a botanist, before he fixes on one or more technical characters, by which to stamp and define such

---

*"Scias Characterem non constituere Genus, sed Genus Characterem" (Linnaeus 1751:119, par. 169).

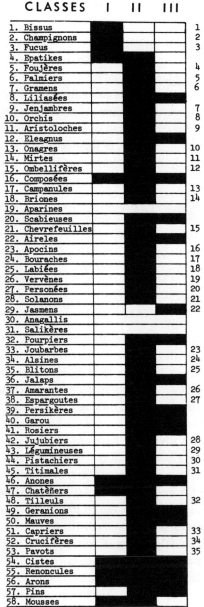

Figure 2.22. Adanson's artificial system number 33, with three classes (vertical columns). The distribution of each of his 58 natural families is indicated by a black box in each class in which the family occurs. Thirty-five families (right) are restricted to one class.

plants as one natural genus. . . . Many persons who can perceive a genus cannot define it. (pp. 276–77)

In retrospect it is not easy to judge in Adanson's favor—that his failure to find a single defining character for any of his 58 natural families is a decisive proof that no such characters exist.

Of the supposedly natural groups of plants recognized by Linnaeus, Adanson, and other early botanists, there is some variation but little in the way of disagreement. A recent comparison of genera attributed to certain families, judged by modern standards of their "correct" systematic position, shows a consistently high performance by one and all (Stafleu 1963:218):

Author	Year	Number of Genera	Number of genera in correct: Family	Number of genera in correct: Order	Percentage of genera in correct: Family	Percentage of genera in correct: Order
Linnaeus	1751	142	118	131	83.1	92.2
B. de Jussieu	1759	172	138	157	80.2	91.2
Adanson	1763	204	170	182	83.3	89.2
A.-L. de Jussieu	1789	225	183	218	81.3	96.8

Jean-Baptiste de Lamarck also focused on artificial systems in their role as helpful, indeed necessary, devices for identifying specimens of plant species with which one is not familiar. He argued that the most efficient artificial system is one in which each division is dichotomous, with the alternatives contradictory and based on single characters easily observed on any part of the plant. These criteria, which are essentially those of Aristotle's "dichotomous division," have remained in use ever since as a guide to making "keys"—the term that replaced "artificial systems." Lamarck's key to the plants of France, arranged as a table, served as the table of contents for one of his books (figure 2.23). More detailed keys are found in each section devoted to particular groups.

Linnaeus' thoughts about natural groups, and the difficulty in finding characters that define them adequately, changed over the years. In 1764, he conceived of four sorts of natural groups, which correspond to the four levels of his hierarchical classifications: orders, genera, species, and varieties. Convinced that such groups, hierarchically arranged, exist in nature, he attempted his own causal explanation of them (adapted from Bremekamp 1953):

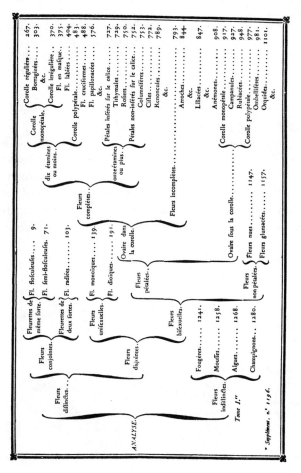

Figure 2.23. Lamarck's key to the major groups of the plants of France. After J.-B. Lamarck (1778), Flore Française, vol. 2 (Paris: Imprimerie Royale), figure on p. 1.

1. In the beginning the thrice exalted Creator covered the medullary substance of the plant with the principles of which the various kinds of cortex consist, and in this way as many individuals were formed as there are now Natural Orders.
2. The vegetable prototypes (par. 1) were mixed with each other by the Almighty, and there are now so many Genera in the Orders as in this way new plants were formed.
3. The generic prototypes (par. 2) were mixed with each other by nature, and in this way in every Genus so many species were formed as at present exist.
4. The Species whose origin was explained (par. 3) were mixed with each other by Chance, and in this way the Varieties arose that here and there are met with. (p. 243)

Some commentators see in Linnaeus' remarks about the mixing of generic prototypes, which produces species, and the mixing of species, which produces varieties, a kind of primitive evolutionism. Indeed, his later remarks, published posthumously, seem to trend in that direction, for he eliminated the influence of the Almighty in all but the origin of orders (Larson 1967:317). But taken together his remarks contain a clear and simple view of the world, easily represented in a branching diagram, which would allow only for four levels of natural groups (figure 2.24).

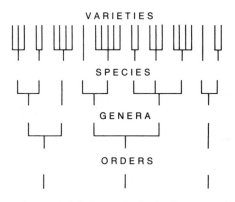

VARIETIES

SPECIES

GENERA

ORDERS

Figure 2.24. Linnaeus' concept of the four sorts of natural groups: orders, genera, species, and varieties. He early (1735) believed that each species was separately created. By 1764, he believed that species arose through nature's mixing of generic prototypes that themselves were the product of the Almighty. Finally, he limited supernatural influence only to the creation of "one plant with reproductive power" for each natural order, with all genera, species, and varieties arising through natural mixing and chance (Larson 1967).

Linnaeus' rigid view, allowing only four levels of natural groups, failed to allow for what, among other things, seemed to be real interrelationships between his natural orders. Rather than increase the number of hierarchical levels, he hit upon the simple expedient of an analogy, that of the geographical map: "All plants show affinities on all sides, like the territories on a geographical map" (translated).* Linnaeus himself never published a map of affinities, but one of his later students (Giseke 1792) made such a map for Linnaeus' natural orders of plants (figure 2.25). Fifty-six orders are represented, each by a circular figure, the diameter of which is roughly proportional to the number of its included genera. Some circles are in contact with one or more other circles, or in more or less close proximity to them, indicating various affinities among them.

From the map itself, a branching diagram may be derived, but it is uninformative with respect to the interrelationships of the orders (figure 2.26). The reason is that any two orders, however close, might have their nearest relationships with orders other than each other. The map, in short, lacks a pathway, with a beginning and an ending, interconnecting all of the territories. In his explanation of the map, however, Giseke states that particular orders are more closely related among themselves than to others. Such is true for the "monocotyledenous province" (a concept dating from Ray 1703), including orders I–XIII, which would in effect constitute a group at a new and higher level in the hierarchy (figure 2.27).

Linnaeus' analogy of a geographical map seems to have been original with him, but there is some similarity between Giseke's (1792) and Morison's (1672) diagrams (figures 2.13, 2.25), and between these and Buffon's (1755) "genealogical tree" of the races of dogs (figure 2.28). Of his tree Buffon writes (1755):

> In order to give a clearer idea of the dog group, their modifications in different climates, and the mixture of their races, I include a figure, or, if one wishes to term it so, a sort of genealogical tree, wherein one can see at a glance all of the varieties. This figure is oriented as a geographical map, and in its construction the relative positions of the climates have been maintained to the extent possible. (p. 255, translated)

---

*"Plantae omnes utrinque affinitatem monstrant, uti Territorium in Mappa geographica" (Linnaeus 1751:27, par. 77).

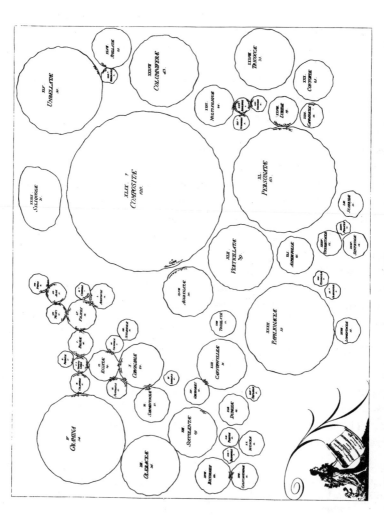

Figure 2.25. Giseke's (1792) "map" or "table of the genealogico-geographical affinities of plants following the natural orders of Linnaeus." Each order is represented by a numbered circle (roman numerals), the diameter of which reflects the number of included genera (arabic numerals). Orders XVI and LIV are omitted. After Giseke (1792).

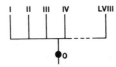

Figure 2.26. A branching diagram derived from Giseke's table (cf. figure 2.25).

Buffon reasoned that climatic effects produced the different races of dogs from a single ancestral race, which he believed to be the shepherd's dog (*chien de berger*):

The shepherd's dog is the source of the tree. This dog, transported into the rigorous climates of the north became smaller and uglier in Lapland, but appears to have been maintained, and even perfected, in Iceland, Russia, and Siberia, where the climate is a little less rigorous and the inhabitants are a little more civilized. These changes have occurred only through the influence of climate. (*Ibid.*)

Giseke's diagram (figure 2.25) lacks a pathway, which is present, however, in Morison's (figure 2.13), from which a branching diagram may be derived (figure 2.29). Because Buffon specified the shepherd's dog as the root of his tree, a branching diagram may be derived from it, too (figure 2.30). In Buffon's tree the pure races are interconnected by solid lines; and the hybrids, by dotted lines. The same components are specified, with or without the hybrid races. Buffon did not provide a formal classification of the races of dogs, but a "classification" may be constructed from his discussion, and, with the hybrid races omitted for simplicity, the "classification" may be represented as a branching diagram (figure 2.31). The "classification" differs from the "tree" in lacking certain components (1 and 9) and including others (11, 12, 13). Interestingly, all of the components, both of the "tree" and the

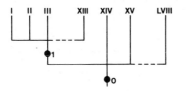

Figure 2.27. A branching diagram derived from Giseke's table, with the added information (component 1) that orders I–XIII constitute the "province of monocotyledons" (Giseke 1792:624).

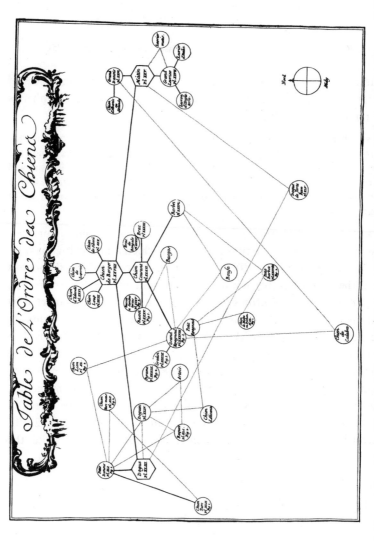

Figure 2.28. The varieties of dogs. After C. de Buffon and L. Daubenton (1755), Histoire Naturelle, Générale et Particulière, avec la Description du Cabinet du Roi (Paris: Imprimerie Royale), vol. 5, plate facing p. 228.

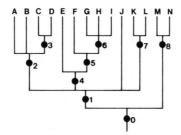

Figure 2.29. A branching diagram derived from Morison (cf. figure 2.13). Species (A–N): A, *villoso*; B, *anthriscus*; C, *oblongo*; D, *brevi*; E, *lutea*; F, *major*; G, *hirsutis*; H, *hirsutioribus*; I, *hirsutissimis*; J, *scandix*; K, *sativum*; L, *sylvestre*; M, *dissimum*; N, *sativum*. Components (0–8): 0, *Umbellae*; 1, *semine rostrato*; 2, *myrrhis annua*; 3, *aspero*; 4, *myrrhis perennis*; 5, *alba*; 6, *minor*; 7, *cerefolium*; 8, *semine rotundo*.

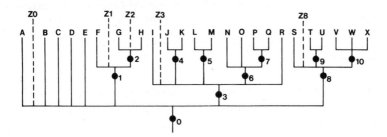

Figure 2.30. A branching diagram derived from Buffon's genealogical tree (cf. figure 2.28). Pure races: A, Shepherd's Dog, *Chien de Berger*; B, Pomeranian Dog, *Chien-loup*; C, Iceland Dog, *Chien d'Islande*; D, Lapland Dog, *Chien de Laponie*; E, Siberian Dog, *Chien de Siberie*; F, Bull Dog, *Dogue*; G, Small Danish Dog, *Petit Danois*; H, Turkish Dog, *Chien-turc*; I, Hound, *Chien courant*; J, Terrier, *Basset*; K, Turnspit, *Basset à jambes torses*; L, Harrier, *Braque*; M, Dalmatian, *Braque de Bengale*; N, Spaniel, *Grand Epagneul*; O, Small Spaniel, *Petit Epagneul*; P, King Charles' Dog, *Gredin*; Q, *Pyrame*; R, Great Water Dog, *Barbet*; S, Irish Greyhound, *Mâtin*; T, Large Danish Dog, *Grand Danois*; U, Albanian Dog, *Chien d'Albanie ou d'Irlande*; V, Large Greyhound, *Grand Levrier*; W, English Greyhound, *Levron d'Angleterre*; X, Italian Greyhound, *Levrier d'Italie*. Hybrid races: Z0—Mastiff Dog, *Dogue de forte race* (F × S); Calabrian Dog, *Chien de Kalabre* (N × T); Alicant Dog, *Chien d'Alicante* ([F × G] × O); Lion Dog, *Chien-lion* (G × N). Z1—Bastard Pug Dog, *Roquet* ([F × G] × G); Pug Dog, *Doguin* (F × G); Artois Dog, *Artois* ([F × G] × [(F × G) × G]). Z2—Turkish Mongrel, *Chien-turc métis* (G × H). Z3—*Bousse* (N × R); Shock Dog, *Chien de Malte ou Bichon* ([O × R] × O); *Burgos* (J × N); Lesser Water Dog, *Petit Barbet* (O × R). Z8—Greyhound With Hair Like a Wolf, *Levrier métis* (S × V). Components: 0, Dogs; 1–2, Unnamed; 3, True hunting dogs, *Vrais chiens de chasse*; 4, Terriers, *Bassets*; 5, Harriers, *Braques*; 6, Spaniels, *Epagneuls*; 7–9, Unnamed; 10, Greyhounds, *Levriers*. English names after C. de Buffon (1812), Natural History, General and Particular, trans. W. Smellie. A new edition, vol. 4 (London: T. Cadell and W. Davies).

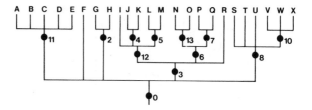

Figure 2.31. A branching diagram derived from Buffon's informal classification of dogs. Components: 0, Dogs; 2, Unnamed—"The Small Danish (G) and Turkish (H) must be joined together" (Buffon, 1755:219, translated; see caption for figure 2.28); 3, True Hunting Dogs; 4, Terriers; 5, Harriers; 6, Spaniels; 7, Unnamed—"The Pyrame (Q) is only a King Charles' Dog" (P; 1755:218, translated); 8, Unnamed—"The Greyhounds (S, V, W, X) and the Large Danish (T) and the Albanian (U) Dogs have, besides a similar form and long snout, the same nature" (1755:218, translated); 10, Greyhounds; 11, Unnamed—"I have grouped together the Shepherd's Dog (A), and the Pomeranian (B), Iceland (C), Lapland (D), and Siberian (E) Dogs because they resemble each other more than they resemble others" (1755:218, translated); 12, Unnamed—"The Hound (I), the Terriers (J, K), and the Harriers (L, M) constitute one race" (1755:226, translated); 13, True Spaniels (large and small).

"classification," are combinable in a single branching diagram (figure 2.32), which may be considered a synthesis of Buffon's views of the interrelationships of the races of dogs. In the synthesis (figure 2.32), the hybrid varieties have their same positions: ZO hybrids attach to the base of component 0; Z1 hybrids, to the base of component 1; Z2, component 2; Z3, component 3; and Z8, component 8.

One rather fanciful diagram was published in 1817 by Dunal (figure 2.33) in an account of nine genera of a certain family of plants ("Anonacées"). In a less elaborate rendition, but in accord with the spirit of the analogy of a map, the nine genera might be likened to towns, and the affinities between them might be likened to roadways (figure 2.34).

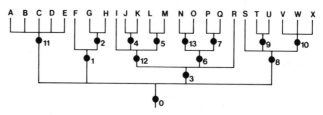

Figure 2.32. A synthesis of Buffon's views of the interrelationships of races of dogs (cf. figures 2.30, 2.31).

Figure. 2.33. Table of affinities of the genera of a family of plants. After Dunal (1817), plate 1.

Exactly what the affinities might represent is difficult to determine from Dunal's account, but he does give two tabular arrangements (figures 2.35, 2.36) and a third arrangement that he calls a linear series (figure 2.37). Despite the numerous affinities (roadways) interconnecting the nine genera (towns), Dunal's table of affinities cannot be represented in an informative branching diagram similar to that derived from Morison's table (figure 2.29). Dunal's map (figures 2.33, 2.34), in short, does

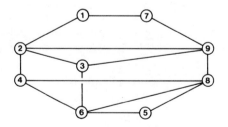

Figure 2.34. Dunal's table of affinities simplified to its essentials (cf. figure 2.33). Genera (numbered according to Dunal's table 3; see below, figure 2.37): 1, *Kadsura*; 2, *Anona*; 3, *Monodora*; 4, *Asimina*; 5, *Porcelia*; 6, *Uvaria*; 7, *Xylopia*; 8, *Unona*; 9, *Guatteria*.

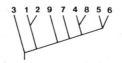

Figure 2.35. A representation of Dunal's first analytical table, with characters omitted (genera numbered as in figure 2.34).

not specify a unique pathway with a beginning and an end, despite his stated objectives:

> Before passing to the history and the description of each genus, I will present here four tables of these genera as I conceive them today. The first two are analytical tables, similar to those that M. Lestiboudois has offerred in his *Botanographie Belgique;* tables of which the purpose is to make clear the differences between the genera. The third is the linear series of the genera, in the order in which I will describe them. The fourth shows the fruit of each genus, arranged in accordance with its natural affinities. This last table is analogous to those published by Giseke, in order to show the multiple affinities of plant families [orders]. (1817:22, translated)

The idea of natural groups arranged in space is seemingly an important one, for it has persisted, with much elaboration, since these early times. One of the best commentaries on this idea is that of A.-P. de Candolle:

> The principal fact that presents itself in this research, the fact around which all other facts are arranged, is that certain beings so resemble each other that they appear to the eye of a naturalist to constitute a distinct group. Such a group, considered to have real existence, resembles other such groups, which form groups among themselves. Thus, the plant kingdom is nothing more than a vast assemblage composed of many subgroups of lower order. It was from this point of view that for the first time Linnaeus, with his usual insight, compared the plant kingdom to a geographical map. This comparison, indicated in his book only by a single phrase, was later developed by Giseke, Batsch, Bernadin de Saint Pierre, L'Heritier, du Petit-Thouard, et al. Even though one might

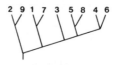

Figure 2.36. A representation of Dunal's second analytical table, with characters omitted (genera numbers as in figure 2.34).

Figure 2.37. A representation of Dunal's linear series of genera, with characters omitted (genera numbered as in figure 2.34).

consider the map a simple metaphor, it is a metaphor so true, and so rich in useful consequences, that it is worth further consideration.

Consider an example of such a map: the classes would correspond to the quarters of the globe; the families, to realms; the tribes, to provinces; the genera, to counties; the species, to villages. If we were to examine such a map we would note its extreme similarity to an ordinary geographical map.

The groups, of course, would be essentially unequal. But a realm (or a family) would be no more or less distinct, no matter what the space it occupies over the surface of the globe (or in the totality of the plant kingdom).

The distance that separates each species, each genus, each tribe, and each family could be actually calculated, if not in an absolute manner, at least in a relative one, and would indicate for the eye the affinities, more or less intimate, that the plants have among themselves. The genera not yet classified according to the natural method would be represented in the form of islands more or less remote from certain continents. In the best known classes we would note that, in certain parts, the genera and species would be close together, the ones next to the others, whereas in certain other parts the genera and species would be widely spaced. Thus, for example, if one would compare the Compositae and the Palmae, both of which certainly are natural groups, one would quickly note that in the first family all of the genera are in contact, so to speak, whereas in the second family they would display remarkable differences, or distances. It is this concept that the linear series, widely used in our books, can never portray, but which is easily obtained by the means of a geographical map. Also, one would note that, as I have already indicated, each genus, or each family, would resemble not only the group that precedes it and the group that follows it, but would have multiple affinities with many other groups. The linear series cannot express these complex affinities, whereas they may be easily represented in a geographical map. It is the more intimate knowledge of these multiple affinities that really constitutes the superiority of one classifier over another.

I will not myself attempt such a map, for that effort still seems premature. Also it could not be well executed until the natural divisions of the dicotyledons were established in a definitive manner. I hope only through these general considerations to remind the classifiers of the goal toward which they should be directed, and to help beginners to understand the natural method. (1844:193–94, translated)

It is well to consider the possibilities for maplike arrangements,

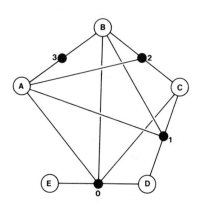

Figure 2.38. Affinities (0–3), or resemblances, among five species (A–E). Species A and B are alike in some respect (3); species A, B, and C, in another respect (2); species A, B, C, and D, in yet another respect (1); and all species, in a final respect (0).

wherein affinities, or resemblances, between entities are represented by lines drawn between them (or simply by spaces that separate them). Consider, for example, five species (A–E) and four affinities (0–3): affinity 0 is common to all five species; 1, to species A–D; 2, to species A–C; 3, to species A–B (figure 2.38). This information could also be represented by a branching diagram in which the affinities serve as structural components (figure 2.39), or, alternatively, by a map having a pathway with a beginning and an ending (figure 2.40). Indeed, various pathways are possible, without alteration of the affinities in their role as structural components (figures 2.41, 2.42).

It is possible, of course, that entities and affinities might be mapped without any informative result, by way of structural components (figures 2.43–2.45). In such cases numerous pathways would be possible, but no informative pathway would be specified.

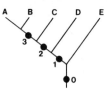

Figure 2.39. The same five species arranged in a branching diagram, of which the structural components are the affinities, or resemblances, among the species (cf. figure 2.38).

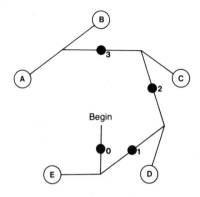

Figure 2.40. Affinities (0–3), or resemblances, among five species (A–E), viewed as if on a map with a pathway specified (cf. figures 2.38, 2.39).

Modern workers have sought in spatial representation a means for resolving natural groups, rather than merely portraying the affinities of groups already recognized by other means. Spatial analysis has been extended into many dimensions (Sneath and Sokal 1973), and various graphic devices have been developed in order to portray spatial affinities (figure 2.46). Yet the overall goal—the resolution of natural groups— has proved elusive through spatial analysis and representation, inasmuch as an informative pathway seems unspecifiable by these considerations alone. Hence, it is perhaps time to consider that Candolle might have been correct in his suggestion that natural groups must first be specified before spatial affinities can be well executed.

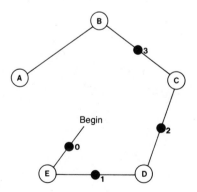

Figure 2.41. Affinities (0–3), or resemblances, among five species (A–E), viewed as if on a map with a pathway specified (cf. figures 2.38–2.40).

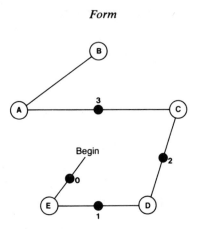

Figure 2.42. Affinities (0–3), or resemblances, among five species (A–E), viewed as if on a map with a pathway specified (cf. figures 2.38–2.41).

Candolle, despite his reluctance to map the entire plant kingdom, did map portions of it (figure 2.47), for example in relation to his classification of the plant family Crassulaceae (1828):

I. Crassulaceae Legitimae
  Isostemoneae
    Polypetalae
      Genera: *Tillaea, Bulliarda, Dasystemon, Septas, Crassula, Globulea*
    Gamopetalae
      Genera: *Curtogyne, Grammanthes, Rochea*

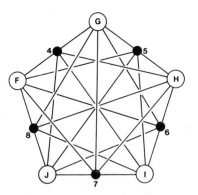

Figure 2.43. Affinities (4–8), or resemblances among five different species (F–J), viewed as if on a map with numerous pathways possible, but no informative pathway specified.

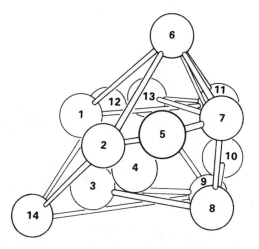

Figure 2.44. Taxonomic model of the Enterobacteriaceae. The species are represented by spheres (balls) connected by rods (sticks) indicating taxonomic distance. After Sneath and Sokal (1973), figure 5–15, p. 260. Copyright © 1973, W. H. Freeman.

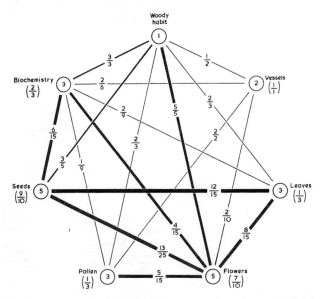

Figure 2.45. Correlations, or affinities, among 22 characters of certain plants. After K. R. Sporne (1977), Some problems associated with character correlations. In K. Kubitzki, ed., Flowering Plants: Evolution and Classification of Higher Categories. Plant Systematics and Evolution, Supplement 1, pp. 33–51 (Vienna and New York: Springer–Verlag), figure 2, p. 44.

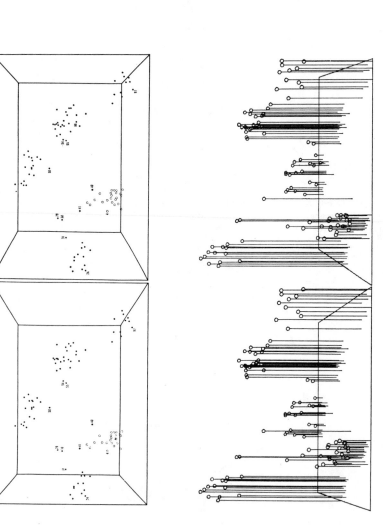

Figure 2.46. Spatial studies of 97 species of bees of the *Hoplitis* complex. The upper pair of diagrams, when viewed in a stereo viewer, shows the species as points in three-dimensional space; the lower pair, as balls on pins, as viewed from a different angle. After F. J. Rohlf (1968), Stereograms in numerical taxonomy. Syst. Zool. 17:246–55; figures 5, 7, pp. 252, 254.

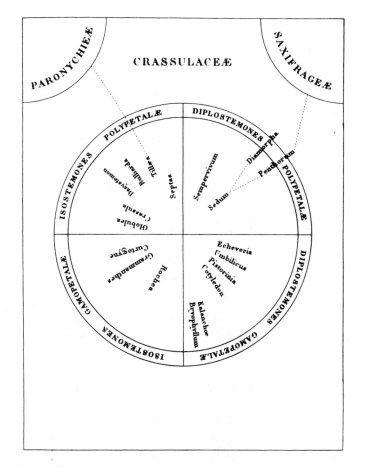

Figure 2.47. Affinities of genera of the plant family Crassulaceae. After Candolle (1828), plate 2.

Diplostemoneae
  Gamopetalae
    Genera: *Kalanchoe, Bryophyllum, Cotyledon, Pistorinia, Umbilicus, Echeveria*
  Polypetalae
    Genera: *Sedum, Sempervivum*
II. Crassulaceae Anomale
    Genera: *Diamorpha, Penthorum*

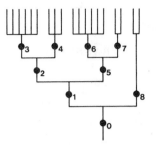

Figure 2.48. A branching diagram derived from Candolle's classification of plants of the family Crassulaceae (cf. figure 2.47). Components: 0, Crassulaceae; 1, Crassulaceae Legitimae; 2, Isostemoneae; 3, Polypetalae; 4, Gamopetalae; 5, Diplostemoneae; 6, Gamopetalae; 7, Polypetalae; 8, Crassulaceae Anomalae.

Most of the information in Candolle's map is easily represented in a branching diagram (figure 2.48), except for the circular arrangement of the genera (Candolle 1828:11, translated):

The central genera follow one another in the order of their affinities in a manner that appears to me very exact. This circular disposition, very adaptable to natural families, does, it seems to me, render the real analogies clearly apparent, and show the complete impossibility of establishing a regularly linear series.

Circular arrangement of genera within families was the novel idea of W. S. Macleay (1819), which was soon extended to embrace a constellation of ideas later known as the "quinarian" approach to classification (Swainson 1834). Basic to this approach is the idea that each natural group is composed of five subgroups, arranged in a circle and exhibiting regular affinities with similarly arranged subgroups of other groups (figures 2.49, 2.50). Some arrangements favored stars rather than circles (figure 2.51). Characteristic of each group is that its subgroups form "aberrant" (subgroups 1–3) and "normal" (subgroups 4–5) assemblages, as, for example, in Macleay's arrangement of fishes (Macleay 1842:198):

Fishes
    Aberrant group
  Ctenobranchii: Gills pectinated
    1. Plagiostomi
    2. Sturiones
    3. Ostinopterygii

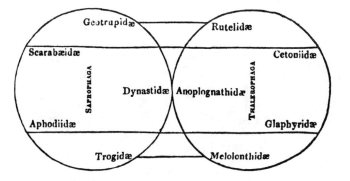

Figure 2.49. Two natural groups (circles) of insects (Saprophaga and Thalerophaga), each including five families circularly arranged. The horizontal lines indicate affinities between families occupying analogous positions in different natural groups (circles). After Macleay (1819), figure on p. 29.

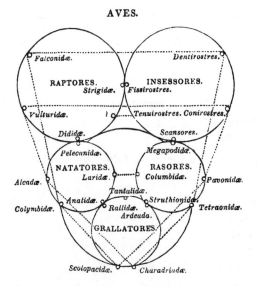

Figure 2.50. The five natural orders (circles) of birds (Insessores, Raptores, Natatores, Rasores, Grallatores), each theoretically including five families circularly arranged. The dotted lines indicate affinities between families occupying analogous positions in different natural orders (circles). The "aberrant" orders are grouped within the large circle, below. After W. Swainson (1837), On the Natural History and Classification of Birds, vol. 2 (London: Longman), figure on p. 200.

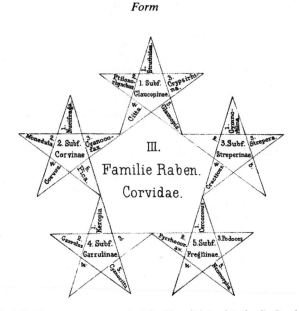

Figure 2.51. Affinities among genera and subfamilies of birds of the family Corvidae. After J. Kaup (1854), Einige Worte über die systematische Stellung der Familie der Raben, Corvidae. J. Ornithol. 2:xlvii–lvi; plate II, figure 10.

Normal group
Actenobranchii: Gills not pectinated
    4. Lophobranchii
    5. Cyclostomi

His divisions of the Ostinopterygii follow the same formula:

Ostinopterygii
    Aberrant group
Acanthopterygii
    1. Balistina
    2. Percina
    3. Fistularina

    Normal group
Malacopterygii
    4. Pleuronectina
    5. Clupeina

So also does his division of the Percina:

Aberrant group
1. Chaetodontidae
2. Percidae
3. Scorpaenidae

Normal group
4. Cirrhitidae
5. Sparidae

And, finally, so does his division of the Scorpaenidae:

Aberrant group
1. *Monocentris*
2. *Trigla*
3. *Scorpaena*

Normal group
4. *Oreosoma*
5. *Gasterosteus*

Contrary to Wilkins' (1668) early attempts at numerical organization, the quinarians believed that quinary subdivision and circular arrangement were real aspects of natural order that had previously gone unnoticed. Macleay was soon criticized, first by Virey (1825) for not paying close enough attention to the earlier development in France of natural methods of classification, and then by Fleming (1829), who advocated a purely dichotomous system (as exemplified by Lamarck's key) as the most natural possible. Macleay responded to his critics in a style seldom injected into scientific discourse (Macleay 1829–30). It is said (Wallace 1855) that in England the death blow to the quinarians was dealt by Strickland (1841):

The plan proposed [by me] is to take any species, A, and ask the question, What are its nearest affinities? If, after an examination of its points of resemblance to all other known species, it should appear that there are two other species, B and C, which closely approach it in structure, and that A is intermediate between them, the question is answered, and the formula BAC would express a portion of the natural system, the survey of which is so far completed. Then take C, and ask the same question. One of its affinities, that of C to A, is already determined; and we will suppose that D is found to form its nearest affinity on the other side. Then BACD will represent four species, the relative affinities of which are determined. By a repetition of this process, supposing our knowledge of the structure of each species to be complete, and our rules for determining the degrees of affinity correct, the whole organized creation might be ultimately arranged in the order of its affinities, and our survey of the natural system would then be finally effected. Now if each species never had more than two affinities, and those in opposite directions, as in the above example, the natural system would form a straight line, as some authors have assumed it to be. But we shall often find, in fact, that a species has only one direct affinity, and in other cases that it has three or more, showing the existence of lateral ramifications instead of a simple line; as shown in this example [figure 2.52], where C, besides its affinity to A and D, has an affinity to a third species E, which therefore forms a lateral ramification.

It was the observation of this fact which led some naturalists to adopt the circular instead of the linear theory, still adhering to the assumption of a symmetrical figure, but changing their notions of its form. Now although we find occasional ramifications in the affinities, and although these ramifications may occasionally anastomose and form a circle, yet it has been shown that the doctrine of a regular figure cannot be sustained . . . . The natural system may, perhaps, be most truly compared to an irregularly branching tree, or rather to an assemblage of detached trees and shrubs of various sizes and modes of growth. [footnote: If this illustration should prove to be a just one, the order of affinities might be shown in museums in a pleasing manner by constructing an artificial tree, whose ramifications should correspond with those of any given family of birds, and by then placing on its branches a stuffed specimen of each genus in their true order]. (pp. 189–190)

Strickland's example of a tree (figure 2.52) does not, unfortunately, specify a pathway. It is, therefore, uninformative in the sense that it conveys little more information than that there are five species interrelated in some way. His tree is not totally uninformative, for it does prohibit some groupings. For example, it specifies that there is no group including B and D and excluding A and C. The tree, in short, allows for as many pathways as it allows of possible beginnings.

Strickland continued:

In order to show that the views here maintained are not chimerical, I will here present one or two *sketch*-maps of different families of birds, though I am well aware that our knowledge of natural history is as yet far too imperfect to pretend to accuracy [figure 2.53]. Such sketches as these can be compared only to the

$$B \text{ -- } A \text{ -- } C \text{ -- } D$$
$$|$$
$$E$$

Figure 2.52. Five species and their affinities as they might appear in the true natural system. After Strickland (1841), figure on p. 190.

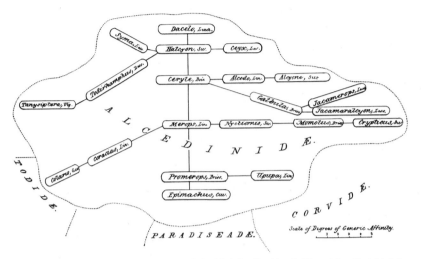

Figure 2.53. Affinities among genera of the bird family Alcedinidae. After Strickland (1841), plate 8.

rude efforts at map-making made by the ancients, of which the Peutinger Table is an example; and it is probably reserved for a distant age to introduce that degree of exactness into natural history which in modern geography is attained by a trigonometrical survey. For the sake of simplicity, in making these sketches I have omitted the consideration of *species,* but assuming that the genera of modern authors consist solely of closely allied species, I have proceeded to group them in what appeared to be their true position in respect of their affinities. In order to place these groups at their true distances, it is necessary to form a *scale of degrees of affinity,* to which the intervals between each genus shall correspond. I am aware that this scale must be, in some measure, arbitrary; but for this there is no remedy. (pp. 190–91)

Strickland's map of the affinities of genera of birds of the family Alcedinidae (figure 2.53) likewise restricts the number of possible pathways, but does not specify an informative one. Wallace later commented (1856):

In such a case an *arrangement* may be possible, but a *classification* may not be so. We must therefore give up altogether the principle of *division,* and employ that of *agglutination* or juxtaposition. (p. 195)

Wallace published various maps of his own in the same style (figure 2.54), and commented:

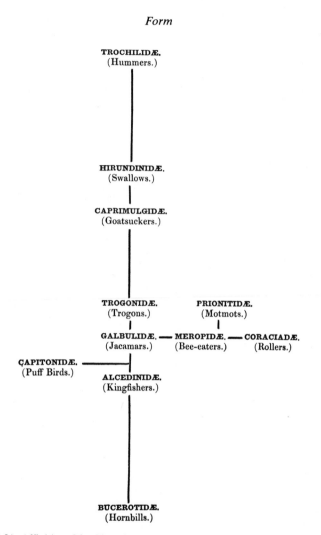

Figure 2.54. Affinities of families of the bird order Fissirostres. After Wallace (1856), figure on p. 205.

The method of representing affinities here adopted we believe to be of the highest value. It is founded on the method suggested by the late Mr. Strickland, and which we believe Dr. Lindley has been the first naturalist to adopt, namely that of placing to the right and left of every family or other group the names of those to which it is most nearly allied. (p. 206)

It is, perhaps, too fine a point to belabor, but Strickland and Wallace, on

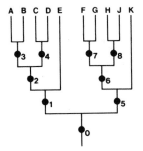

Figure 2.55. Ten species interrelated by way of components 0–8.

the one hand, and the quinarians on the other, were to some extent arguing at cross purposes. For the quinarians, circular affinities were those among natural groups already arranged in a classification. For Strickland, and Wallace too, it proved difficult to recognize major groups of birds; hence their concept of affinity was more inclusive, embracing a range of possibilities from which, hopefully, some order (or "*classification*" in Wallace's sense) might emerge.

These cross purposes may be illustrated by an example. Consider a group of 10 species with relations (and components) as shown in figure 2.55. A quinarian could assert, correctly, that there are two subgroups each of five species (figure 2.56). An opponent could assert, equally correctly, that there is a network of affinities (figure 2.57) bearing little resemblance to two groups of five species each. Yet this network could transform decidedly in that direction with the addition of an informative pathway with the appropriate beginning. The result could be a branching diagram with components 2–4 and 6–8.

Finally, it is always possible to divide a complex group into five subgroups without artificiality. For example, the scheme of Linnaeus'

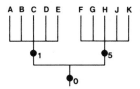

Figure 2.56. A quinarian arrangement of the ten species (cf. figure 2.55).

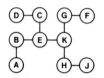

Figure 2.57. A map-like arrangement of the ten species (cf. figure 2.56).

sexual system could be reduced to five groups of plants (figure 2.18; components 4, 16, 19, 24, 28). If the groups were natural to begin with, they would be no less natural if they were the only ones recognized in a classification.

Some of the above points can be appreciated in a diagram of Milne-Edwards (figure 2.58), for which a branching diagram may be specified (figure 2.59). Within the confines of this classification of the major groups of vertebrate animals, lines of affinity are drawn between sharks (Chondroptérygiens) and whales (Cétacés), between frogs (Anoures) and turtles (Chéloniens), between herbivorous marsupials and rodents (Rongeurs), etc. Such lines of affinity cross the major divisions of the classification. At best, they represent a residue of resemblance unsubsumed in the components of the classification. For a group such as Strickland's birds (figure 2.53), there are present affinities of this sort plus all of those that would be subsumed in a classification if it had been worked out. It is no wonder, therefore, that Strickland felt that his affinities are more interesting than those of the quinarians. Yet had he worked out a classification of his birds, such affinities as he might have been able to draw across the divisions of his classification would doubtlessly consist of the same sort of affinities drawn by the quinarians.

Milne-Edwards' classification is notable for its grouping fishes and amphibians under the name "Vertébrés anallantoidiens" ("Anamniota" is a more modern term)—a grouping, based on a negative character, that has since been generally abandoned.

The quinarian controversy eventually passed away without much in the way of a resolution. Perhaps it marked the end of an era, for systematics was soon to be influenced by new sorts of information gleaned from studies of development. Embryological studies had a long history in continental Europe before their impact was felt in the English-speaking world. One early statement, remarkable for its clarity, was published by William Carpenter (1841):

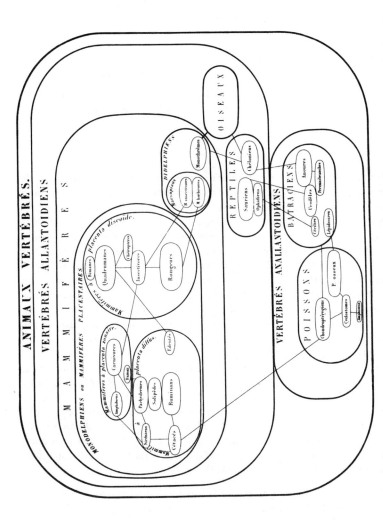

Figure 2.58. Natural affinities of vertebrate animals. After M. Milne-Edwards (1844), Considérations sur quelques principes relatifs à la classification naturelle des animaux. Ann. Sci. Nat., ser. 3, 1:65–99; plate.

Form

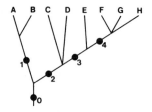

Figure 2.59. The branching diagram implied by Milne-Edwards' classification (cf. figure 2.58). A, Poissons; B, Batraciens; C, Reptiles; D, Oiseaux; E, Didelphiens; F, Mammifères a placenta diffus; G, Mammifères à placenta zonaire; H, Mammifères à placenta discoide. Components: 0, Animaux vertébrés; 1, Vertébrés anallantoidiens; 2, Vertébrés allantoidiens; 3, Mammifères; 4, Monodelphiens.

Allusion was just now made to the correspondence which is discernible between the transitory forms exhibited by the embryos of the higher beings, and the permanent conditions of the lower. When this was first observed, it was stated as a general law, that all the higher animals in the progress of their development pass through a series of forms analogous to those encountered in ascending the animal scale. But this is not correct; . . . the correspondence is much closer between the embryonic Fish and the foetal Bird or Mammal, than between these and the adult Fish. . . . The view here stated may perhaps receive further elucidation from a simple diagram [figure 2.60, left]. Let the vertical line represent the progressive change of type observed in the development of the foetus, commencing from below. The foetus of the Fish only advances to the stage F; but it then undergoes a certain change in its progress towards maturity, which is represented by the horizontal line FD. The foetus of the Reptile passes through the condition which is characteristic of the *foetal* Fish; and then, stopping short at the grade R, it changes to the perfect Reptile. The same principle applies to Birds and Mammalia; so that A, B, and C,—the *adult* conditions of the higher groups,—are seen to be very different from the *foetal*, and still more from the *adult*, forms of the lower; whilst between the embryonic forms of all the classes, there is, at certain periods, a very close correspondence, arising from the law of gradual progress from a general to a special condition, already so much dwelt upon. (pp. 196–97)

A similar diagram (figure 2.60, right) was published anonymously a few years later by Robert Chambers, who commented that "this diagram shews only the main ramifications; but the reader must suppose minor ones, representing the subordinate differences of orders, tribes, families, genera, etc., if he wishes to extend his views to the whole varieties of being in the animal kingdom" (1844:212–13).

From such views of divergence in ontogenetic time it was but a short

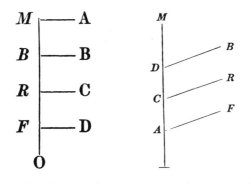

Figure 2.60. Branching diagrams of the main groups of vertebrates (Fishes, Reptiles, Birds, Mammals). Left, after Carpenter (1841), figure on p. 197. Right, after Chambers (1844), figure on p. 212.

step to views of divergence in evolutionary time. Prior to the origin of evolutionary theory, it was thought that, in a natural classification, species ought to be grouped together because they share, or ought to share, "natural affinity" (a "natural" rather than "artificial" general group similarity). But there was no clear answer to the question: "what is natural affinity?" At that time, two possible answers were available, both somewhat vague: (1) the plan of creation, (2) the true affinities that species have, one to another. The plan of creation could only be guessed at by study of the ways species resemble and differ from one another. And the notion of true affinity (or resemblance) was always open to the question: "what is 'true' as opposed to 'false' affinity?" To some extent, all things truly resemble one another, and in that sense have affinities with one another. Nevertheless, the principle of natural affinity proved useful for a time. Organisms were carefully studied in order to discover their true or natural affinities. Agreement was reached on many points. For example, comparison of a bee with a hummingbird would have left no doubt, despite the presence of "wings" in both organisms, that the bee has natural affinity with insects, and the bird with vertebrates. Bees resemble insects in numerous other ways, and hummingbirds resemble other birds and vertebrates generally. To group together bees and hummingbirds, because both fly and visit flowers, would have been considered "artificial"—even though the group might have had practical usefulness in describing where the organisms might be found and what they might be doing there.

The concept of natural affinity among species was nature-oriented, and was generally believed to reflect real relations that existed in nature as a result of creation. These relations, or affinities, could not be directly observed but could be discovered by study of the resemblances and differences between species. To make a judgment of affinity on the basis of resemblance of only one character, or only a few characters, was considered unreliable, for an artificial rather than natural group might result. Characters that seem to display true resemblance, or natural affinity, were termed "homologous" characters. And characters that seem to display false resemblance or affinity were termed "analogous." Wings of different birds were considered "homologs." Wings of birds and bees were considered "analogs." The job of the taxonomist, or classifier, was somehow to distinguish between homologs and analogs, and in doing so to discover natural affinities. Once discovered, the natural affinities could be made the basis of a "natural system" of classification that would harmonize with the plan of creation, unknown though that plan may be.

Such concepts do not really differ greatly from modern ones, although the language in which the modern concepts are expressed, and the concepts themselves, are more precise. An important element of modern concepts was added as a result of evolutionary theory, initiated by Lamarck at the beginning of the nineteenth century, and further developed by Darwin in the mid-nineteenth century. In his book on the origin of species, Darwin stated: "Community of descent is the hidden bond which naturalists have been unconsciously seeking, and not some unknown plan of creation, or the enunciation of general propositions, and the mere putting together and separating objects more or less alike" (1859:420).

After scientists accepted the principle of evolution, natural affinity was viewed as the result of evolution rather than creation. But the idea of natural affinity, and the way to discover it, changed hardly at all. Groups thought to be natural were still thought to be natural, and for the same reasons: the observed similarities and differences between species. To the question, "what is a group?" a pre-evolutionary scientist might have responded: "a *natural* group—a group with natural affinity." Pressed further by the question, "what is natural affinity?" one might have added: "the plan of creation." To the same question ("what is a group?"), an early evolutionist might have similarly responded: "a *natural* group—a group with natural affinity." Pressed further, however, by "what is natural affinity?" one would have added: "propinquity of descent." Particular groups need not have changed at all from pre- to post-evolutionary times. In a practical sense, therefore, the ideas of "plan of creation" and "propinquity of descent" were, for all intents and purposes, initially synonymous. They were not to remain so, at least not entirely.

The idea of "natural affinity" in the sense of "propinquity of descent" early and easily lent itself to graphic representation in the form of a branching diagram—soon to be called a "family tree," "phylogenetic tree," "phyletic tree," etc. An early genealogical diagram of this kind was published by Duchesne in 1766 (figure 2.61), another by Lamarck in 1809 (figures 2.62, 2.63). A similar diagram, the first of many published in this style during the next 130 years, was published by Agassiz in 1844. Interestingly, Agassiz's diagram (figure 2.64) was not intended to reflect evolutionary relationships, for Agassiz was not an evolutionist (Patterson 1977). Another, entirely theoretical, diagram was published by Darwin in 1859 (figure 2.65). The first comprehensive diagram pertaining to real organisms was published by Haeckel in 1866 (figure 2.66). In Haeckel's figure, there is a clear correspondence between groups of organisms and the lines of the diagram—the "branches" of the "tree."

Most modern diagrams of this kind are patterned after Agassiz's, Darwin's, or Haeckel's (relative to them, Duchesne's and Lamarck's are upside down). But a few are patterned on Duchesne's and Lamarck's (figure 2.67). In actuality, it makes no difference. A tree may be upside down, rightside up, or turned sideways (figure 2.68). It is still the same tree, with the same branches in the same relative positions. The branches of the tree, of course, are intended to portray lines of descent extending through time. As such, the trees are often viewed as simple representations of history. The bifurcations particularly are so viewed—as representing the division of an ancestral species into descendant species that subsequently evolve, or diverge, in their own fashions through the eons of time. Some branches might change, or evolve slowly, through time. Others might divide, or bifurcate, again and again, producing numerous terminal twigs representing species alive today. Others might become extinct.

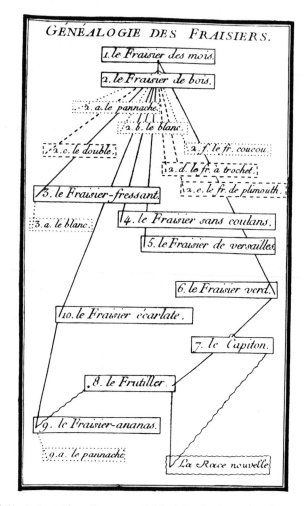

Figure 2.61. A branching diagram, published by Duchesne in 1766, showing the relationships between the various types of strawberries, as he conceived them. After A. N. Duchesne (1766), Histoire Naturelle des Fraisiers (Paris:Didot), plate facing p. 228.

With the advent of evolutionary theory and, especially, phyletic trees, scientists began to think about groups in a different way. Asked, "what is a group?" they began to respond: "a branch of a phyletic tree." Pressed by, "what is a phyletic tree?" they could answer: "a diagram of phyletic relationships." Pressed still further by, "what is phyletic relationship?" they could add: "propinquity of descent." To some persons, the answers seemed increasingly vague, leading further and further into the realm of theory, away from things that could be touched and observed. It seemed that something had happened to the tangible similarities and differences displayed by different species. Where were the homologs and analogs? These, too, became redefined in terms of evolutionary theory. Homology became "resemblance due to common ancestry." Analogy became "resemblance due only to common function." Yet the characters considered homologous could still be demonstrated as an ultimate justification for the enterprise of evolutionary classification. And in most cases, they were the same homologous characters, recognized, presumably, in the same way and for the same reasons. They could be felt and observed just as clearly as in the old days of pre-evolutionary biology. Thus it was that evolutionary theory settled, albeit somewhat uneasily, over the enterprise of biological classification. The job of the

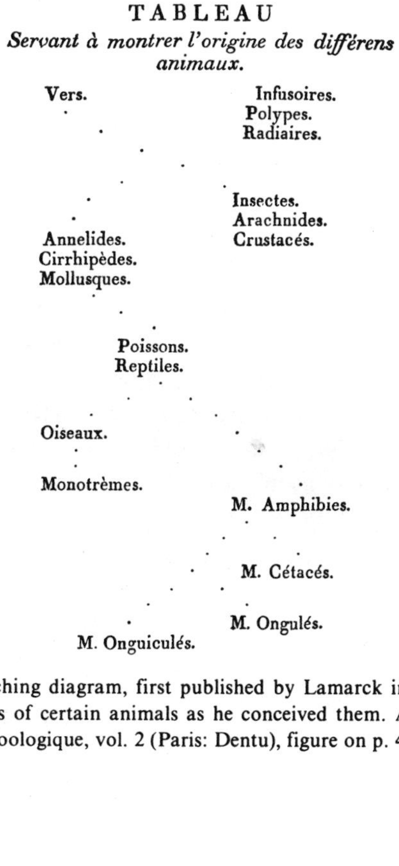

Figure 2.62. A branching diagram, first published by Lamarck in 1809, showing the evolutionary relations of certain animals as he conceived them. After J.-B. Lamarck (1809), Philosophie Zoologique, vol. 2 (Paris: Dentu), figure on p. 463.

TABLE

SHOWING THE ORIGIN OF THE VARIOUS ANIMALS.

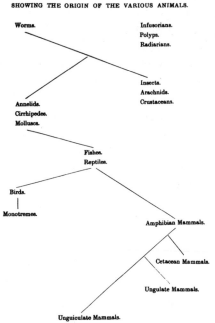

Figure 2.63. Lamarck's branching diagram as it appeared in the only English translation of his work. After J.-B. Lamarck (1914), Zoological Philosophy; trans., with an introduction, by H. Elliot (London: Macmillan; reprint, New York: Hafner, 1963), figure on p. 179.

taxonomist was still to distinguish homologous from analogous characters. But one could think about homologs, if one wished to do so, in a different way—as the indicators of common descent rather than the plan of creation. Once determined, they could be used to construct a phyletic tree. Once constructed, a tree could be dismembered to make a classification. The only problem was where to apply the ax. The same tree could be chopped up in many different ways, and it often was.

STABILITY OF CLASSIFICATION

Stability in classification has always been considered desirable, but there has always been general agreement that stability must give way to genuine scientific advance. Evolutionary theory was an advance not to be denied entry into the area of biological classification. But it entered at its own risk, promising as it did to produce stable classification based on the principle of propinquity of descent. Like most revolutionary regimes, it initially promised a lot, and subsequently gave only a little in the way of stability. But consideration of the stability problem was postponed by the development of the "New Systematics" of the 1930s and 1940s—with its emphasis on the species as the real unit of evolution. During that period, the job of the taxonomist became, so it was said, to detect evolution at work—by careful scrutiny of, and experimentation upon, the real units of evolution, the evolving species. In the context of the tunnel vision of the time, the problem of stability evaporated; no one really cared about it.

Not until after the Second World War did the stability problem

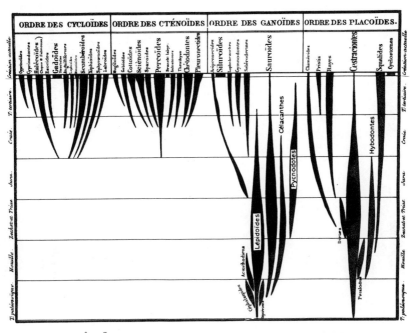

Figure 2.64. A branching diagram, published by Agassiz in 1844, showing the genealogy of the class of fishes. After L. Agassiz (1833–1844), Recherches sur les Poissons Fossiles, vol. 1 (Neuchatel), plate facing p. 170.

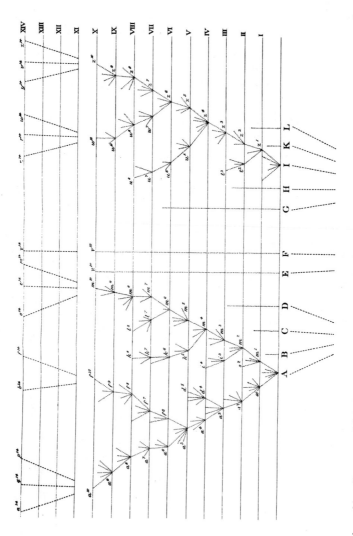

Figure 2.65. A branching diagram, published by Darwin in 1859, theoretically showing the subsequent development and diversification of species A–L, through time periods I–XIV. After Darwin (1859), plate facing p. 117.

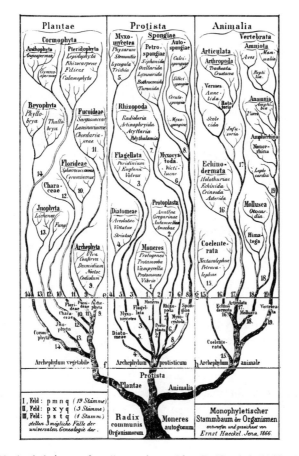

Figure 2.66. A phyletic tree for all organisms. After E. Haeckel (1866), Allgemeine Entwicklungsgeschichte der Organismen (Berlin: Reimer), plate 1.

emerge for serious consideration. By that time, evolutionary classification, bolstered by the seemingly impressive accomplishments of the "New Systematics," already had a lengthy history of nearly 100 years. Numerous studies had been conducted. Numerous phyletic trees had been constructed. And the ax of the classifier had been applied by numerous pairs of hands to numerous trees representing, supposedly, the history of numerous groups of plants and animals. The more popular groups had been studied and restudied by generations of evolutionists. For the popular groups, it became possible to study the

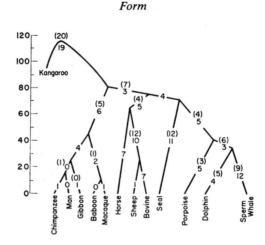

Figure 2.67. A tree of the interrelationships of 13 different mammals, based on molecular structure of proteins (myoglobins). After M. Goodman and G. W. Moore (1974), Phylogeny of hemoglobin. Syst. Zool. 22:508–32; figure 3, p. 514.

changes wrought by succeeding generations of taxonomists. The phyletic trees changed through time and so also, it seemed, did the chopping techniques of the classifiers. The changes did not always seem to represent genuine scientific advances. This conclusion was reached independently by several different scientists of different countries. The revolutionary regime of evolutionary classification had a problem, and the problem did not go away. It grew.

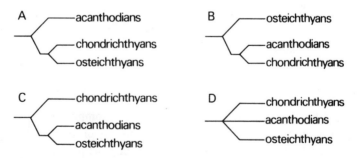

Figure 2.68. Alternative cladograms, showing the possible interrelationships between three groups: acanthodians (fossil fishes known only from the Paleozoic Era), chondrichthyans (chimaeras, sharks, and rays), and osteichthyans (bony fishes). After R. Miles (1973), Relationships of acanthodians. In P. H. Greenwood, R. S. Miles, and C. Patterson, eds., Interrelationships of Fishes. Supplement no. 1 to the Zoological Journal of the Linnean Society (London), 53:63–103 (New York and London: Academic Press), figure 1, p. 65.

EVOLUTIONISM

The various attempts to respond to the problem in book-length may be mentioned here. One response, best and uniquely represented by Simpson (1961), was to try to state clearly the principles of evolutionary classification. Simpson's book stemmed from his long experience and well-known expertise in paleontology, but perhaps more than anything else merely summarized the practice of the last 100 years. In doing so, his book marked and perhaps hastened the end of evolutionary theory as traditionally applied to classification. It was a "Bible" written retrospectively, so to speak, near the end rather than the beginning of an intellectual tradition. Simpson doubtlessly performed a great service by bringing the theory, such as he was able to formulate it, into the open, where it could be inspected and criticized by the world at large. The taxonomic world lost no time in reacting, and by reacting, taxonomy moved from one era into another.

PRACTICAL TAXONOMY

Another response, best represented by Blackwelder (1967; also Boyden 1973), emphasized that evolutionary theory of classification is, after all, only theory; and that taxonomists might better stick to the facts rather than expend their energies in futile theorizing about ancestors and the like. In doing so, taxonomists would produce natural classifications, just as they always had done in the past, because, if enough facts are available, no artificial classification could ever result from legitimate taxonomic enterprise. There is merit, of course, in Blackwelder's argument. Evolutionists did tend to regard their interpretations as facts rather than theories, particularly in regard to phyletic trees—especially if the trees were "documented" by "fossil evidence." Perhaps no evolutionist ever went so far as to claim his tree was completely factual, but numerous evolutionists did claim that their trees were as close to the truth as available evidence would permit, and it is difficult to see much difference between the one claim and the other. No doubt Blackwelder is correct in pointing out also that taxonomy has a practical side, which has endured, and will continue to endure, theoretical upheavals from time to time. He argues, in effect, that taxonomy is a practical matter, independent of theory. But his argument leaves something to be desired by persons who believe that taxonomy, and science generally, involve both facts and theory—data and their interpretation. Blackwelder attempts no refutation of the idea (e.g., Popper 1959) that facts are dependent upon theory—an idea that would probably strike him as absurd. The criticisms of Blackwelder (and Boyden), therefore, seem directed mainly toward evolutionists—of whom there were many— insensitive to theory, its uses in science, and its limitations. But neither Blackwelder nor Boyden develops these themes. Their books are not contributions to a general theory of classification, but rather retreats from general theory and theorizing, at least within the traditional evolutionary context.

PHENETICS

Another, and in some ways more impressive, response—best represented by Sokal and Sneath (1963; also Jardine and Sibson 1971; Sneath and Sokal 1973; Clifford and Stephenson 1975)—was a mathematical and computer-oriented approach, sometimes called "numerical taxonomy." Here it will be discussed in its "phenetic" aspect, as "phenetic taxonomy." It, too, was a movement away from evolutionary theorizing. It focused on the problem of "natural affinity" as evidenced by actual similarities and differences. Specimens could be observed and their characters noted. If coded quantitatively, the characters could be added together and averaged. Species could be discriminated and defined quantitatively. Once defined, they could be compared, and a numerical value could be calculated to represent the similarity between two species. By appropriate techniques, species could be grouped, the groups could be compared, and the values of similarity computed. By adopting the strategy of grouping together most-similar species and, subsequently, most-similar groups, a classification could result.

The results of initial test cases of phenetic taxonomy were encouraging because the groups formed by the computer in many cases were the same groups that nonmathematical taxonomists had defined a few or many years before. The hope was that the computer might one day take over, preserving the best practices of taxonomy, while eliminating the worst—particularly the arbitrariness that seemed to underlie, and sometimes to dominate, the traditional nonmathematical approaches to classification.

An early procedure of phenetic taxonomy, one that has persisted ever since, was to portray the results of phenetic analyses by means of a branching diagram (figure 2.69)—a diagram with many of the same

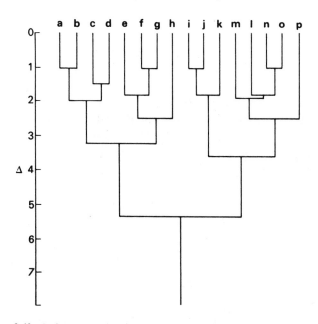

Figure 2.69. A phenogram showing the pattern of overall similarity between species a-p. The ordinate, in units 0-7, is a scale of decreasing similarity. After Sneath and Sokal (1973), figure 5-7, p. 231. Copyright © 1973, W. H. Freeman.

properties as a phyletic tree. It was argued that phenetic trees and phyletic trees need not really differ except in the ways they are interpreted by humans. For a given group, both trees were constructed from the same data base: the observed similarities and differences of organisms. But it was argued that a phenetic tree was a better, or at least more empirical, interpretation of the data, because it was mathematically exact, and because it did not encourage its interpreters to go beyond the facts into the speculative realm of evolutionary history. Because of the potential repeatability of mathematical procedures, as programmed in the computer, phenetic taxonomy offered a promise of ultimate stability. Tell the computer how to construct a tree, and it will construct the same tree from the same data time and time again. Tell the computer how to construct one tree, and it will construct all trees the same way. Tell the computer how to chop one tree, and it will chop it the same way over and over again if need be. Tell it how to chop one tree, and it will chop all trees in the same way. Or so it seemed.

Phenetic taxonomy soon developed problems of its own. And the question to be asked by the computer ("how do I chop the tree?") was seldom answered in an explicit way (figure 2.70). Instead, new mathematical methods were devised to construct better phenetic trees. And soon, it seemed, there were more permutations of phenetic techniques than there ever were trees, phenetic or otherwise. Each new technique, of course, had the potentiality of producing a different tree from the same data. And so it did. Thereupon arose a new question for the computer to ask: "what is the best procedure to construct a tree?" No answer was forthcoming. And, for want of a definable goal, phenetic taxonomy became lost in a maze of its own technical inventiveness. And there it remains today, with a very uncertain future.

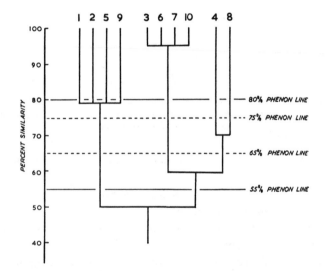

Figure 2.70. A phenogram that, according to Sokal and Sneath, is to be chopped into pieces (phenons) by drawing horizontal lines. Their legend reads: "Dendrogram to show the formation of phenons (for explanation, see text)." Their text reads: "An example of the delimitation of phenons can be seen in Figure [2.70]. Drawing a horizontal line across the dendrogram at a similarity value of 75% creates four 75-phenons, 1, 2, 5, 9; 3, 6, 7, 10; 4; and 8 .... A second phenon line at 65% forms three 65-phenons. The advantage of phenons is that it is obvious that they are arbitrary and relative groups. This is not true of the Linnean nomenclatural scheme. If some investigator felt that the taxa in Figure [2.70] should be divided into two instead of three groups, the phenon line would have to be drawn at a similarity value between 50% and 60%; or he might feel that the two phenon lines were too close together and did not summarize the main relations very fairly, as a result of which he might draw the line at the 80% level." After Sokal and Sneath (1963), figure 9–1 and pp. 251–53. Copyright © 1963, W. H. Freeman.

Whatever their other theoretical accomplishments, evolutionary and phenetic taxonomists did prove that a branching diagram was useful—or at least interesting—in at least two ways. Indeed, one might argue that a major accomplishment of evolutionary taxonomy was the invention of the branching diagram and its subsequent widespread use, however successful, to depict propinquity of descent. Similarly, a major accomplishment of phenetic taxonomy might be considered the use of the branching diagram to depict not propinquity of descent but propinquity of phenetic relationship ("overall" similarity). As a result, it became possible to view the branching diagram in a more general way: as a device to portray propinquity (general group similarity—be it phyletic, phenetic, or whatever). Suddenly, the branching diagram assumed a new significance, independent of evolutionary or phenetic theory. And it became possible to explore the stability problem at a new and more general level by investigating the relation between the branching diagram and classification, independent of any statement or preconception of the nature of propinquity (general group similarity). This possibility was not exploited until after another, and perhaps most important, response to the stability problem.

PHYLETICS

Relationship This response is best represented by Hennig (1950, 1966; also Crowson 1970; and, to some extent, Ross 1974), although its basic elements had been clearly stated many years before (Mitchell 1901; Rosa 1918, 1931). Hennig was concerned with the apparent inadequacy of evolutionary theory as applied to classification. Unlike Blackwelder or Sokal and Sneath, he neither retreated from the inadequacy nor proposed an alternative to the evolutionary, or phylogenetic, system. He confronted the inadequacy and tried to rectify it within its own context. Its basic cause, he observed, was ambiguity in the concept of "natural affinity," which in the evolutionary context means "propinquity of descent" or "common ancestry." He recognized that all organisms may reasonably be assumed to be related through common ancestry. He concluded, therefore, that "common ancestry" by itself is insufficient to define a useful concept of phyletic relationship (in the sense of general group similarity). Hence, he defined phyletic relationship unambiguously, and more restrictively, as degree of common ancestry, such that a species C is more closely related to a species D than to a species B or A, if

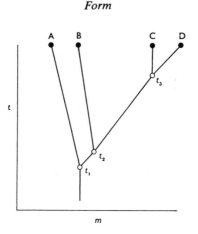

Figure 2.71. A phyletic tree illustrating Hennig's concept of relationships of four hypothetical species (A, B, C, D); *t* is time, and *m* is a measure of overall similarity. Species C and D are more closely related to each other than to species A and B. Species B is more closely related to species C and D than to species A. After Brundin (1966), figure 3, p. 16.

C and D have a common ancestor that is not also an ancestor of B or A (figure 2.71). Immediately, much past evolutionary taxonomy came into sharper focus. It became possible to distinguish between precise and ambiguous phyletic trees, and precise and ambiguous statements of relationships. And it was immediately obvious that much of the "best" existing evolutionary taxonomy is interpretable according to Hennig's definition of relationship—which retrospectively explained, perhaps, why the "best" evolutionary taxonomy had been so adjudged.

Monophyly   His concept of relationship caused Hennig to examine other concepts of evolutionary taxonomy, for example, the concept of monophyly. Within the context of evolutionary taxonomy, a group was said to be "natural" or "monophyletic" if its included species were believed to have a common ancestor. He pointed out that any group whatever is "monophyletic" under that definition. He consequently defined the concept unambiguously, and more restrictively, such that a group is "monophyletic" if its included species have an ancestor in common only to themselves (figure 2.72). It therefore became possible to require that the phyletic tree, however chopped, yield groups that conform to this general and unambiguous criterion of monophyly. And it was immediately obvious that many of the "best-founded" groups in existing evolutionary classification conformed to this criterion—which retrospectively explained, perhaps, why the "best-founded" groups had been so adjudged.

Synapomorphy   Hennig considered also the nature of evidence of relationship, and he concluded that only shared advanced, or derived, characters, which he called "synapomorphies," constitute evidence (figure 2.73). The concept of synapomorphy was unambiguous, and more restrictive, relative to the traditional concept of homology, which it replaced in Hennig's system; for all homologous resemblances may be considered synapomorphies at one or another level of phyletic relationship. It therefore became possible to scrutinize, with reference to the concept of synapomorphy, the "evidence" of relationship—the homologies—accumulated during the history of evolutionary taxonomy. Some of this "evidence" was immediately recognized to consist not of

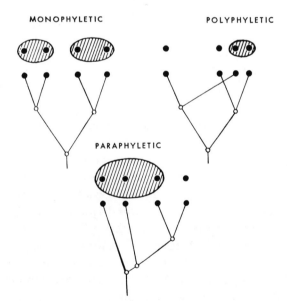

Figure 2.72. Diagrams illustrating Hennig's concept of monophyly. Black dots represent species; circles, common ancestors. The species of a monophyletic group have an ancestor in common only to themselves. Nonmonophyletic groups may be either paraphyletic (based on symplesiomorphy) or polyphyletic (based on convergence). Modified from Hennig (1969), figure 1, p. 18.

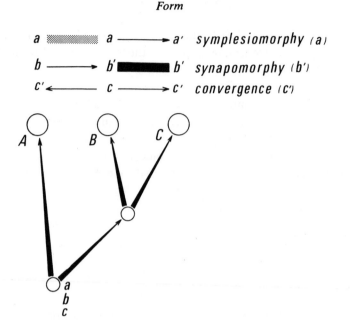

Figure 2.73. Diagram illustrating Hennig's concept of resemblance, which may be divided into synapomorphy, symplesiomorphy, and convergence. A, B, C are species; a, b, c are characters that change through time into advanced characters a', b', c', After Hennig (1966), figure 44, p. 147. Copyright © 1966, Board of Trustees of the University of Illinois.

shared advanced characters ("synapomorphies"), but rather of shared primitive characters ("symplesiomorphies"), which are really synapomorphies associated with the wrong level of relationship. An example may illustrate the point: the chimpanzee and orang-utan, in having more body hair than humans, are not, therefore, evidenced to be more closely related to each other than to humans, for the presence of hair is evidence only that all three species are mammals; what might seem to be a homologous resemblance between the chimpanzee and orang-utan is symplesiomorphic, i.e., retention of a primitive character of mammals— a mammalian synapomorphy. But it was also immediately obvious that much of the "best," or "most important," evidence consisted of synapomorphies—which retrospectively explained, perhaps, why the "best" evidence had been so adjudged.

Hennig's concepts of relationship, monophyly, and synapomorphy initially seemed to comprise a potent and revolutionary critique of traditional evolutionary taxonomy, at least as it was presented in the summary of Simpson. But in many ways, Hennig's concepts—in contrast to Simpson's—allowed a deeper and more critical understanding of past taxonomic efforts, particularly why some succeeded better, or were more convincing, than others. Hennig's critique consequently began to look less like a revolutionary manifesto, and more like an incisive exposition of the best of past taxonomic practice and theory. Indeed, it now seems an attempt, generally successful, to tighten the entire theoretical structure of evolutionary taxonomy. And the question is moot whether Simpson's or Hennig's is really the more traditional exposition of it.

Cladograms A unique feature of Hennig's concepts is that they are largely understandable and, indeed, definable only with reference to a branching diagram, which plays a role also in specifying the nature of classification as he understands it. In effect, he recommends that degree of common ancestry be the one and only general group similarity. Or, in other words, all taxa should be monophyletic groups (figure 2.74). Because all monophyletic groups can be specified in the structure of a phyletic tree, all that a classification can do is to mirror that structure, in part or completely. This concept of classification, and its relation to a phyletic tree, has often been misunderstood, perhaps at times even by Hennig himself. The idea, in effect, is that a proper phyletic tree requires no chopping in order to be converted into a classification. Conceptually, therefore, this proper tree differs from the traditional tree of evolutionary taxonomy.

The idea of a phyletic tree—one requiring no chopping to be converted into a classification—seems to have originated as an outgrowth of Hennig's own practical taxonomic experience, as seems the case also for Mitchell and Rosa. Neither Hennig, Mitchell, nor Rosa developed the idea to its full generality, and there is some inconsistency in Hennig's diagrams (e.g., 1966: figure 15). In any case, Hennig's critics, sensing the apparently new and unique nature of his phyletic tree, labeled it a "cladogram," and Hennig's ideas collectively as "cladism." These terms unfortunately were intended to have an explicitly evolutionary significance pertaining to the actual branching, or speciation events, of phylogeny. Although such events are an implication of

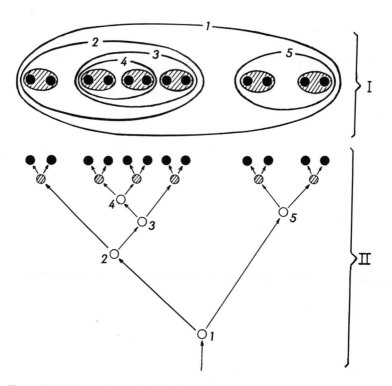

Figure 2.74. Diagram illustrating Hennig's concept of the relation between a phyletic tree and the taxa of a classification. II.1–5 are hypothetical ancestors whose descendant species (black dots) form monophyletic groups I.1–5. After Hennig (1966), figure 18, p. 71. Copyright © 1966. Board of Trustees of the University of Illinois.

Hennig's diagrams, they are an implication, also, of phenetic trees, which might, therefore, be considered cladograms in that sense. The term "cladistic" would also apply to a classification designed to reflect the branching structure of a phenetic tree, for the usual phenetic tree is similar to Hennig's in requiring no chopping in order to be converted into a classification. Indeed, phenetic taxonomists have come to regard their phenetic trees as "classifications," for the trees in themselves specify all groups that from the phenetic viewpoint might legitimately be recognized as taxa (figure 2.75).

Synapomorphy Patterns The majority of the monophyletic groups specified by Hennig's type of trees (cladograms) might be considered evolutionary in nature, reflecting actual speciation events—branchings of the historical process. And so might the groups specified by phenetic trees. But *all* of Hennig's groups correspond by *definition* to patterns of synapomorphy. Indeed, Hennig's trees are frequently called synapomorphy schemes. The concept of "patterns within patterns" seems, therefore, an empirical generalization largely independent of evolutionary theory, but, of course, compatible with, and interpretable with reference to, evolutionary theory. The concept rests on the same

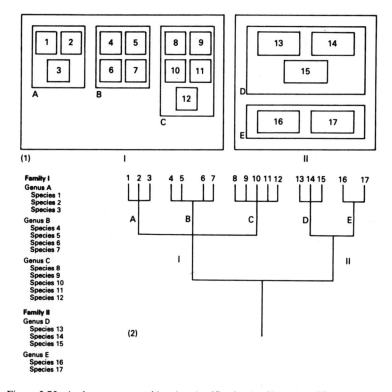

Figure 2.75. A phenogram considered a classification by Heywood. His legend reads: "Two forms of presenting the same hierarchical classification of two families, five genera and seventeen species listed: (1) a box-within-box arrangement; (2) a tree-like phenogram." After V. H. Heywood (1973), Chemosystematics—an artificial discipline. In G. Bendz and J. Santesson, eds., Chemistry in Botanical Classification, pp. 41–54 (London and New York: Academic Press); figure 1, p. 46. Copyright © 1973, Academic Press.

empirical basis as all other taxonomic systems (the observed similarities and differences of organisms). But the concept is not wholly independent of evolutionary theory, for one of its basic elements (nature of evidence) is synapomorphy, or shared advanced character. The other basic elements, namely relationship (what is evidenced) and monophyly (what is resolved), are definable only with reference to the branching diagram, and carry no necessary evolutionary connotation. Indeed, the concept of synapomorphy may be definable purely as an element of pattern—a unit of resolution, so to speak. If so, Hennig's system would be understandable not merely as the theory of "phyletic" taxonomy but as the general theory of taxonomy of whatever sort. The general properties of Hennig's system—the basic elements and their logical interrelations as exhibited by the branching diagram—are, perhaps, the more interesting properties of this system.

Trees and Classification   The relation between a phyletic tree and a classification has traditionally been conceived as a relation between a temporal continuum (the tree) which is "known" to one degree or another (figure 2.76), and a hierarchy of discontinuous taxa to be constructed on the basis of what is "known." The problem of constructing a classification has traditionally been conceived as one involving the division (chopping) of a temporal continuum into discrete pieces (taxa). How the division is accomplished was considered always to involve a certain arbitrariness (in applying the ax). Hennig's branching diagram, differing as it does from a traditional phyletic tree, is no more nor less continuous than the classification that mirrors its structure. It is, therefore, of general interest to determine if Hennig's type of branching diagram is a more appropriate representation of information concerning relationships (general group similarity of whatever sort) than the traditional tree—be it phyletic, phenetic, or otherwise. This subject will be considered in further detail below. But it may be mentioned here that the traditional conception of the tree as a temporal continuum probably has been the root cause, perhaps totally unnecessary, of most arbitrariness in past classification. If so, Hennig's type of tree, or cladogram, holds the key to a general solution to the stability problem.

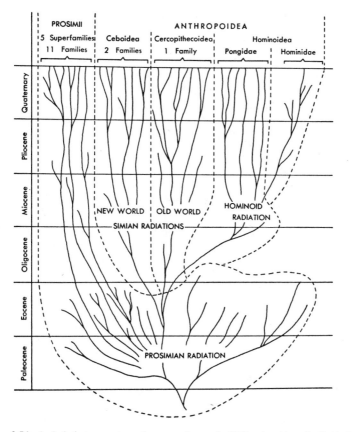

Figure 2.76. A phyletic tree portrayed as a continuum (solid lines), arbitrarily divided (by dashing) into two main groups (Prosimii and Anthropoidea) that are recognized as taxa. After Simpson (1961), figure 28, p. 213.

Trees and Cladograms Some differences between a traditional phyletic tree and a synapomorphy scheme may be appreciated with reference to figures 2.77 through 2.80. Assume that an actual phylogeny is shown in figure 2.77, and that it has been sampled by collection of specimens, and discovery of 13 diagnosable species (both fossil and recent), at four different times, A, B, C, and D. A synapomorphy scheme for the 13 species, if it could be constructed in its complete detail, would be that of figure 2.78 (a phenetic tree constructed for 13 species would

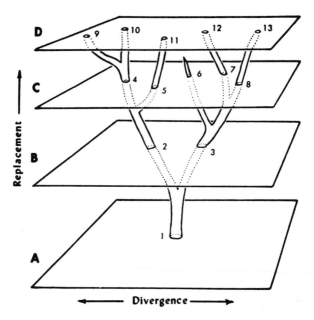

Figure 2.77. An assumed real history of a group of species evolving through time (vertical axis), known from samples of specimens at four different time levels (A, B, C, D). The specimens comprise 13 samples, each definable and identifiable as a species by observation of its characters. After N. D. Newell (1956), Fossil populations. In P. C. Sylvester-Bradley, ed., The Species Concept in Paleontology, pp. 63–82 (London: Systematics Association, pub. no. 2); figure 2, p. 68.

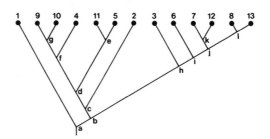

Figure 2.78. A cladogram for species 1–13, derived from figure 2.77; a–l are branch points.

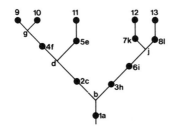

Figure 2.79. A first-order tree, derived from the cladogram of figure 2.78.

have the same general form, with the 13 species at terminal positions on the tree). The scheme (figure 2.78) may be absolutely correct in what it says about "patterns within patterns" of synapomorphy. But the scheme need not, and in this hypothetical case cannot, be correct in estimating the actual branching that took place. The scheme overestimates the number of branch points by 240 percent and in itself provides no indication of which points are artifacts in a historical sense. Viewed as a synapomorphy scheme, the diagram may be assumed to be correct. Viewed as a phyletic tree, the diagram may be assumed to approximate, to one degree or another, an unknown historical reality ("known" in this hypothetical case to be that of figure 2.77).

The scheme (figure 2.78) includes 13 species (1–13) and 12 branch points (a–l). The sense of each branch point is that the species distal to it have synapomorphic resemblance. Species (8, 13) distal to point l, for example, presumably exhibit some advanced characters shared only among themselves. Viewed simply as a synapomorphy scheme, the diagram has no more sense than the above. Viewed as a traditional tree, the diagram has an *additional* sense: that each branch point represents a

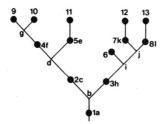

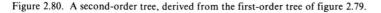

Figure 2.80. A second-order tree, derived from the first-order tree of figure 2.79.

historical event (a speciation). Although the difference lies mainly in viewpoint, the difference will be emphasized in what follows by distinguishing between two types of diagrams: cladograms (e.g., a synapomorphy scheme) and trees (e.g., the traditional tree):

Cladogram: branch point = synapomorphic resemblance
Tree: branch point = synapomorphic resemblance + speciation event

It may be assumed that a cladogram, if constructed correctly and completely, includes in its branching structure all of the branch points (speciation events) of the actual phylogeny, or evolutionary history, of its species. The speciations of figure 2.77, for example, are included in figure 2.78 as points b, d, g, i, and j. It may be assumed also that a cladogram may include additional branch points, which would be artifacts in a historical context. The additional points would accurately represent synapomorphic resemblance, but they would not represent speciations. The cladogram may, nevertheless, be considered a starting point for arriving at an accurate tree—a best estimate of actual history. All that is necessary to convert a cladogram into an accurate tree is to eliminate the artifactual branch points.

Elimination of a branch point may be accomplished if a given species is removed from its terminal position and placed at a branch point. For example, branch point a may be eliminated by placing species 1 at the branch point. Elimination of a branch point may be accomplished, in effect, by reducing the length of the line extending from point a to species 1 until species 1 and point a are coincident. In a historical context, placing species 1 at branch point a designates species 1 as the ancestor of species 2–13. Elimination of branch points from a cladogram, therefore, is equivalent to designating species as ancestors. And a cladogram, therefore, is a branching diagram in which no ancestors are designated.

Designation of ancestors may involve one or more criteria, such as the following: (1) the ancestor should be a species, rather than a group of species; (2) the ancestor should be known only from fossils. It is possible to designate groups as ancestors, but that possibility is not needed for this example, in which all ancestors are in fact species. That an ancestor should be known only from fossils follows from the fact that any branch point implied by two contemporaneous species will not be eliminated by the discovery of fossil specimens of either or both species. In any case, the above list of criteria is not definitive; rather it is suggestive of some criteria that would permit ancestors to be designated in reference to the hypothetical example of figure 2.78.

With consideration of the above criteria, all "possible ancestors" may be designated, and the cladogram (figure 2.78) may be transformed into a tree (figure 2.79). This tree is a first-order approximation of the actual historical reality (figure 2.77), for it includes only the minimum number of speciation events, as represented by branch points. The minimum number can be independently calculated by counting the maximum number of contemporaneous species (5 in this case) and subtracting one; if all the species are contemporaneous, which would be true if all species were recent species, or all species were fossil species from the same time horizon, the minimum number and the maximum number are the same, one less than the number of species.

The first-order approximation sets one limit (the minimum) to the number of branch points; the cladogram sets the other limit (the maximum). Provided the cladogram correctly reflects the "patterns within patterns" of synapomorphy, the actual historical reality (figure 2.77) must be described either by figures 2.78 or 2.79, or must lie somewhere between figures 2.78 and 2.79. A first-order approximation (figure 2.79) has been derived from the cladogram (figure 2.78) by a process of designating "all possible ancestors." A second-order approximation may be achieved by rejecting one or more of the "possible ancestors," if there are reasons to do so.

Rejection of "possible ancestors" may involve criteria such as the following: (1) an ancestor should be primitive (relative to all supposed descendants) in all known characters; (2) an ancestor should occur earlier in time than all of its descendants. Failure to meet either criterion could be considered grounds for rejection. If applied correctly to figure 2.79, these criteria are sufficient to reject species 6 as a "possible ancestor" (species 6 should not be primitive in all characters; in any event, species 6 is contemporaneous with species 7 and 8). And the second-order approximation may be achieved (figure 2.80).

Trees may, therefore, be classified either as first-order or second-order trees:

Tree: branch point = synapomorphic resemblance + speciation event
    A. First-order tree: all "possible ancestors" designated (minimum speciations)

B. Second-order tree: all rejectable ancestors rejected (final estimate of speciations)

That a first-order or second-order tree may be achieved does not mean that the tree is correct—even though in this case figure 2.80 exactly duplicates the branchings of the hypothetical reality (figure 2.77). It means only that, if the synapomorphy scheme is correct, a correct approximation might be achieved by way of a two-step process. But the two-step process does not guarantee a correct result even if the process is correctly applied to a correct synapomorphy scheme. About all that can be said for the process is that its result is about as close to the reality as allowed by available evidence and methods. But how close is that? If figure 2.77 were not available, how might the truth of figure 2.80 be evaluated? Are there really only five branch points? Might there not be six, seven, eight, or even twelve? The weakness in the process lies in the second step: rejecting "possible ancestors." Have enough been rejected? Rejection requires information that may be considered in relation to a set of criteria. Even though the criteria may be sound, available information may be insufficient. Is there a way to test the sufficiency of the information? Have too many "possible ancestors" been rejected? Are the criteria adequate for rejection? These questions will be postponed for consideration in the next chapter, for they do not help to clarify the nature of cladograms, trees, or classification, the subjects now at hand.

The distinction drawn here between a cladogram and a tree is one that Hennig did not emphasize. But he did emphasize that a cladogram, as a summary of "patterns within patterns" of synapomorphy, is of fundamental importance for evolutionary theorizing (see above) and for classification (see below). Its importance as a basis for both, but especially for classification, stems from its structural properties and, ultimately, from what its structural properties represent: the observed "patterns within patterns." Unlike a traditional tree (e.g., figures 2.66 and 2.76), the cladogram is not a temporal continuum, wherein one species must be separated from another, and one group from another, only arbitrarily—in a manner analogous to chopping a tree into sections of trunk and branches. As indicated, however, a cladogram can function as a basis for estimating a historical reality by a two-step process. A third step may be added: the transformation of the second-order tree (figure 2.80) into a third-order tree—a temporal continuum unperturbed by sampling discontinuities (figure 2.81).

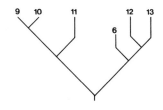

Figure 2.81. An ideal tree (third-order tree), derived from the second-order tree of figure 2.80.

Ideal Trees Such an unperturbed continuum is, of course, the ideal tree of the traditional evolutionist, for it is the ultimate estimation of the historical reality. And, for this example, it is—as demonstrated by comparison between figures 2.77 and 2.81. For the evolutionary taxonomist, it is the type of concept one strives to achieve prior to constructing a classification, for according to that view it is the best type of concept on which to base classification. And according to that belief, the only obstacle to stable classification—as stable as classification could ever be—is the lack of information required to arrive at a true concept of the historical continuum in its natural unperturbed state.

It is to Hennig's credit to have indicated that the ideal tree (figure 2.81) is really a derivative concept of a cladogram (figure 2.78); that the ideal tree, although it requires more information and accessory criteria (the three-step process) than the cladogram, actually obscures the information summarized by, and the empirical basis of, the cladogram; that the ideal tree is prone to chronic misunderstanding about sufficiency of information (see above); and that, as a basis for classification the ideal tree is inferior to the cladogram, for the ideal tree imposes an additional requirement inviting arbitrary action from the taxonomist (chopping the tree).

Although Hennig indicated these properties of cladograms, he may not have foreseen their full importance. Other writers commenting on his work have considered the ideal tree as the ideal basis for classification in Hennig's sense (figure 2.82). Griffiths (1974), for example, proposed to chop the ideal tree in a nonarbitrary way, not by chopping through the branches, but through the branch points. This proposal has been widely discussed, and the implication has been noted (e.g., by Mayr 1974:109) that, according to this view, an ancestral species ceases to exist once it has divided into two or more descendant species (figure

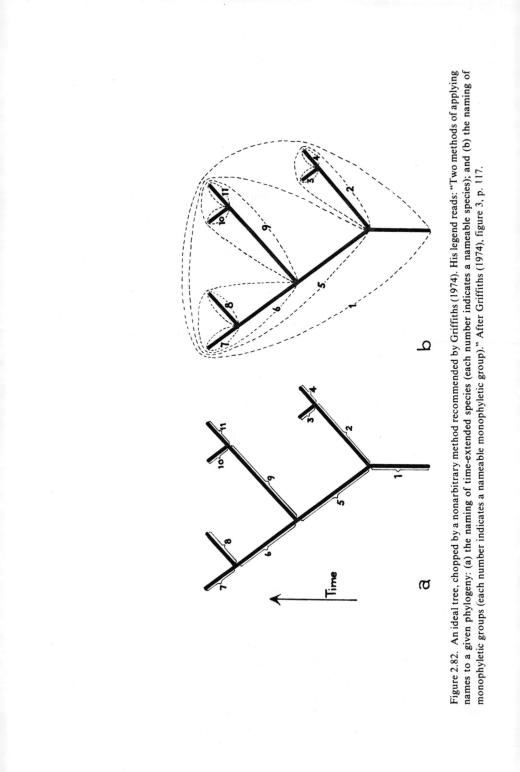

Figure 2.82. An ideal tree, chopped by a nonarbitrary method recommended by Griffiths (1974). His legend reads: "Two methods of applying names to a given phylogeny: (a) the naming of time-extended species (each number indicates a nameable species); and (b) the naming of monophyletic groups (each number indicates a nameable monophyletic group)." After Griffiths (1974), figure 3, p. 117.

2.82a). This proposal, applied to figures 2.77 through 2.81, would mean that species 5, 7, and 8 are not "real" species—a conclusion that may be interesting from a theoretical standpoint, but which contradicts the previous empirical determination that species 5, 7, and 8 exist. The conclusion depends upon a correct outcome of the three-step process of ideal tree construction. Interestingly, the outcome may be incorrect, and the conclusion false, even though the cladogram from which the tree is ultimately derived is absolutely correct in reflecting patterns of synapomorphy. What then is the purpose of imposing a conclusion that may be false on the empirical basis of the conclusion—to modify the data to fit the theory? The only purpose would be consistency in the use of the ideal tree. But the ideal tree need not be used at all.

Considered as a basis for forming groups (figure 2.82b), the ideal tree, even if chopped through the branch points, yields some or all of the same groups as the cladogram (see below). In any event, the ideal tree does not yield any groups different from those yielded by the cladogram. The ideal tree is not, therefore, the ideal basis for classification. It imposes a certain species concept, but otherwise performs as a cladogram, although not, perhaps, in as detailed a manner.

The cladogram, in effect, satisfies the concern about futile theorizing of Blackwelder, Sokal and Sneath, and other taxonomists, for that theorizing, or at least a large part of it, belongs to the three-step process of deriving a tree from a cladogram.

Synapomorphy and Homology   What is synapomorphy, exactly? As indicated above, synapomorphy is a restriction of the traditional concept of homology. But what is homology? It has been a key concept of comparative anatomy for over one hundred years. As a concept, it has had a tortuous and disputatious history. There have been numerous attempts to define it and numerous attempts to specify exactly how to recognize it. There are even different kinds of it: special homology, serial homology, etc. Fortunately, statements of homology are easy to recognize: for example, paired fins of fishes are homologous to the arms and wings of land vertebrates. But what does the word "homologous" mean? It may mean different things to different persons. But perhaps to all biologists it means at least "comparable." But what does "comparable" mean? Are not all things "comparable"? Potentially yes, but actually biologists do not compare arms with eyeballs or noses (although there is no rule that biologists may not do so). They compare, for example, arms of a human, or wings of a bird, with the front legs of a frog or the pectoral fins of fishes. To a biologist, these organs display interesting similarities and differences, interesting perhaps only because a biologist can generalize about them. And by stating that they are homologous, one means at least that general statements can be made about them. If one could generalize about noses and arms, they might be called homologous, too. A statement of homology may be put into an alternative form:

(1) Homologous Paired Appendages
        Pectoral fins
        Arms (forelegs)
        Wings

Although fins, arms, and wings may be observed to differ in some respects, they may also be observed to have common properties, such as form, mode of embryonic development, internal structure of muscles, nerves, blood vessels, skeleton, movement, function, chemical composition, etc. To the extent that they have common properties, the common properties would define the concept of "Homologous Paired Appendages." But the common properties would not define the term "homologous." Perhaps all the term need mean is that there are known, or might be discovered, some common properties of, for example, fins, arms, and wings, such that uniting these three types of structures as "Homologous Paired Appendages" might possibly serve as a useful basis for discussion—a basis perhaps more useful than any other, and a discussion more interesting than, for example, a discussion of the similarities and differences among wings, noses, and eyeballs. To make a homology statement, therefore, is to assume, suppose, or expect that such is the case.

Homology and Classification   A homology statement is an attempt to generalize and, in a sense, to classify. Stating that certain structures are homologous is much the same as stating that certain species belong to the same genus. In the above discussion, there are really four homology statements, which comprise a set with three subsets. And one might suppose that the four sets, or more properly the set with its three subsets, might form a basis for classification of the organisms that possess the homologous structures:

(2) Homology Statements   Classification
Set: Homologous Paired Appendages ..... Taxon: Vertebrata
    Subset: Pectoral Fins .... Subtaxon: Pisces
    Subset: Arms .... Subtaxon: Tetrapoda
    Subset: Wings .... Subtaxon: Aves

The assumption, or expectation, underlying the classification is that other congruent homologies (generalizations) are known, or may be found, to correspond to the four groups (Vertebrata, Pisces, Tetrapoda, Aves) defined by the four homology statements. Fishes (Pisces), for example, live in the water and breathe with gills. Four-legged animals (Tetrapoda) crawl or run about on the land and breathe with lungs. And birds (Aves) have feathers and fly. Insofar as these and other generalizations could be made about these groups, the groups could be said to be based on, or to exemplify, homologies, or homologous resemblance. Such has been the traditional use and significance of homology.

With increased knowledge, the assumption, or expectation, of congruent generalization either is, or is not, realized in a given case. Increased knowledge of vertebrates, for example, has revealed two basic types of reproduction. Eggs are either laid in the water, where they are kept moist; or they are laid out of the water, in which case they soon

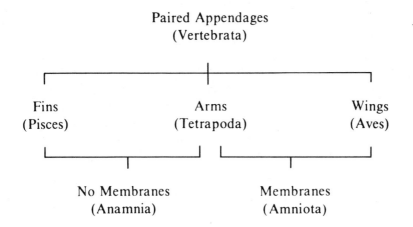

develop a complex of membranes that in effect create an aquatic habitat for the developing embryo. Fishes and some tetrapods (amphibians) exhibit the former type of reproduction; other tetrapods (reptiles and mammals) and birds exhibit the latter. This pattern of incongruent generalities is problematical, for it contradicts the original assumption, or expectation, underlying the classification based on appendages.

Another type of problem contradicting the original assumption, or expectation, is sometimes created by the discovery of previously unknown organisms. The Devonian fossil *Eusthenopteron*, for example, has a "pectoral fin" with the internal skeleton of an arm, rather than a fin (figure 2.83). And the Jurassic fossil *Archaeopteryx* has a feathered "wing" with well developed digits ("fingers") suggestive of an arm, rather than a wing (figure 2.84). In some ways, therefore, *Eusthenopteron* has homologous resemblance with Pisces, and in other ways with Tetrapoda; and *Archaeopteryx,* in some ways with Tetrapoda, and in others with Aves.

These two types of problems, which may be called incongruence and intermediacy, offer various approaches to a solution. One may accept a traditional classification, and simply declare that the contradictions to it are exceptions: "*Eusthenopteron* may have the skeleton of an arm in its fin, but it's still a fish." Or one may develop a phenetic philosophy and seek to maximize the number of generalizations by choosing among contradictory classifications, or parts of classifications: "*Eusthenopteron* has more fishlike characters than tetrapod characters and is, therefore, a fish, not a tetrapod." Hennig suggested that the problem is with the nature of the concept of homology itself, rather than with truly contradictory information.

Synapomorphy and Classification   In a phyletic context, Hennig referred to a set of homologous characters as a "transformation series," which could be further analyzed into relatively primitive and relatively advanced characters. And although all characters are primitive at some level, and advanced at another, only at the advanced level do characters define congruent sets of ultimate utility in classification. At that advanced level, the characters become synapomorphies in Hennig's terminology. With respect to vertebrates, wings may be considered modified arms; arms, modified fins; fins, modifications of a more

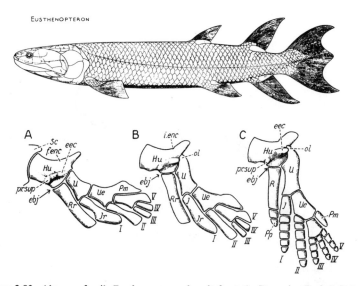

Figure 2.83. Above, a fossil, *Eusthenopteron foordi*, from the Devonian Period. Below, the skeleton of the left arm, as seen from the side. A, *Eusthenopteron*; B, hypothetical intermediate; C, primitive tetrapod. *Hu*, humerus; *I*, intermedium ray; *Pm*, postminimus; *Pp*, prepollex; *Sc*, endoskeletal shoulder girdle; *U*, ulna; *Ue*, ulnare; *ebj*, elbow joint; *eec*, ectepicondylar process; *f. enc*, entepicondylar foramen; *i. enc*, notch corresponding to *f. enc*; *ol*, olecranon process of ulna; *pr. sup*, supinator process; *I, II, III, IV, V*, digits I–V. Above, after E. Jarvik (1960), Théories de l'Évolution des Vertébrés (Paris: Masson), figure 12, p. 42. Below, after E. Jarvik (1964), Specializations in early vertebrates. Ann. Soc. Roy. Zool. Belgique 94:11–95; figure 27, p. 78.

primitive body wall; and embryonic membranes, modifications of a more primitive type of development. Viewed from this perspective, the two transformation series are not contradictory, and the problem of incongruence evaporates:

Ordinary Body Wall
Modified Body Wall (Fins) .... Vertebrata
    Ordinary Fins
    Modified Fins (Arms) .... Tetrapoda
        Ordinary Development
        Modified Development (Membranes) .... Amniota
            Ordinary Arms
            Modified Arms (Wings) .... Aves

Viewed from this perspective, *Eusthenopteron*, despite some primitive

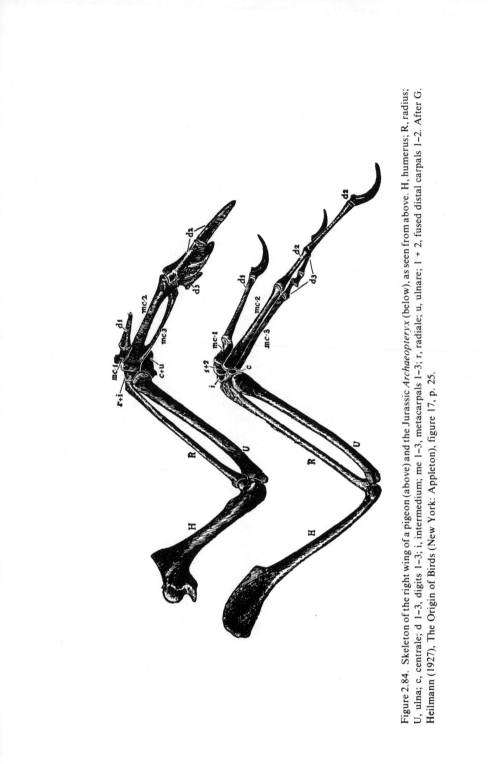

Figure 2.84. Skeleton of the right wing of a pigeon (above) and the Jurassic *Archaeopteryx* (below), as seen from above. H, humerus; R, radius; U, ulna; c, centrale; d 1–3, digits 1–3; i, intermedium; me 1–3, metacarpals 1–3; r, radiale; u, ulnare; 1 + 2, fused distal carpals 1–2. After G. Heilmann (1927), The Origin of Birds (New York: Appleton), figure 17, p. 25.

characters (vertebrate synapomorphies), is a tetrapod; and *Archaeopteryx,* despite some primitive characters (tetrapod synapomorphies), is a bird. From this perspective, the "fishlike" characters (vertebrate synapomorphies) of *Eusthenopteron* indicate only that *Eusthenopteron,* like all tetrapods, is a vertebrate. And the "reptilelike" characters (tetrapod synapomorphies) of *Archaeopteryx* indicate only that *Archaeopteryx,* like all birds, is a tetrapod.

Synapomorphy as a Subset   What is a synapomorphy, exactly? Hennig defined synapomorphy as a shared advanced character, or apomorphy. It is contained as an element of a statement such as: vertebrate arms are an advanced character relative to fins. Therefore, organisms with arms share that apomorphy and are synapomorphic in that respect. Statements of this sort (arms are modified fins) abound in the literature of comparative anatomy, even in its pre-evolutionary period of the seventeenth and eighteenth centuries. Why should this have been so? Early comparative anatomists might have contented themselves with apparently less argumentative or less theoretical statements, such as: of vertebrate appendages there are two kinds, fins and arms.

(3) Appendages   (4) Fins
    Fins   Ordinary Fins
    Arms   Modified fins (Arms)§

Both kinds of statements contain an element of homology (both statements specify a set that includes fins and arms). One statement, (3), apparently includes only an element of homology: fins and arms comprise a set (of appendages). The other statement, (4), includes an element of homology plus an obvious element of synapomorphy §: arms are modified fins. Statement (3) is arguable insofar as the reality of the set might be disputed. Statement (4) is arguable in the same sense and in another sense as well: the reality of the set might be accepted, but a person might contend that fins are modified arms.

(5) Arms
    Ordinary Arms
    Modified Arms (Fins)§

To some extent, this difference in arguability is only a problem of words, for the two statements are more similar than they might seem.

Consider two types of structures, X and Y, which are stated to be homologous and, therefore, to form a set S:

(6) Set S: a, b, c
       Subset X: a, b, c
       Subset Y: a, b, c, d, e§

If the properties of X are a, b, and c, and the properties of Y are a, b, c, d, and e, the defining properties of the set S are, therefore, a, b, and c. Set S is synonymous with subset X. And Y is a synapomorphy. Y is modified X.

The statement that X and Y are homologous means that there is a set S that includes X and Y as subsets. The statement that Y is modified X means that with respect to the defining characters of X, X is synonymous to S.

The concepts of homology and synapomorphy have, therefore, an empirical base. And the concepts are interrelated. Homology implies generality (that there is a set that includes . . .), and synapomorphy implies relative, or restricted, generality (that there is a subset included in . . .). This empirical basis was, of course, evident to pre-evolutionary biologists, who considered the "modifications" as modifications of ideas, or types (archetypes), rather than modifications of historical descent (evolutionary modifications in the modern sense). Pre-evolutionary biologists invoked the Plan of Creation in order to explain the "modifications," but in reality it was the pattern of "modifications" (synapomorphies) that made manifest the Plan of Creation. And it was the same pattern (or set of patterns within patterns) that early evolutionists sought to explain by the theory of descent, but in reality it was the pattern of synapomorphies that made manifest the theory of descent. In the world of taxonomy, the two explanations were equivalent in an important sense: there could only be one Creation and, alternatively, there could only be one process of historical descent. There could be, therefore, only one natural system—one hierarchical pattern of synapomorphies that would embrace all living things.

In some cases, statements of homology and, especially, synapomorphy, may seem problematical.

(7) Set S: b, c, d
       Subset X: a, b, c, d§
       Subset Y: b, c, d, e§

Given the information of statement (7), that there are structures X and Y with properties a, b, c, d, and b, c, d, e, respectively, one may conclude that there is a Set S, defined by b, c, and d, but there is no basis for stating that X is modified Y, or that Y is modified X. However, one might assert that both X and Y are modified S—that X and Y are both synapomorphies.

To consider synapomorphy as restricted, or relative, generality makes possible further abstract analysis, which will not be pursued here. The present argument is intended to show that synapomorphy has the same empirical basis as homology, that both concepts are interdependent, that they may be considered without reference to evolutionism, and that an evolutionary element of interpretation may be added to them without necessarily changing their empirical basis. Homology does not exist independent of synapomorphy, any more than a set exists independent of its members (subsets).

We may ask why early anatomists viewed nature as a hierarchical system of patterns within patterns (or sets within sets). We may ask, but we may not be able to answer, at least not definitively. Why they chose to do so was, perhaps, the belief that nature was best described in that fashion. To some extent, the belief may have had an empirical basis. A heart, for example, may be empirically considered a modified blood vessel by anyone who observes the embryological process of vascular development.

Ontogenetic Transformations　The process of embryological development has often been considered important in relation to homology statements. Homologous structures, so it is said, should exhibit a common pattern of development. But the patterns of embryological differentiation surely played a role also in the resolution of the major synapomorphies of early, even pre-evolutionary, biologists. For only in the embryological process could structures actually be observed to transform from one state of organization to another. It is no coincidence that the word "evolution" originally referred to the ontogenetic process.

In modern time, it has never been doubted that life has evolved from the simple to the complex, at least with regard to its levels of cellularity. Unicellular organisms have always been considered primitive with respect to multicellular organisms. An important reason for this belief, one might suppose, is the observation that multicellular organisms typically exhibit a unicellular stage that, through a process of development, actually transforms itself into a multicellular adult. An adult, without doubt, is a modified egg (or, more properly, a modified zygote). That the multicellular stage might, therefore, be considered an advanced character need not stem from a belief in the biogenetic law—that the ontogenetic process recapitulates phyletic history. Rather, it stems from the greater generality of the unicellular stage.

(8) Set: Unicellular Organisms
  Ordinary Unicellular Organisms (unicellular throughout life)
  Modified Unicellular Organisms (with multicellular adult)§

The significance of ontogenetic character transformations is sometimes dismissed as tending toward conundrums. For example, how can a zygote be "homologous" to an adult? Or, better, how can an amoeba be homologous to an adult *Homo sapiens*? But such questions are not meaningless, for they relate to the nature of cellularity and its taxonomic correlate, the group of cellular organisms (the Eukaryota)—a subject that, unfortunately, cannot be discussed further here. What is important is to note that ontogenetic character transformations played some role in the resolution of synapomorphies during the early history of taxonomy; and that these early resolutions, largely empirical in nature, provided a basis for subsequent resolution of finer details in the patterns of synapomorphy displayed by the living world.

The sense of synapomorphy, then, is that one character is less general, or more restricted, than its homologs, relative to some empirical source of reference.

Synapomorphy and General Group Similarity   If some number of species, A, B, C, etc., are said to be synapomorphic in some respect, let us say in having wings rather than ordinary arms, or fins, or an unmodified body wall without appendages, or no body wall at all (as in a unicellular organism), the species A, B, C, etc., are synapomorphic relative not to some but to all other organisms. Thus a statement of synapomorphy implies comparison, and therefore homology, between certain characters of a specified group and certain characters of all other organisms, characters such as organs, more inclusive parts of other organisms, or even organisms as wholes. Among vertebrates, paired appendages such as wings, arms, and fins may be compared among themselves, but how may such structures be compared to parts of a limbless amphioxus, flatworm, coelenterate, or protozoan? Such comparisons are possible, but only at more general (more inclusive) levels. For example, the body wall of vertebrates, including its paired appendages, may be compared to that of amphioxus or a flatworm, which lack paired appendages, with the resolution of paired appendages (or a body wall with paired appendages) as a synapomorphy:

(9) Body Wall
    No Appendages
    Appendages§

Comparison of the body wall of these organisms would resolve the mesoderm (or a body wall with mesoderm) as a synapomorphy:

(10) Body Wall
    No Mesoderm
    Mesoderm§
        No Appendages
        Appendages§

A comparison between metazoans and protozoans would resolve multicellularity (or a life cycle with multicellular adult) as a synapomorphy (statement 8). One may argue, therefore, that there is an unbroken chain of homology from any structure, such as a vertebrate organ (the wing), to an entire protozoan, by way of increasingly more inclusive levels of biological organization. Such seems to be the nature of homology (and synapomorphy): a homology statement (synapomorophy) seems to involve, ultimately, all of life while designating, perhaps, only one or a few restricted subsets of it.

So in reality, to assume homology between two structures (or other types of characters) is to imply at least two synapomorphies: one synapomorphy for the restricted subset, and one synapomorphy for the inclusive set (cf., statement 6):

(11) Fins§
    Ordinary Fins ⎫
    Modified Fins§ ⎬ Homology Statement
                   ⎭

And there is a possibility that three synapomorphies may be implied (cf., statement 7):

(12) Fins§
     Modified Fins A§ ⎫
     Modified Fins B§ ⎬ Homology Statement
                       ⎭

Although two or more synapomorphies may be implied, they need not all be explicitly recognized or explicitly stated. But even when no restricted subset can be understood as a synapomorphy, the inclusive set may be so understood explicitly (cf., statement 3):

(13) Fins§
     Fins A ⎫
     Fins B ⎬ Homology Statement
             ⎭
or

(14) Appendages§
     Fins ⎫
     Arms ⎬ Homology Statement
           ⎭

It is impossible, therefore, to make a homology statement without also making an explicit synapomorphy statement. Or, in other words, all sets may be conceived as restricted subsets (synapomorphies) of still other sets.

Of course, one may make synapomorphy statements of one sort or another without much effect. Although the statements may have an empirical base, they are not, therefore, in themselves automatically true, or even generally significant. To interpret a synapomorphy in an evolutionary context, as a shared advanced character, does not guarantee it actually to be such. What, then, is the significance of synapomorphy statements, and how is their significance to be judged—according to what standard of relevance?

There are, perhaps, many possible standards of relevance. One, perhaps the most important, is generality. Indeed, if a synapomorphy statement has any real significance, it may consist of the generalization that certain species are synapomorphic not merely in one respect, but also in other respects that are as yet unknown. This generalization may be stated as a prediction: that a certain group of species, observed to be synapomorphic in one respect, with further study will be observed to be synapomorphic in other respects not yet discovered; or, stated alternatively, that no synapomorphies will be discovered that are shared by only some species of the group and one or more species outside the group. This prediction, in effect, is that newly discovered synapomorphies may be more inclusive, or less inclusive, but in any case will not be incongruent with those already discovered. This prediction may be called the hypothesis of congruence, or the hypothesis of general synapomorphy. A cladogram, therefore, may be considered a hypothesis of general synapomorphy for the species or groups included in the cladogram in the pattern specified by the branching structure. The cladogram of figure 2.71, for example, is a hypothesis of general synapomorphy between (1) C and D, (2) B, C, and D, and (3) A, B, C, and D, of which the pattern may be summarized thus:

(14) ABCD
      BCD
      CD

To predict that some pattern of synapomorphy is general is not the same as proving it to be so. Such a prediction is, however, open to testing with new samples of information. And the testing can be performed in an unbiased way if synapomorphies can be resolved from new information independently of previous resolutions of synapomorphies.

As a general group similarity, synapomorphy, therefore, is no more nor less than generality: what can be stated to be generally true for groups of species. Synapomorphy is not a new concept, for it includes all valid generalizations that have heretofore been made about groups of species. The word "valid" is used here in what might be a peculiar sense. It might seem perfectly "valid," for example, to state that bees and hummingbirds both have wings and visit flowers. And that generalization might seem, therefore, a "valid" synapomorphy in the sense of generalization. But the sense of synapomorphy is that species truly synapomorphic are synapomorphic generally (with respect to other features as well). Available information indicates that bees and hummingbirds are not generally synapomorphic—a conclusion that may be tested by sampling the available knowledge of these organisms, or by accumulating new information.

A synapomorphy statement is, therefore, a hypothesis of general synapomorphy. Rejected hypotheses of general synapomorphy (or congruence) are not valid synapomorphy statements: they are merely invalid, or rejected, hypotheses of generality. They are commonly called convergences (figure 2.73).

To the extent that synapomorphy statements can be made, tested, and corroborated, a pattern of congruent synapomorphies results. Congruence of generalizations about the living world has been perceived, in one degree or another, for hundreds, perhaps thousands of years. And it has been variously interpreted; as revealing, for example, the Plan of Creation or the phylogenetic history of life. That congruence, once it has been perceived (in whatever degree), may be interpreted in one context or another, or that it may seem more or less interesting or important (depending on the viewpoint of the interpreter), has no necessary connection with the pattern of congruence, or even with the manner in which the pattern is discovered. This is not to say that the interpretation can play no helpful role in discovery of the pattern, but only that the interpretation may have no necessary role. Despite the numerous writings of evolutionists, to the effect that evolutionism is inseparable from the taxonomic process (of pattern recognition), it is doubtful that any necessary role has so far been demonstrated. And the question of the role, if any, that evolutionism plays in the perception of pattern remains open to investigation. It will not be further pursued here because this account is focused on the logical structure of pattern, and so far at least, this structure seems understandable in terms largely independent of evolutionary theory. In their most general sense, therefore, synapomorphies are merely the defining characters of a group.

Synapomorphy and Evolutionism As for modern statements of synapomorphy, they seem distinctive only in having an explicitly evolutionary connotation. It is true, of course, that by having this evolutionary connotation, statements of synapomorphy are brought into relation with other statements of modern science, e.g., those that pertain to interrelationships of species and those that pertain to the processes and mechanisms of genetics, heredity, and evolutionary change. And the possibility is created that these statements might alter, or otherwise have an impact on, the empirical basis of synapomorphy resolution. Yet this seems to have happened only to a limited extent, if at all. It is not yet generally possible, for example, to resolve synapomorphies by study of the genetic processes that produce them. And it seems doubtful if the study of process and the study of pattern need ever merge more than they do now.

There is, of course, a danger that, once coupled with evolutionary theory, the taxonomic enterprise might lose contact with some portion of its empirical base. But such need not be the case. Indeed, the empirical base of taxonomy seems clearer today than ever before, largely because of the efforts of Hennig and others to state clearly the principles of phylogenetic taxonomy. The resulting clarification in the concepts of taxonomy was not to be confined to the area of phylogenetic interpretation, for it became extended to a more general level as, it may be hoped, this book demonstrates.

Thus, the cladogram, as a summary of the pattern of synapomorphy, also satisfies, or nearly so, the need for empiricism stressed by Blackwelder, Sokal and Sneath, and many other taxonomists (not to mention humanity at large), for the "patterns within patterns" are largely, or ultimately, empirical in nature. It is true that the concept of synapomorphy, as commonly defined, may be considered to have a nonempirical element (the "advanced" part of "shared advanced character"); however, as indicated above, synapomorphy may be alternatively viewed in purely empirical terms. But, more importantly, patterns of synapomorphy, if they emerge repeatedly, as they have, from independent studies of different samples of characters of different levels of different structural and functional systems, have themselves a generality that is empirical enough. The concept of synapomorphy, even in an evolutionary context, could in no way impose upon nature the generality displayed by the "patterns within patterns" that taxonomic procedure has already uncovered. There is no reasonable alternative to accepting this structured knowledge (the patterns of synapomorphy) as a real manifestation of the living world. What else could be accepted in its place? It is not, therefore, an artifact of taxonomic procedure.

It will not be argued here that construction of repeatable cladograms (repeated observation of the same patterns of synapomorphy) depends upon infallibly correct resolution of synapomorphies, as interpreted in an evolutionary sense. Probably the system of taxonomic enterprise permits a wide margin of error in resolving individual synapomorphies. Perhaps all that is necessary is that a taxonomist guess right more often than wrong, and that taxonomists jointly keep busy enough to maintain a useful signal-to-noise ratio.

Summary In summary of the above concepts elaborated by Hennig (1950, 1966) and by other authors in response to the stimulus of his work, it may be said that his major accomplishment was perhaps the invention of the cladogram (in a phyletic context) and the realization that its structure is of fundamental importance to classification. Hennig also used the cladogram to define the concepts of relationship, synapomorphy, and monophyly, which logically related these elementary concepts in a unified system. Thus, it would seem that the job of taxonomists has not changed much as a result. We still have to discriminate characters and recognize homologs (synapomorphies), so as to discover the "patterns within patterns" manifested by the living world—which was always our job and which will remain so, as long as we and the living world continue to exist, and as long as we have the curiosity to investigate it. As classifiers, we have been relieved of some responsibilities, futile perhaps, imposed by early enthusiasts of evolutionism. And our ax, with our license to use it, has been taken away. Hopefully, these changes will be for the better.

EVOLUTIONISM REVISITED: GRADISM

The last major response to the stability problem, and the last to be considered here, is that of Mayr (1969), which was written after the other responses and was, to some extent, an attempt to respond to them, too. However, Mayr's book is primarily a revision of an earlier work stemming from the period of the "New Systematics" (Mayr, Linsley, and Usinger 1953), and only secondarily an attempt at a modern exposition of taxonomic theory. Mayr, nevertheless, tried to synthesize the diverse taxonomic theories into a new system, based on the concept of genetic similarity (as his preferred general group similarity). He called his system "evolutionary taxonomy," claimed to distinguish it from "phenetic" and "cladistic" theories, and claimed to trace its origin to Darwin. Mayr's concept of genetic similarity and its significance typifies the clarity of the concepts of his system. According to Mayr, genetic similarity cannot be measured directly, but must be inferred by a process of "weighting similarity." "Weighting" is defined as *a method for determining the phyletic information content of a character*" (p. 218). With respect to this method, Mayr states:

The scientific basis of a posteriori weighting is not entirely clear, but difference in weight somehow results from the complexity of the relationship between genotype and phenotype. (p. 218)

He states also:

To undertake a successful a posteriori weighting of characters requires a thorough knowledge of the history of previous classifications of a given group and an ability to make value judgments. Yet no clearly better method has so far been found. (p. 219)

The vagueness of these concepts makes them difficult to discuss. Mayr concludes that:

A classification based on phyletic weighting has numerous advantages. It is the only known system that has a sound theoretical basis, it has greater predictive value than other kinds of classifications, it stimulates a character-by-character comparison of organisms believed to be phylogenetically related, and it encourages the study of additional characters and character systems in order to improve the soundness of the classification and hence its information content and predictive value. Finally, it leads to the discovery of interesting evolutionary problems. (p. 86)

However, he provides no evidence to support the truth of the first two claims (both of which have been subsequently refuted by Farris 1980) or to refute the view that the last three claims are true of most or all kinds of classifications.

Mayr recommends use of a branching diagram (figure 2.85), which, according to his view, successfully combines both "phenetic" and "cladistic" aspects and therefore represents a new synthesis of taxonomic

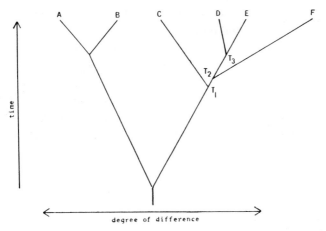

Figure 2.85. A branching diagram of the type recommended by Mayr (1969). His legend reads: "The phyletic dendrogram (phylogram) of the evolutionary taxonomist. The slower a line evolves, the more it approaches the vertical; the more rapidly it evolves (as T_2 to F), the more it approaches the horizontal." After Mayr (1969), figure 10–18, p. 256. Copyright © 1969, McGraw Hill, Inc. Used with the permission of McGraw-Hill Book Company.

theory. Elsewhere, Mayr considers these aspects as "variables," and he states that the purpose of the "evolutionary taxonomist" is to "maximize simultaneously in his classification the information content of both types of variables" (1974:95).

With respect to evolutionary theorizing, Mayr believes that "as complete as possible a reconstruction of phylogeny must precede the construction of a classification" (Mayr 1974:95). This "complete reconstruction" would seem, therefore, to be that of the traditional evolutionist, as represented by the ideal tree (figure 2.81). If so, Mayr's recommendations for the formation of taxa would amount to rules of thumb that might be applied during the chopping procedure. Indeed, Mayr (1969:97) contends that "taxa ranked in higher categories represent the main branches of the phylogenetic tree." As for the chopping procedure, Mayr maintains the traditional view that the procedure "cannot be carried out without an element of arbitrariness" (1969:98).

What Mayr's various claims add up to is difficult to say. Perhaps he has done as well as anyone to characterize his own work: "The time has not yet come to present a well-balanced methodology of macrotaxonomy" (1969:199). Although many of Mayr's comments amount to a restatement of traditional evolutionism, rather than a critical search for elements of a general theory, the thrust of Mayr's argument centers on an appraisal of the total biology of organisms. For this reason, Mayr believes that his system is more meaningful biologically. Be that as it may, one of Mayr's key concepts, representing to some extent a departure from traditional evolutionism, is that of general adaptive level, or grade—a concept Mayr adopted from the writings of Julian Huxley. For this reason, Mayr's system may be termed "gradism." An attempt will be made below to investigate the nature of gradistic theory and practice.

3

SYSTEMATIC PATTERNS: COMPONENT ANALYSIS

CLADOGRAMS, TREES, AND COMPONENTS

It has been stated above that the cladogram contains information sufficient for specifying groups and subgroups to be recognized as taxa in classifications, and that the information resides in the structure of the cladogram. The unit of information of this kind may be called a component, corresponding to a statement of general synapomorphy. Given two species, A and B, for example, a component in the form "AB" is a statement of general synapomorphy: that species A and B share apomorphy, and are therefore related. What follows is an analysis of the possible components of cladograms and other branching structures—what may be called "component analysis." For convenience, the term "taxon" is used to indicate an entity that may be a single species, or a group of species.

COMPONENT ANALYSIS OF TWO TAXA

The sense of a branching, or dendritic, diagram *seems* immediately obvious. The sense of the alternative diagrams of figure 3.1, for example, could be said to be that taxon A is ancestral to B, or that taxa A and B are separated by some unspecified phenetic or gradistic distance. By way of introduction, the phyletic viewpoint is adopted.

EQUIVALENT TREES

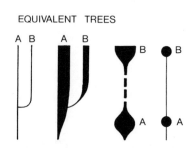

Figure 3.1. Branching diagrams of various styles, showing taxon A as ancestral to taxon B.

Consider an analysis of two taxa (figure 3.2). There are only three ways that two taxa (A and B) might be phyletically interrelated: (1) taxon A is ancestral to B (tree 3.2.1); (2) taxon B is ancestral to A (tree 3.2.2); (3) an unknown taxon X is ancestral to both A and B (tree 3.2.3). But there is another way of stating these three possibilities, and that way consists of what may be called "component analysis." Tree 3.2.1, for example, has two components: (1) taxa A and B are related by common ancestry (the general component); (2) the common ancestor [of A and B] is A (the unique component). Similarly, tree 3.2.2 has two components: (1) taxa A and B are related by common ancestry (general); (2) the common ancestor is B (unique). And finally, tree 3.2.3 has two components: (1) taxa A and B are related by common ancestry (general);

ANALYSIS OF TWO TAXA			
TREES	COMPONENTS		GROUPING
	GENERAL	UNIQUE	
1 ●B ●A	AB	A	AB 1
2 ●A ●B	AB	B	AB 1
3 A B ●● Y X	AB	X	AB 1
CLADOGRAM			
A B ∨ Y 1	AB		AB 1

Figure 3.2. Component analysis of two taxa, A and B, showing the three possible trees, one possible cladogram, and one possible grouping (classification).

(2) the common ancestor is neither A nor B but an unknown taxon X (unique).

The general component is called "general" because it occurs as a component of all three trees (3.2.1–3); the unique components, in contrast, are unique to each tree ("A" to tree 3.2.1; "B" to tree 3.2.2; "X" to tree 3.2.3). The general component ("AB") defines, and is the sole component of, a dendritic structure that in this context may be called a "phylogram" (cladogram 3.2.1). The phylogram is defined by the general component: "taxa A and B are related by common ancestry," which is the total information content of the phylogram. This component, however, has itself two parts, which may be called "cladistic" and "phyletic," respectively. The cladistic part is "Taxa A and B are related"; and the phyletic part is "by common ancestry." If the phyletic part is omitted, only the cladistic part remains: "Taxa A and B are related" (by some unspecified relation). In fact, figure 3.2 does not specify the nature of the relation between taxa A and B. For that reason, the trees in themselves are not phyletic trees; they are simply trees. And the cladogram in itself is not a phylogram; it is simply a cladogram.

A cladogram, therefore, may be defined as a branching, or dendritic, structure, or dendrogram, illustrating an unspecified relation (general synapomorphy) between certain specified terms that in the context of systematics represent taxa. If the relation is considered common ancestry, there is reason to call the structure a phylogram; or, alternatively, if the relation is considered phenetic or gradistic similarity (however measured), there is reason to call the structure a phenogram or gradogram, respectively. In either case, the structure itself need not change—only the interpretation of its significance.

Because it is defined by a general component, a cladogram denotes a set of trees. A cladogram, therefore, is not a tree. A tree may be defined as a branching, or dendritic, structure, or dendrogram, having one or more *general* as well as *unique* components (or combination of components). A cladogram, in contrast, is a dendritic structure having only one or more general components. Conceived as a set, a cladogram does not resemble any one of its member-trees more than another: cladogram 3.2.1 does not resemble tree 3.2.3 more than trees 3.2.1 and 3.2.2. It resembles all of its member-trees equally. And the resemblance consists only of the general component shared by all member-trees and the cladogram. A cladogram, therefore, is not *a* tree; it is a *set* of trees.

A cladogram is a branching structure joining certain terms (representing taxa) that are related by some unspecified relation. In itself, a cladogram conveys no sense of phylogeny, common ancestry, phenetic resemblance, gradistic resemblance, ecological resemblance, or any other relation that might conceivably join the terms (representing taxa). Of course, it may seem odd to think of a phenogram, for example, as a type of cladogram. But a phenogram does, after all, have a cladistic (branching) aspect that as *branching* is no different than the branching of a phylogram. And the implication of a phenogram, like that of a phylogram, is that certain groups (represented by the components) exist, an implication that can be investigated independently of any implied phenetic (or phyletic) relations.

There is only one way that two taxa might be grouped (classified): as subtaxa of some inclusive taxon. In other words, the two taxa must be classified together: in figure 3.2 the grouping AB is the only one possible. Not surprisingly, its terms ("AB") are those of the general component. Thus, there is a one-to-one correspondence between the grouping and the cladogram (the general component of the trees). Put another way, the groups (inclusive taxa) of a classification correspond on a one-to-one basis to the components of a cladogram (the general components of the member-trees of the set denoted by the cladogram).

COMPONENT ANALYSIS OF THREE TAXA
Other properties of cladograms may be considered in relation to a component analysis of three taxa (fig. 3.3). For three taxa, there are four possible cladograms (primary, or fully resolved, cladograms 1–3, and tertiary, or unresolved, cladogram 1) and, naturally, four corresponding classifications (groupings 1–4). The four cladograms, and the groupings, differ among themselves in their components. The cladograms comprise what may be called three "primary cladograms" and one "tertiary cladogram" (and the corresponding groups comprise what might be called three "primary classifications" [groupings 1–3] and one "tertiary classification" [grouping 4]). Each primary cladogram is a set of six "primary trees," and the tertiary cladogram is a set of four "tertiary trees."

The component analysis of a primary tree (in a phyletic context) is as follows. For tree 3.3.1 there are two general components: (1) taxa A, B, and C are related by common ancestry (a "tertiary component"); (2) taxa

PRIMARY TREES	COMPONENTS				GROUPINGS
	GENERAL		UNIQUE		
	Tertiary	Primary	Tertiary	Primary	
ANALYSIS OF THREE TAXA					
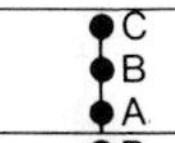 1 — C, B, A	ABC	BC	A	B	ABC BC [1]
2 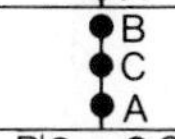 — B, C, A	ABC	BC	A	C	ABC BC [1]
3 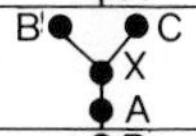 — B', C, X, A	ABC	BC	A	X	ABC BC [1]
4 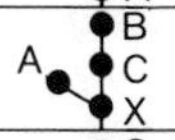 — B, C, X, A	ABC	BC	X	C	ABC BC [1]
5 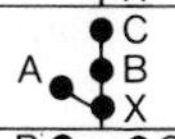 — C, B, X, A	ABC	BC	X	B	ABC BC [1]
6 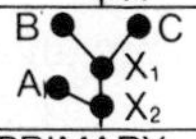 — B, C, X₁, X₂, A	ABC	BC	X_2	X_1	ABC BC [1]
PRIMARY CLADOGRAMS					
1 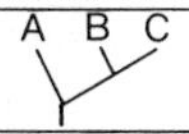 — A B C	ABC	BC			ABC BC [1]
2 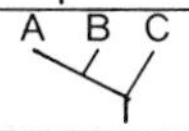 — A B C	ABC	AB			ABC AB [2]
3 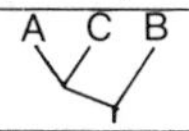 — A C B	ABC	AC			ABC AC [3]
TERTIARY TREES					
1 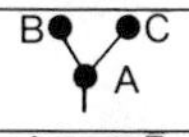 — B, C, A	ABC		A		ABC [4]
2 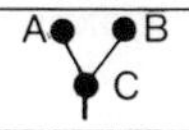 — A, B, C	ABC		C		ABC [4]
3 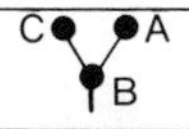 — C, A, B	ABC		B		ABC [4]
4 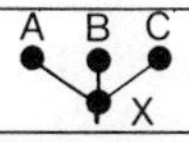 — A B C, X	ABC		X		ABC [4]
TERTIARY CLADOGRAM					
1 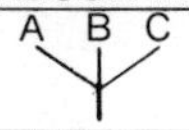 — A B C	ABC				ABC [4]

Figure 3.3. Component analysis of three taxa, A, B, C.

B and C are related by common ancestry (a "primary component"). There are two unique components: (1) the ancestor [of ABC] is A ("tertiary component"); (2) the ancestor [of BC] is B ("primary component"). The six primary trees (3.3.1–6) form a set denoted by primary cladogram 3.3.1. There is a similar set of six different trees denoted by primary cladogram 3.3.2, and another set of six different trees denoted by primary cladogram 3.3.3.

Each cladogram corresponds to a classification (grouping). Clado-gram 3.3.1 (and trees 3.3.1–6), for example, corresponds to grouping 3.3.1, according to which there is an inclusive taxon (ABC), including two subtaxa: (1) including only A (implied and not listed); and (2) including BC.

There are in addition four "tertiary trees" (3.3.1–4), each having only one general component (ABC) and one unique component (A, B, C, or X). These form a set denoted by the tertiary cladogram, of which the corresponding classification (grouping 3.3.4) is simply an inclusive taxon (ABC) subdivided (by implication) into three noninclusive subtaxa (A, B, and C).

In the three-taxon analysis, both components and trees are sorted into primary and tertiary types. Each primary tree has a maximum number of general components (the maximum number = the number of terms [representing taxa] minus one, or $N-1$). The more inclusive component is the tertiary; and the less inclusive, the primary. For each general tree-component, there is a corresponding unique tree-component. For primary tree 3.3.1, for example, the general tertiary component (ABC) corresponds to the unique tertiary component (A); and the general primary component (BC), to the unique primary component (B).

The unique components of primary tree 3.3.6 (X_1, X_2) are understood to be different $(X_1 \neq X_2)$. If they are the same $(X_1 = X_2)$, the tree is a tertiary tree (tertiary tree 3.3.4), defined by only one general tertiary component (ABC) and one unique tertiary component (X).

It is true of the three-taxon analysis, as for the two-taxon analysis, that there is a one-to-one correspondence between classifications (groupings) and cladograms (sets of trees).

COMPONENT ANALYSIS OF FOUR TAXA

An analysis of four taxa (figures 3.4–3.9) is almost excessively complex. There are twelve sets (primary cladograms 3.4.1, 3.5.1–12), each of

ANALYSIS OF FOUR TAXA							
PRIMARY TREES	COMPONENTS						GROUPINGS
	GENERAL			UNIQUE			
	TERTIARY	SECONDARY	PRIMARY	TERTIARY	SECONDARY	PRIMARY	
1 (D C B A)	ABCD	BCD	CD	A	B	C	ABCD BCD CD [1]
2 (C D B A)	ABCD	BCD	CD	A	B	D	ABCD BCD CD [1]
3 (D C B X / A)	ABCD	BCD	CD	X	B	C	ABCD BCD CD [1]
4 (C D B X / A)	ABCD	BCD	CD	X	B	D	ABCD BCD CD [1]
5 (D C X / B, A)	ABCD	BCD	CD	A	X	C	ABCD BCD CD [1]
6 (C D X / B, A)	ABCD	BCD	CD	A	X	D	ABCD BCD CD [1]
7 (D C X_1 X_2 / B, A)	ABCD	BCD	CD	X_2	X_1	C	ABCD BCD CD [1]
8 (C D X_1 X_2 / B, A)	ABCD	BCD	CD	X_2	X_1	D	ABCD BCD CD [1]
9 (C D / X B A)	ABCD	BCD	CD	A	B	X	ABCD BCD CD [1]
10 (C D / X_1 X_2 B A)	ABCD	BCD	CD	A	X_2	X_1	ABCD BCD CD [1]
11 (C D / X_1 B X_2 A)	ABCD	BCD	CD	X_2	B	X_1	ABCD BCD CD [1]
12 (C D / X_1 X_2 X_3 B A)	ABCD	BCD	CD	X_3	X_2	X_1	ABCD BCD CD [1]
PRIMARY CLADOGRAM (A B C D) 1	ABCD	BCD	CD				ABCD BCD CD [1]

Figure 3.4. Component analysis of four taxa, A, B, C, D.

ANALYSIS OF FOUR TAXA (CONT'D)				
PRIMARY CLADOGRAMS	GENERAL COMPONENTS			GROUPINGS
	Tertiary	Secondary	Primary	
A B C D — 1	ABCD	BCD	CD	ABCD BCD CD — 1
A C B D — 2	ABCD	BCD	BD	ABCD BCD BD — 2
A D B C — 3	ABCD	BCD	BC	ABCD BCD BC — 3
D A B C — 4	ABCD	ABC	BC	ABCD ABC BC — 4
D B A C — 5	ABCD	ABC	AC	ABCD ABC AC — 5
D C A B — 6	ABCD	ABC	AB	ABCD ABC AB — 6
C D A B — 7	ABCD	ABD	AB	ABCD ABD AB — 7
C A D B — 8	ABCD	ABD	BD	ABCD ABD AB — 8
C B A D — 9	ABCD	ABD	AD	ABCD ABD AD — 9
B C A D — 10	ABCD	ACD	AD	ABCD ACD AD — 10
B A C D — 11	ABCD	ACD	CD	ABCD ACD CD — 11
B D A C — 12	ABCD	ACD	AC	ABCD ACD AC — 12

Figure 3.5. Component analysis of four taxa (cont'd).

twelve primary trees (3.4.1–12); and in addition, another three sets (primary cladograms 3.6.13–15), each of nine primary trees (3.6.145–153). There are four sets (secondary, or partially resolved, cladograms 3.7.1–4), each of eight secondary trees (3.7.1–8); and in addition, another six sets (secondary cladograms 3.8.5–10), each of nine secondary trees (3.8.33–41). Finally, there is a set (tertiary cladogram 3.9.1) of five tertiary trees (3.9.1–5). The second suite of secondary cladograms (3.8.5–10) are interpreted here (trees 3.8.33–41) in their most restrictive sense; a less restrictive interpretation will be discussed below.

Some aspects of the above analyses are summarized in figure 3.10. Apparent is a one-to-one correspondence between classifications (groupings) and cladograms. Because of the large number of trees, there is no correspondence between them and classifications, except by way of their general components, which form sets (cladograms).

Comparison of the members of a given set of trees shows diversity that in a phyletic context is the cause of controversy, e.g., over the nature of ancestors in general, and the identity of particular ancestors. It may be kept in mind, however, that the cladogram (or phylogram) resembles equally all of the member-trees of its set, as defined by its general components. Of some importance, then, is that cladistic analysis operates at the level of generality of cladograms, not of trees. For many purposes, such as classification, cladistic analysis renders superfluous the analytical morass that often engulfs efforts to resolve trees.

Within the phyletic context, cladistic analysis has great appeal because of its generality. However, its generality is sometimes misunderstood, and cladistic analysis is sometimes confused with tree analysis (and cladograms, such as primary cladogram 3.4.1, are sometimes confused with trees having unknown unique components, such as primary tree 3.4.12). Trees and cladograms are not incompatible; neither is tree analysis incompatible with cladistic analysis. The point is simply that cladograms and cladistic analysis have a generality greater than that of trees and tree analysis. And cladograms and classification stand in a one-to-one relation which trees and classification do not share.

It has been shown that a given cladogram denotes a set of trees. In addition, some "cladograms" denote sets of other cladograms (quotes are used to indicate this quality of some cladograms). Consider a four-

ANALYSIS OF FOUR TAXA (CONT'D)					
PRIMARY TREES	COMPONENTS				GROUPING
	GENERAL		UNIQUE		
	Tertiary	Primary	Tertiary	Primary	
145 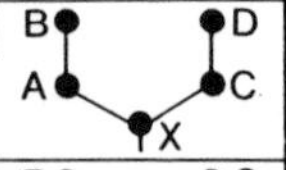	ABCD	AB CD	X	A C	ABCD AB CD — 13
146 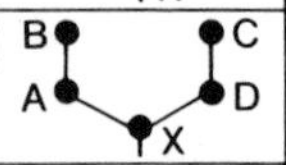	ABCD	AB CD	X	A D	ABCD AB CD — 13
147 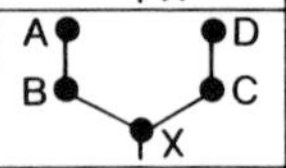	ABCD	AB CD	X	B C	ABCD AB CD — 13
148 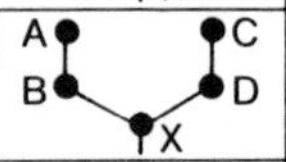	ABCD	AB CD	X	B D	ABCD AB CD — 13
149 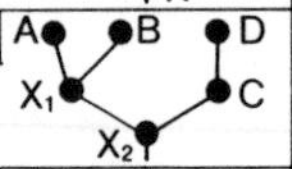	ABCD	AB CD	X_2	X_1 C	ABCD AB CD — 13
150 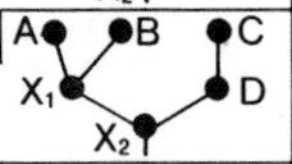	ABCD	AB CD	X_2	X_1 D	ABCD AB CD — 13
151 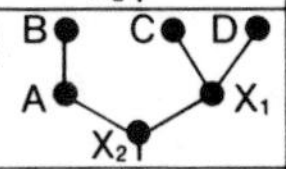	ABCD	AB CD	X_2	A X_1	ABCD AB CD — 13
152 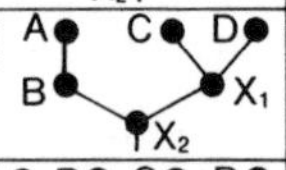	ABCD	AB CD	X_2	B X_1	ABCD AB CD — 13
153 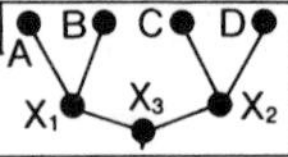	ABCD	AB CD	X_3	X_1 X_2	ABCD AB CD — 13
PRIMARY CLADOGRAMS					
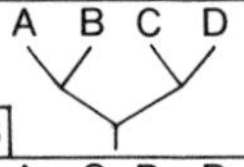 13	ABCD	AB CD			ABCD AB CD — 13
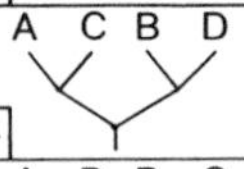 14	ABCD	AC BD			ABCD AC BD — 14
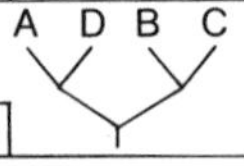 15	ABCD	AD BC			ABCD AD BC — 15

Figure 3.6. Component analysis of four taxa (cont'd).

ANALYSIS OF FOUR TAXA (CONT'D)							
SECONDARY TREES	COMPONENTS						GROUPINGS
	GENERAL			UNIQUE			
	TERTIARY	SECONDARY	PRIMARY	TERTIARY	SECONDARY	PRIMARY	
1 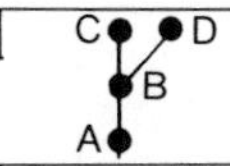 C● ●D / B / A●	ABCD	BCD		A	B		ABCD BCD 1
2 B● ●C / D / A● 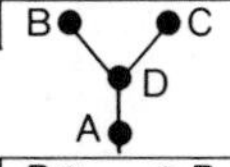	ABCD	BCD		A	D		ABCD BCD 1
3 D● ●B / C / A● 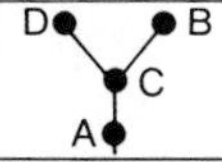	ABCD	BCD		A	C		ABCD BCD 1
4 B● C● ●D / X / A● 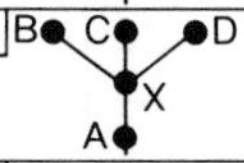	ABCD	BCD		A	X		ABCD BCD 1
5 C● ●D / B / A●●X 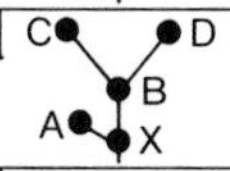	ABCD	BCD		X	B		ABCD BCD 1
6 B● ●C / D / A●●X 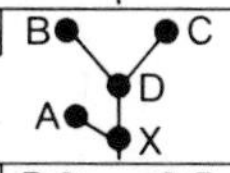	ABCD	BCD		X	D		ABCD BCD 1
7 D● ●B / C / A●●X 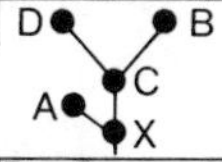	ABCD	BCD		X	C		ABCD BCD 1
8 B● C● ●D / X₁ / A●●X₂ 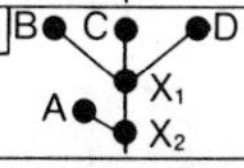	ABCD	BCD		X_2	X_1		ABCD BCD 1
SECONDARY CLADOGRAMS							
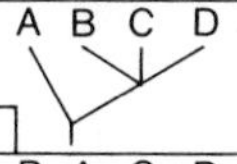 A B C D 1	ABCD	BCD					ABCD BCD 1
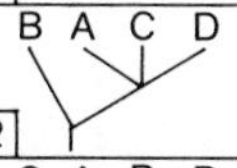 B A C D 2	ABCD	ACD					ABCD ACD 2
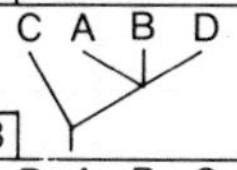 C A B D 3	ABCD	ABD					ABCD ABD 3
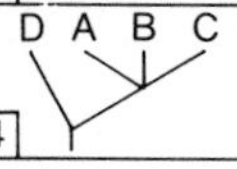 D A B C 4	ABCD	ABC					ABCD ABC 4

Figure 3.7. Component analysis of four taxa (cont'd).

ANALYSIS OF FOUR TAXA (CONT'D)							
SECONDARY TREES	COMPONENTS						GROUPING
	GENERAL			UNIQUE			
	TERTIARY	SECONDARY	PRIMARY	TERTIARY	SECONDARY	PRIMARY	
33 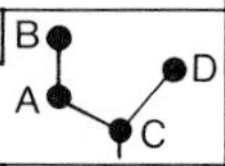	ABCD		AB	C		A	ABCD AB · 5
34 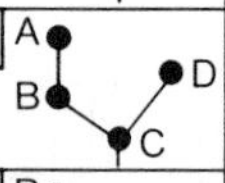	ABCD		AB	C		B	ABCD AB · 5
35 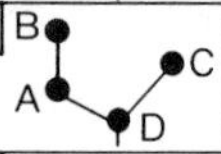	ABCD		AB	D		A	ABCD AB · 5
36 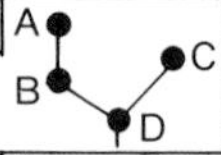	ABCD		AB	D		B	ABCD AB · 5
37 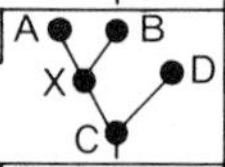	ABCD		AB	C		X	ABCD AB · 5
38 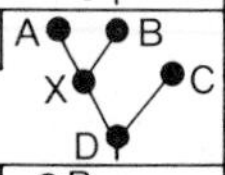	ABCD		AB	D		X	ABCD AB · 5
39 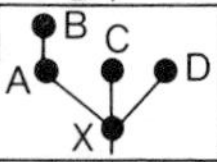	ABCD		AB	X		A	ABCD AB · 5
40 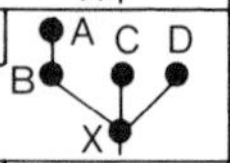	ABCD		AB	X		B	ABCD AB · 5
41 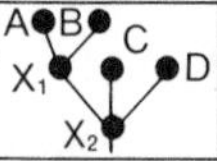	ABCD		AB	X_2		X_1	ABCD AB · 5
SECONDARY CLADOGRAMS							
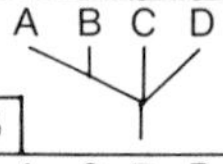 5	ABCD		AB				ABCD AB · 5
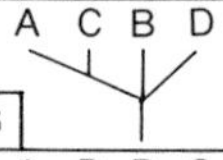 6	ABCD		AC				ABCD AC · 6
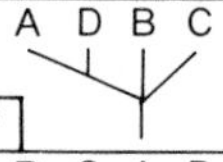 7	ABCD		AD				ABCD AD · 7
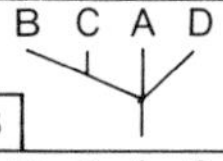 8	ABCD		BC				ABCD BC · 8
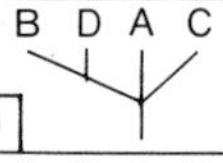 9	ABCD		BD				ABCD BD · 9
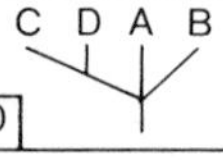 10	ABCD		CD				ABCD CD · 10

Figure 3.8. Component analysis of four taxa (cont'd).

ANALYSIS OF FOUR TAXA							
TERTIARY TREES	COMPONENTS						GROUPING
	GENERAL			UNIQUE			
	Tertiary	Secondary	Primary	Tertiary	Secondary	Primary	
1 B C D A	ABCD			A			ABCD 1
2 A B C D	ABCD			D			ABCD 1
3 A B D C	ABCD			C			ABCD 1
4 A C D B	ABCD			B			ABCD 1
5 A B C D X	ABCD			X			ABCD 1
TERTIARY CLADOGRAM							
A B C D 1	ABCD						ABCD 1

Figure 3.9. Component analysis of four taxa (concluded).

taxon tertiary cladogram (3.9.1). It is defined by the tertiary component ABCD, and it denotes a set of five tertiary trees (3.9.1–5). It does so because the tertiary component ABCD is the general component of all five tertiary trees. Yet the tertiary component ABCD is a general component, also, of all secondary cladograms (3.7.1–4, 3.8.5–10), secondary trees (3.7.1–8, 3.8.33–41), primary cladograms (3.4.1, 3.5.1–12, 3.6.13–15), and primary trees (3.4.1–12, 3.6.145–153). The four-taxon tertiary "cladogram" (cladogram 3.9.1), therefore, denotes a set of all other four-taxon cladograms and four-taxon trees that contain the same tertiary component (ABCD).

The subsets of a four-taxon "cladogram" are shown in figure 3.11. The set of all four-taxon cladograms and trees is called a "structure set." The set's notation is a tertiary "structure" (a tertiary "cladogram" with no terms, such as A, B, etc.), of which the node is a circle. The circle denotes a set of five structures: one tertiary (unresolved), two secondary (partially resolved), and two primary (fully resolved). Each of the structures is defined by a "clade type." Of the two secondary structures,

TAXA	TREES				CLADOGRAMS				GROUPINGS
	Primary	Secondary	Tertiary	Total	Primary	Secondary	Tertiary	Total	
2	3			3	1			1	1
3	18		4	22	3		1	4	4
4	171	86	5	262	15	10	1	26	26

Figure 3.10. Summary of component analyses of two, three, and four taxa.

(1) is defined by a tertiary structure to which a primary structure has been added, whereas (2) is defined by a tertiary structure to which a secondary structure has been added. Of the two primary structures, (1) is defined by a tertiary structure to which two different primary structures have been added, whereas (2) is defined by a tertiary structure to which both a secondary and a primary structure have been added. Each of the five structures denotes a set of one or more cladograms, each of which denotes a set of trees. The first of the two secondary structures (defined by a tertiary plus primary clade type) is here again treated in its most restrictive sense.

That a given tertiary or secondary "cladogram" can be a set of other cladograms means that the corresponding classification can be a set of classifications. Interesting as this may be in itself, there are implications

	STRUCTURE SET 				
	TERTIARY	SECONDARY		PRIMARY	
STRUCTURES (sets of cladograms)		1 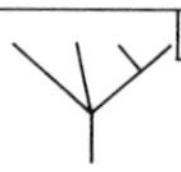	2 	1 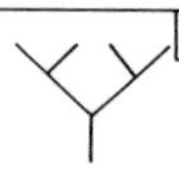	2 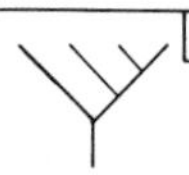
CLADE TYPES	T	T+P	T+S	T+P$_1$+P$_2$	T+S+P
CLADOGRAMS (all possible)	1	6	4	3	12
TREES (all possible)	5	54	32	27	144

Figure 3.11. Set-logic of 4-taxon analysis.

pertaining to information content. Any given cladogram or classification has an actual information content, specified by its components. If it is a nonprimary cladogram (or classification), it has in addition a potential information content, potentially specifiable by one or more components of its member cladograms (the member cladograms of the set it denotes). For example, tertiary cladogram 3.9.1 (four taxa) is defined only by one component (ABCD). Potentially, it includes two other components (a secondary and a primary, or two primary components), among the various secondary and primary components of cladograms 3.7.1–4, 3.8.5–10, 3.5.1–12, and 3.6.13–15.

That cladograms (and classifications) have an actual and a potential information content means that they can (1) represent the current state of knowledge of relationships at any given time; and (2) evolve in relation to increased knowledge. Put another way, cladograms and classifications evolve as actual information content increases (and potential information content diminishes).

CLADOGRAMS AND INFORMATION

Comparative studies of organisms result in the accumulation of information. Summaries of information in the form of cladograms pose certain problems, stemming from a basic question: how is information integrated into a coherent summary? The simplest case involves two species, let us say a lamprey and a shark (figure 3.12). Certain features may be found in one species and not the other, in both species, or in neither species: a lamprey, for example, has an elongate body (character 1), a single nostril (character 2), a notochord (character 3), etc.; a shark has a more robust body (character 4), two nostrils (character 5), a vertebral column (character 6), etc.; both species have eyes (character 7), cranial nerves (character 8), kidneys (character 9), etc.; neither species has lungs (character 10), a spoken language (character 11), and a bony skeleton (character 12). This information may be summarized in the form of a table (table 3.1). It is easy to see that the information may be integrated into a general form (table 3.2). The corresponding cladogram is figure 3.12, which poses no problems in this case, because it includes only two species (for two species there is only one possible cladogram).

Form

Figure 3.12. A cladogram for two species.

The cladogram does pose a problem, however, in its exact relations with the information of table 3.2: how is the information (character–types) to be understood as represented in the cladogram? Consider character-type A: where is it represented in the cladogram? There are three possibilities: (1) the line extending to the lamprey (figure 3.13.1); (2) the line extending to the shark (figure 3.13.2); (3) the line extending to both species (figure 3.13.3).

Possibility 3.13.3 may be seen immediately to be unsatisfactory, for it implies something true for both species, whereas character-type A asserts that the species differ. Possibility 3.13.2 may also seem unsatisfactory, but the reason is not so obvious. For example, consider the statement, that it is true that the shark does not have the characters found only in the lamprey: namely, an elongate body (character 1), a single nostril (character 2), a notochord (character 3), etc. The statement is true for what it denies to be true for the shark. Possibility 3.13.1 seems intuitively preferable, because it is simple and direct, without the complicated phrasing of possibility 3.13.2, namely that "it is true that the shark does not have...." On intuitive grounds, therefore, one might regard the information (character–types) to be represented in the cladogram as specified in figure 3.14.1, which omits the negative occurrences of character-types. Adding the negative occurrences, however, does not change the nature of the cladogram (figure 3.14.2).

An example with three species poses an additional problem, because there are four possible cladograms. As a third species may be added the lancelet with certain features unique to it, solenocytes (character 13), atriopore (character 14), endostyle (character 15); and certain features

Table 3.1. Characters and Their Occurrence in the Lamprey and Shark

Species					Characters							
	1	*2*	*3*	*4*	*5*	*6*	*7*	*8*	*9*	*10*	*11*	*12*
Lamprey	+	+	+	−	−	−	+	+	+	−	−	−
Shark	−	−	−	+	+	+	+	+	+	−	−	−

Table 3.2. Character-Types and Their Occurrence in the Lamprey and Shark (cf. table 3.1)

	Character-Types			
Species	*A*	*B*	*C*	*D*
Lamprey	+	−	+	−
Shark	−	+	+	−

common to all three species, gill slits (character 16), bilateral symmetry (character 17), and a dorsal nerve tube (character 18). The information may be organized in a general form (table 3.3). The four possible cladograms may be immediately assessed with reference to the positive, rather than the negative occurrences of the character-types (figure 3.15), as represented by single lines. Cladogram 3.15.1 includes all character-types, whereas cladograms 3.15.2–4 do not include character-type D; and cladograms 3.15.2–3 include a line for a character-type (?) unrepresented in the information (table 3.3). On this basis, cladogram 3.15.1 may be accepted as a true summary of the information (table 3.3), and the other cladograms may be rejected.

The cladograms may alternatively be compared with reference only to the relevant positive occurrences, namely character-type D, which distinguish one cladogram (3.15.5) from the others (3.15.6–8). On this basis, too, cladogram 3.15.5 may be accepted as a true summary of the relevant information (table 3.3), and the other cladograms may be rejected.

A full analysis of the four possibilities, with reference both to positive

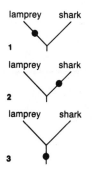

Figure 3.13. Three possibilities for representation of information in a cladogram for two species.

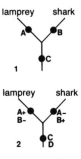

Figure 3.14. Representation of character-types (A, B, C) in a cladogram for two species. 1: Positive occurrences. 2: Positive and negative occurrences (cf. table 3.2).

and negative occurrences of character-types, is given in figure 3.16. As before, only one cladogram (3.16.1) includes all positive occurrences of character-types, as represented by single lines; the other cladograms (3.16.2–4) do not allow for character-type D+ except as multiple occurrences. Also, cladograms 3.16.2–3 include a line for a character-type represented only by a negative occurrence (B– in 3.16.2; C– in 3.16.3); cladogram 3.16.1 also includes a line representing a negative occurrence (A–) but the line also represents a positive occurrence (D+).

A criterion by which to judge the different possible cladograms is not as obvious in the full analysis (figure 3.16) as in the analysis only of positive occurrences (figure 3.15). But regarded as an integration of information, a cladogram may be judged according to how much information it does in fact integrate, or, in other words, how concentrated a summary it is. In the above full analysis (figure 3.16), all of the original information (table 3.3) is related to each of the four possibilities. One way to judge each possibility is simply to count the number of character occurrences (table 3.4). Cladogram 3.16.1 includes five positive, and seven negative, occurrences; cladograms 3.16.2–3, six

Table 3.3. Character-Types and Their Occurrence in the Lancelet, Lamprey, and Shark (cf. tables 3.1, 3.2)

| Species | Character-Types | | | | | |
	A	B	C	D	E	F
Lancelet	+	–	–	–	+	–
Lamprey	–	+	–	+	+	–
Shark	–	–	+	+	+	–

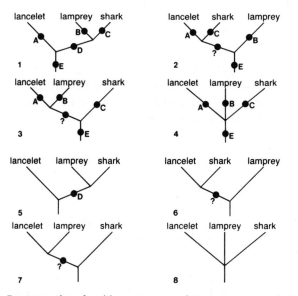

Figure 3.15. Representation of positive occurrences of character-types (A, B, C, D, E) in cladograms for three species. 1–4: All positive occurrences representable by single lines. 5–8: Positive occurrences (D) that differ among the cladograms (cf. table 3.3).

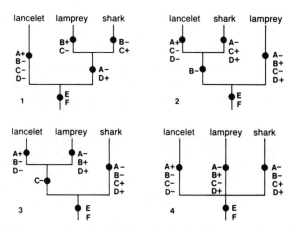

Figure 3.16. Representation of positive (A+, B+, C+, D+, E) and negative (A-, B-, C-, D-, F) occurrences of character-types in cladograms for three species (cf. figure 3.15 and table 3.3).

positive and seven negative; cladogram 3.16.4, six positive and eight negative; the original information (table 3.3) includes eight positive and 10 negative.

The difference in occurrences indicates the degree to which the various cladograms integrate, or generalize, the original information (table 3.5, left). In having the lowest number of occurrences, cladogram 3.16.1 may be judged the most efficient integration. Why such is the case is demonstrated by further analysis. Firstly, the occurrences common to all cladograms (character–types E and F) may be eliminated as irrelevant to the differences among cladograms (table 3.5, center). It is apparent that cladogram 3.16.4 integrates the information no better than table 3.3, and in a manner common also to cladograms 3.16.1–3. Secondly, the occurrences still common to cladograms 3.16.1–3 (character-types A, B, and C, each of which contributes three positive and five negative occurrences among cladograms 3.16.1–3 such that each cladogram receives three positive and five negative occurrences) may be eliminated, as irrelevant to the differences among cladograms 3.16.1–3 (table 3.5, right). It is apparent that the cladograms differ with respect only to character-type D, which appears fewer times in cladogram 3.16.1 than in cladograms 3.16.2–4. The full analysis thus yields the same result as the analysis based only on positive occurrences (figure 3.15), namely that cladogram 3.16.1 (3.15.1) is a true summary of the information (table 3.3), and that the other cladograms may be rejected. Adding the negative occurrences, in short, accomplishes nothing, which is a way of saying that negative occurrences are uninformative with respect to the problem of evaluating the different possible cladograms that might summarize, more or less efficiently, a given sample of information.

But there is more, which derives from the inference to be drawn from a

Table 3.4. Positive (+) and Negative (–) Character-Types and Their Occurrences among Cladograms (cf. figure 3.16 and table 3.3)

Clado-grams	Positive Occurrences						Negative Occurrences						Total
	$A+$	$B+$	$C+$	$D+$	$E+$	Total	$A-$	$B-$	$C-$	$D-$	$F-$	Total	Occurrences
3.16.1	1	1	1	1	1	5	1	2	2	1	1	7	12
3.16.2	1	1	1	2	1	6	2	1	2	1	1	7	13
3.16.3	1	1	1	2	1	6	2	2	1	1	1	7	13
3.16.4	1	1	1	2	1	6	2	2	2	1	1	8	14
table 3.3	1	1	1	2	3	8	2	2	2	1	3	10	18

Table 3.5. Positive (+) and Negative (–) Character-Types and Their
Occurrences among Cladograms (cf. figure 3.16 and table 3.4)

Cladograms	All Character-Types			E and F Eliminated			A, B, and C Eliminated		
	+	–	Total	+	–	Total	+	–	Total
3.16.1	5	7	12	4	6	10	1	1	2
3.16.2	6	7	13	5	6	11	2	1	3
3.16.3	6	7	13	5	6	11	2	1	3
3.16.4	6	8	14	5	7	12	2	1	3
table 3.3	8	10	18	5	7	12	2	1	3

cladogram such as 3.16.1—namely that, with cladogram 3.16.1 as a basis for inference, there exists a group including the lamprey and shark and excluding the lancelet. The inference based on cladogram 3.16.2 (that there is a group including the lancelet and shark and excluding the lamprey) and the inference based on cladogram 3.16.3 (that there is a group including the lancelet and lamprey and excluding the shark) conflict with each other, and with the inference based on cladogram 3.16.1. One may anticipate that, of the three inferences, one of them might be true and others false. If cladogram 3.16.1 is accepted as a basis for inference and the others rejected, the acceptance amounts to asserting that cladogram 3.16.1 is true, and the rejection amounts to asserting that cladograms 3.16.2 and 3.16.3 are false. If the decisive factor for acceptance is the occurrence of character-type D+, then the correlated occurrence of character-type A- is coincidental; and the occurrences of character-types B-(cladogram 3.16.2) and C-(cladogram 3.16.3) are also coincidental. In other words, the occurrence of negative characters, if considered informative, is falsely informative in all except one (two of three) of the possible cladograms, if one among them is true.

The above examples are unproblematical in the sense that all positive occurrences are combinable in one cladogram. Real information typically includes positive occurrences that conflict with one another (table 3.6). The four possible cladograms may be compared with

Table 3.6. Character-Types and Their Occurrence in Three Species

Species	Character-Types						
	A	B	C	D1	D2	E	F
1	+	–	–	–	+	+	–
2	–	+	–	+	+	+	–
3	–	–	+	+	–	+	–

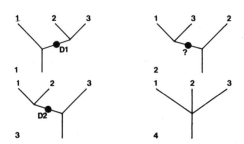

Figure 3.17. Positive occurrences (D1, D2) that differ among cladograms for three species (cf. table 3.6).

reference to the relevant occurrences of the character-types (figure 3.17). Two cladograms (3.17.1 and 3.17.3) integrate the information to the same degree. Although cladogram 3.17.2 might be rejected, no single primary cladogram can be accepted as an efficient summary; the only single cladogram that could be accepted is the tertiary cladogram 3.17.4.

A totally uninformative set of information, which allows for all possible character-types, is presented in table 3.7. Three cladograms integrate the information to the same degree (figure 3.18). No primary cladogram can be rejected; no single primary cladogram can be accepted as an efficient summary; the only single cladogram that could be accepted is tertiary cladogram 3.18.4.

Various procedures have been advocated for the purpose of combining conflicting information into a summary that can be represented in a cladogram. The procedures are sometimes called "clustering procedures" or "clustering algorithms," many of which in recent years have been discussed and compared in the journal *Systematic Zoology*. Clustering procedures vary among themselves and give varying results, any one of which may be used as a basis for inference. That a clustering procedure may be applied to conflicting information, and might yield an unambiguous result in the form of a single cladogram, does not guarantee that the result, or the inference based upon it, is true.

Table 3.7. Character-Types and Their Occurrence in Three Species

Species	A	B	C	D1	D2	D3	E	F
1	+	−	−	−	+	+	+	−
2	−	+	−	+	+	−	+	−
3	−	−	+	+	−	+	+	−

Character-Types

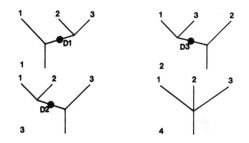

Figure 3.18. Positive occurrences (D1, D2, D3) that differ among cladograms for three species (cf. table 3.7).

Some information with conflicting positive occurrences is contained in table 3.8, and the relevant information is related to the four possible cladograms in figure 3.19 (cladograms 3.19.1-4). A full analysis of positive and negative occurrences (table 3.9) shows that cladogram 3.19.1 is the most efficient summary, and that the decisive factor is the occurrence of character-type D4+ (cf. cladograms 3.19.5-8) which in effect replicates character-type D1+ of the previous example (cf. table 3.7, cladogram 3.18.1). In terms of its efficiency, cladogram 3.19.1 is only 1 positive occurrence more efficient than cladograms 3.19.2-3.

This example illustrates that, for a given sample of information containing conflict in positive occurrences, various cladograms are possible, which differ more or less among themselves in their efficiency in integrating the information. Although one cladogram may be judged most efficient according to a stated criterion, and may be accepted as the best cladogram in that sense, there is no guarantee that, if the cladogram is used as a basis for inference, the inference will be true. Suppose, nevertheless, that cladogram 3.19.1 is used as a basis for inference, namely that there exists a group containing species 2 and 3 and excluding species 1. If so, additional samples of information should tend to give the same result. Suppose also that an additional sample of information is taken, and that the sample is similar enough to the

Table 3.8. Character-Types and Their Occurrence in Three Species

| Species | Character-Types | | | | | | | | |
	A	B	C	D1	D2	D3	D4	E	F
1	+	−	−	−	+	+	−	+	−
2	−	+	−	+	+	−	+	+	−
3	−	−	+	+	−	+	+	+	−

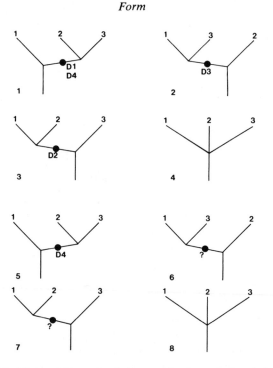

Figure 3.19. Positive occurrences in cladograms for three species. 1–4: All positive occurrences (D1, D2, D3, D4). 5–8: Positive occurrences that differ among cladograms (cf. table 3.8).

information of table 3.8, such that the sample is most efficiently summarized by a cladogram that duplicates cladogram 3.19.1. One may ask, what is the probability of achieving the same cladogram due to chance alone? Because there are three informative cladograms possible (3.19.1–3), the chance is 1 in 3, which is not a particularly low

Table 3.9. Positive (+) and Negative (–) Character-Types and Their Occurrences among Cladograms (cf. figure 3.19 and table 3.8)

Cladograms	All Character-Types			A, B, C, E, and F Eliminated			D1, D2, and D3 Eliminated		
	+	–	Total	+	–	Total	+	–	Total
3.19.1	10	10	20	6	4	10	1	1	2
3.19.2	11	10	21	7	4	11	2	1	3
3.19.3	11	10	21	7	4	11	2	1	3
3.19.4	12	11	23	8	4	12	2	1	3
table 3.8	14	13	27	8	4	12	2	1	3

probability. However, additional sampling with the same result would eventually produce a significantly low probability due to chance alone. A significantly low probability would offer some basis on which to judge the truth of the inference.

The conclusion to be drawn is not that a particular clustering procedure necessarily leads to the truth, as if truth depends upon mere conformity to procedure. The conclusion is that a particular clustering procedure yields a result that might be true or false, and that a judgment as to its truth or falsity may be considered an independent problem that can be investigated in other ways.

This conclusion should not be taken to imply that all clustering methods are in any sense equally valid. As shown by Farris (1977, 1979), for example, certain phenetic procedures give anomalous results in some circumstances. He supplied a hypothetical set of information for eight species, in which certain characters were multiplied by various factors (table 3.10). With or without the factors, the distribution of positive occurrences, which do not conflict among themselves, yields cladogram 3.20.1 as the most efficient summary. Without the factors, the same result is achieved by a certain phenetic procedure (UPGMA); but with the factors, the phenetic procedure yields no informative cladogram as a summary (figure 3.20.2). The phenetic procedure, in short, seems to behave as if it were clustering according to negative, rather than positive, occurrences. If, as argued above, negative occurrences are mainly false if considered informative, any clustering procedure sensitive to negative occurrences may produce anomalous results for that reason alone.

Conflict among positive occurrences, such as found in tables 3.6–3.8,

Table 3.10. Character-Types and Their Occurrence in Eight Species
(modified from Farris 1977: table 1)

Species	A	B	C	D	E	F	G	H	I	J	K	L	M	N
1	+	−	−	−	−	−	−	−	+	−	−	−	+	−
2	−	+	−	−	−	−	−	−	+	−	−	−	+	−
3	−	−	+	−	−	−	−	−	−	+	−	−	+	−
4	−	−	−	+	−	−	−	−	−	+	−	−	+	−
5	−	−	−	−	+	−	−	−	−	−	+	−	−	+
6	−	−	−	−	−	+	−	−	−	−	+	−	−	+
7	−	−	−	−	−	−	+	−	−	−	−	+	−	+
8	−	−	−	−	−	−	−	+	−	−	−	+	−	+
Factor	5	1	5	1	5	1	5	1	3	1	3	1	1	1

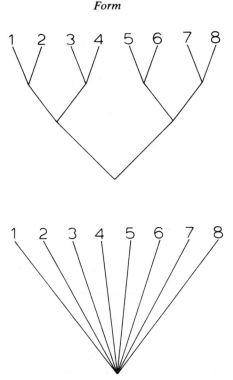

Figure 3.20. Two cladograms for eight species (cf. table 3.10). After Farris (1979), figure 1, p. 201.

may be considered in relation to the inferences for which cladograms serve as the basis. For example, if cladogram 3.19.1 is true (if there exists a group including species 2 and 3 and excluding species 1), then cladograms 3.19.2–3 are false (it is false that there exists a group including species 1 and 3 and excluding species 2; it is false, also, that there exists a group including species 1 and 2 and excluding species 3). It follows that character-types D1 and D4 are true, in the sense that they define a real group; and that character-types D2 and D3 are false, in the sense that they define unreal groups. Given a case of conflict among positive occurrences, therefore, one or more occurrence is false; that is, one or more positive occurrence is not a real occurrence.

Under the assumption that conflict among positive occurrences indicates one or more mistake in their designation as positive occurrences, the conflict may be investigated with this indication in mind. A concrete example may be considered with reference to certain features

Table 3.11. Characters and Their Occurrence in the Lancelet, Lamprey, and Shark

	Characters																				
Species	*1*	*2*	*3*	*4*	*5*	*6*	*7*	*8*	*9*	*10*	*11*	*12*	*13*	*14*	*15*	*16*	*17*	*18*	*19*	*20*	*21*
Lancelet	−	−	−	−	−	−	−	−	−	−	−	−	+	+	+	+	+	+	+	+	+
Lamprey	+	+	+	−	−	−	+	+	+	−	−	−	−	−	−	−	+	+	+	+	+	+
Shark	−	−	−	+	+	+	+	+	+	−	−	−	−	−	−	−	+	+	+	−	−	−

shared by the lancelet and lamprey: an eel-like body (character 19), dorsal and caudal fins confluent (character 20), and a body wall with a smooth exterior surface, uninterrupted by paired fins (character 21). The suite of characters relating to this problem is summarized in table 3.11, and their character-types are summarized in table 3.12, which repeats the structure of the hypothetical example of table 3.6.

If character-type D1 is true, and character-type D2 is false, the implication is that characters 7, 8, and 9 are true, and that characters 19, 20, and 21 are false. "False" does not mean that the lancelet and lamprey truly do not have an eel-like body (character 19), dorsal and caudal fins confluent (character 20), and a body wall with a smooth exterior surface, uninterrupted by paired fins (character 21), for most assuredly they do. "False" means that the characters are not positive occurrences that define a real group including the lancelet and lamprey and excluding the shark.

How might the characters be found to be false positive occurrences? One way would be to discover that the features of the lancelet and lamprey (thought to constitute characters shared by them) are fundamentally different in some way, in which case they would no longer conform to character-type D2, but rather to separate characters of character-types A and B. And it would be correct to assert that characters 19, 20, and 21 are falsely considered as type-D2 characters.

Table 3.12. Character-Types and Their Occurrence in the Lancelet, Lamprey, and Shark

	Character-Types						
Species	*A*	*B*	*C*	*D1*	*D2*	*E*	*F*
Lancelet	+	−	−	−	+	+	−
Lamprey	−	+	−	+	+	+	−
Shark	−	−	+	+	−	+	−

Another way would be to discover the characters in the shark. If all three were discovered there, the characters would no longer conform to character-type D2, but rather to character-type E. And it would be correct to assert that characters 19, 20, and 21 are falsely considered as type-D2 characters.

Still a third way to find the characters to be false positive occurrences would be to discover that they are not positive occurrences because they are either (a) negative occurrences, in which case they would no longer conform to character-type D2, and they would raise a question about the existence and distribution of the corresponding positive occurrences, which might or might not be discoverable in the shark, or (b) not characters at all. In either case it would be correct to assert that characters 19, 20, and 21 are falsely construed as type-D2 characters.

These various ways are diagrammed in figure 3.21. Each way involves a change in the judgment of the generality of a character. In the first case (figures 3.21.1–2) a character is found to be less general than previously supposed, in the sense that it is found to consist of two separate characters (of types A and B) rather than one (of type D2). In the second case (figures 3.21.3–4) a character is found to be more general than previously supposed, in the sense that it is found to occur in three species (character-type E) rather than in two species (character-type D2). In the third case (figures 3.21.5–6) a character is again found to be less general than previously supposed, in the sense that it is found to occur in one species (character-type C) rather than in two species (character-type D2). In the final case (figures 3.21.7–8), the character is found to have no generality at all, in the sense that it cannot be represented in the cladogram in such a way as to be informative.

What of the characters shared by the lancelet and lamprey? Character 19, "an eel-like body," is a statement that might be questioned on the grounds that "an eel-like body" is a mere gestalt perception, and that the shapes of the lamprey and lancelet are actually two quite different characters that are readily distinguishable from each other (and from the shape of eels) if studied in greater detail. Character 20 "dorsal and caudal fins confluent," is said to occur in the embryonic shark, if in fact the dorsal and caudal fins are represented by the embryonic fin fold (figure 3.22); and the character might well be considered of greater generality (character-type E). Character 21, "a body wall with a smooth

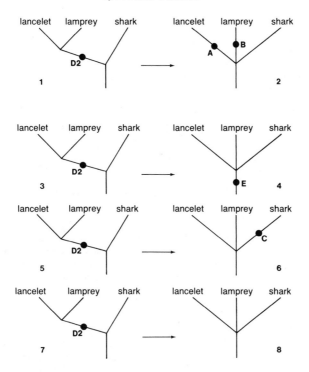

Figure 3.21. Possibilities for reinterpretation of positive occurrences of character-types in cladograms for three species. 1–2: An apparently positive occurrence (D2) reinterpreted as two less general positive occurrences (A, B). 3–4: An apparently positive occurrence (D2) reinterpreted as a more general positive occurrence (E). 5–6: An apparently positive occurrence (D2) reinterpreted as a negative occurrence corresponding to a positive occurrence (C) in the third species. 7–8: An apparently positive occurrence (D2) reinterpreted as no occurrence, either positive or negative.

exterior surface uninterrupted by paired fins," is a statement that is hardly distinguishable from "paired fins absent"; and the character might well be considered a negative occurrence.

Any character, of course, should be open to question and reinterpretation. What of the characters that conflict with those above? Character 7 (eyes), character 8 (cranial nerves), and character 9 (kidneys) each refers to organs that have been extensively studied and found to have (1) a coherent structure that may be identified and studied at gross, histological, and ultrastructural levels; (2) an assignable

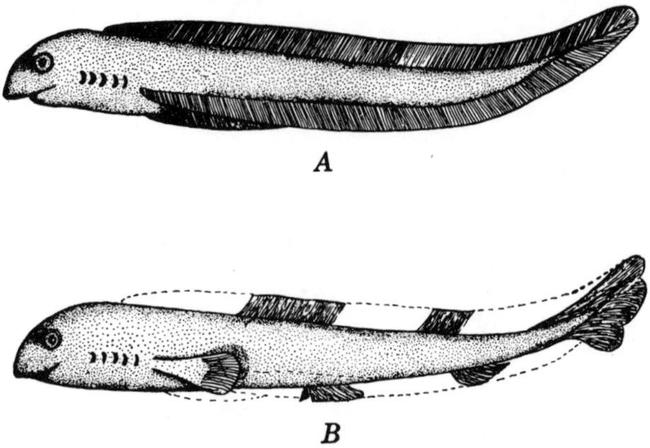

Figure 3.22. The relation of fins to embryonic fin folds according to the fin-fold theory. A: Fin folds as exemplified by embryonic sharks. B: Adult shark fins as remnants of fin folds. After C. K. Weichert (1959), Elements of Chordate Anatomy (New York: McGraw-Hill), figure 10.35, p. 280. Copyright © 1959, McGraw-Hill. Used with permission of McGraw-Hill Book Company.

function as elucidated by physiological studies at behavioral, neural, hormonal, and biochemical levels; and (3) a coherent embryological development that itself may be studied either structurally or physiologically at one or another level. For these reasons characters 7, 8, and 9 seem immune to the kind of reinterpretation applied to characters 19, 20, and 21. There is little reason, therefore, to view the conflict between character-types D1 and D2 as problematical. Character-type D2, as represented by characters 19, 20, and 21, can be rejected in its entirety.

Consider again the sample of information in table 3.8. The use of a clustering procedure sensitive to positive occurrences leads to the acceptance of cladogram 3.19.1. The inference is that character-types D1 and D4 are true and character-types D2 and D3 are false. It follows that the characters represented by types D2 and D3 should be amenable to reinterpretation, whereas the characters represented by types D1 and D4 should not be amenable to reinterpretation, or at least not with an equal facility. In a general sense, therefore, a clustering procedure is merely a means of predicting which positive occurrences, of those that conflict among themselves, are likely to be real and which are not. To test that prediction, additional information is required. A clustering procedure is unnecessary for recognizing conflicting positive occurrences, which may be studied directly without any predictions about which of them are likely to be real and which are not.

To the extent that conflicting positive occurrences can be studied and reinterpreted, conflicting occurrences disappear—if not in fact, at least in one's best judgment. If all conflict is resolved, such that all positive occurrences are combinable in a single cladogram, the choice of the most efficient summary is unproblematical: it is that cladogram that includes all positive occurrences as single lines. Yet as long as there is conflict among positive occurrences, there is a problem that may be investigated: namely, of the conflicting occurrences, which are real and which not? This residual problem cannot be solved, except perfunctorily, through the use of a clustering procedure. Its solution is possible only through study of organisms and new knowledge of, or new insight into, their real characteristics.

TREES AND INFORMATION

As summaries of information, cladograms do not in themselves imply a notion of evolution, or historical descent of species or other taxa. Nevertheless, cladograms, and branching diagrams generally, may be viewed from an evolutionary perspective, and viewed as such they may be termed phyletic trees. Considered as phyletic trees, branching diagrams pose certain problems, stemming from a basic question: how are characters represented in a phyletic tree?

The simplest case involves two species, let us say, again, a lamprey and a shark and certain information about them (figure 3.23 and table 3.2). Two general types of phyletic trees are possible, and each poses a problem in its exact relations with the information of table 3.2. Consider character-type A (characters unique to the lamprey): how are characters of type A represented on the various types of trees? For the bifurcating type of tree (figures 3.23.1–4) there are three possibilities: (1) the line extending to the lamprey (figure 3.13.1); (2) the line extending to the shark (3.13.2); (3) the line extending to both species (figure 3.13.3). Possibilities (1) and (2) imply that characters of type A evolved subsequent to the last occurrence of the species ancestral to the lamprey and shark (subsequent to the bifurcation); possibility (3) implies that the

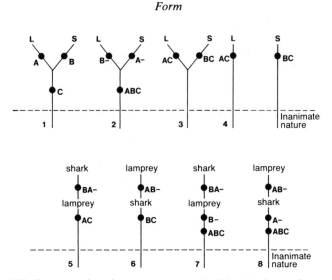

Figure 3.23. Representation of character-types (A, B, C) in phyletic trees for two species, lamprey and shark (cf. table 3.2).

characters evolved prior to the last occurrence of the ancestral species (prior to the bifurcation).

Because characters of type A occur only in the lamprey and not in the shark, possibility (2) may immediately be seen to be unsatisfactory. According to possibility (1), therefore, the characters unique to the lamprey evolved with the lamprey and not before (3.23.1); according to possibility (3), the characters evolved before, and were retained by the lamprey but were lost, or transformed, during the evolution of the shark (figure 3.23.2).

Consider characters of type C (present both in the lamprey and shark). The characters might have evolved either before (figures 3.23.1–2) or after (figures 3.23.3–4) the last occurrence, if any, of the species ancestral to both.

For trees without a bifurcation (figures 3.23.5–8), there are various possibilities for representing character-types. If the lamprey is considered ancestral to the shark (figures 3.23.5 and 3.23.7), characters of type A might have evolved early, and later have been lost or transformed. If the shark is considered ancestral to the lamprey (figure 3.23.6), characters of type A might have evolved only later; or, alternatively (figure 3.23.8), they might have evolved early, then have been lost, or transformed, only to reappear again later. Further possibilities in endless numbers would involve one or more additional cycles of still earlier evolution and loss, or transformation.

With the numerous possibilities for representing characters on trees even in the simplest possible example, how might the numerous (actually limitless) possibilities be evaluated? One way is simply to count the numbers of evolutionary gains and losses specified by each possibility (table 3.13). The possibilities differ among themselves in what they imply about gains and losses. If the true numbers of gains and losses were known, or assumed to be known, the possibilities could be judged according to how well they estimated the true numbers. But no true numbers of gains and losses can be derived from table 3.2. If the different possibilities are to be judged, therefore, they can be judged only in comparison among themselves. But such comparison itself poses a problem, namely, the criterion of judgment.

One possible criterion is parsimony. If this criterion is applied broadly, so as to include both evolutionary gains and evolutionary losses, the members of each class of trees, bifurcating and non-bifurcating, may be compared among themselves. A comparison of trees 3.23.1–4 shows that the character-types are represented more parsimoniously by tree 3.23.1 than by trees 3.23.2–4 (table 3.13); a comparison of trees 3.23.5–8 shows that the character-types are represented more parsimoniously by trees 3.23.5–6 than by trees 3.23.7–8; a comparison of trees 3.23.1 and 3.23.5–6 shows that the character-types are represented more parsimoniously by tree 3.23.1 than by trees 3.23.5–6. Interestingly, tree 3.23.1 corresponds to the cladogram, accepted on intuitive grounds

Table 3.13. Evolutionary Gains and Losses Implied by Various Placements of Character-Types in Phyletic Trees (cf. figure 3.23 and table 3.2)

Trees	Evolutionary Gains				Evolutionary Losses				Total Gains and Losses
	A	B	C	Total	A	B	C	Total	
3.23.1	1	1	1	3	0	0	0	0	3
3.23.2	1	1	1	3	1	1	0	2	5
3.23.3	1	1	2	4	0	0	0	0	4
3.23.4	1	1	2	4	0	0	0	0	4
3.23.5	1	1	1	3	1	0	0	1	4
3.23.6	1	1	1	3	0	1	0	1	4
3.23.7	1	2	1	4	1	1	0	2	6
3.23.8	2	1	1	4	1	1	0	2	6
table 3.2	?	?	?	?	?	?	?	?	?

as the best summary, in its representation of character-types (figure 3.14.1).

That the representation of one tree (3.23.1) may be found to be more parsimonious than that of other trees (3.23.5–6) is no proof that the one is true, or that the others are false, as statements about historical ancestry and descent. All that may reasonably be claimed is that, with respect to certain data (table 3.2), one representation (3.23.1) is truly more parsimonious than the others (3.23.5–6). Or, in other words, parsimony may be used as a criterion governing mode of representation of characters in phyletic trees. According to this mode, characters are properly represented in a tree at their most parsimonious level of generality. In more concrete terms, a character is represented by one line of a tree better than by two lines.

For a given set of character-types, one may ask if there is one tree in which the set is most parsimoniously represented. Consider the character-types of table 3.14. Two different trees are equally parsimonious in representing the character-types, for each tree implies two evolutionary gains and no losses (figure 3.24). But the trees differ in their complexity: tree 3.24.1 includes three lines; tree 3.24.2, only two. That one tree (3.24.2) represents the character-types by lines fewer than those of the other tree (3.24.1) is also an aspect of parsimony. Tree 3.24.2 is more parsimonious than tree 3.24.1 by one line, even though each tree implies the same number of evolutionary gains and losses of character-types. The most parsimonious phyletic tree, therefore, may be understood in a general sense as that tree implying the fewest evolutionary gains and losses with the fewest lines.

For two taxa, how many different phyletic trees are possible? This question is not easily answered. One may ask instead, how many different sets of character-types are possible for two taxa? Eight different sets exhaust the logical possibilities (table 3.15). For each set a most parsimonious "tree" is specifiable (figures 3.25.1–8); in that sense

Table 3.14. Character-Types and Their Occurrence in the Lamprey and Shark (cf. figure 3.24)

	Character-Types		
Species	A	B	C
Lamprey	–	–	+
Shark	–	+	+

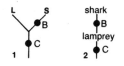

Figure 3.24. Parsimonious representation of character-types in two phyletic trees with different numbers of lines (cf. table 3.14).

there are eight possible "trees." If hybridization is considered, as represented in trees with reticulation, an infinite number of trees is possible, differing in the complexity of their reticulation. All such reticulate trees would be equally parsimonious in their implied evolutionary gains and losses if the character-sets are parsimoniously represented in the trees. But each reticulate tree would be less parsimonious in its number of lines than the corresponding nonreticulate tree. Compare, for example, trees 3.25.1 and 3.25.9. Each tree implies three evolutionary gains and no losses; tree 3.25.1 includes three lines, and tree 3.25.9 includes seven lines. Tree 3.25.1 is more parsimonious than tree 3.25.9 by four lines.

Given a tree most parsimonious in its representation of a certain set of

Table 3.15. Eight Sets of Character-Types for Two Taxa (lamprey and shark) and the Corresponding Most Parsimonious "Trees" (MPT) (cf. figure 3.25)

		Character-Types			
Set	Species	A	B	C	MPT
1	Lamprey	+	−	+	3.25.1
	Shark	−	+	+	
2	Lamprey	−	−	+	3.25.2
	Shark	−	+	+	
3	Lamprey	+	−	+	3.25.3
	Shark	−	−	+	
4	Lamprey	−	−	+	3.25.4
	Shark	−	−	+	
5	Lamprey	+	−	−	3.25.5
	Shark	−	−	−	
6	Lamprey	−	−	−	3.25.6
	Shark	−	+	−	
7	Lamprey	+	−	−	3.25.7
	Shark	−	+	−	
8	Lamprey	−	−	−	3.25.8
	Shark	−	−	−	

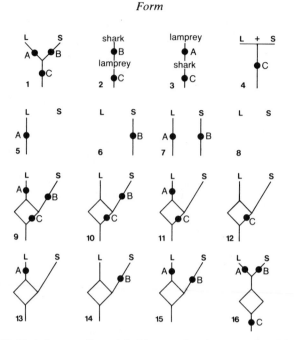

Figure 3.25. Phyletic trees. Trees 1–8: Most parsimonious trees for different sets of character-types (cf. table 3.15: sets 1–8). Trees 9–15: A phyletic tree with reticulation and parsimonious representation of seven sets of character-types (cf. figures 3.25.1–7 and table 3.15: sets 1–7). Tree 16: A phyletic tree with reticulation, additional lines, and parsimonious representation of one set of character-types (cf. figures 3.25.1, 3.25.9, and table 3.15: set 1).

character-types, what inferences may be derived from it? Consider set 3.15.8—no character-types known for the lamprey or shark, and no character-types known to be shared by them. One might infer that the absences are true absences or, alternatively, that the absence of the character-types is a mere artifact of sampling. Consider sets 3.15.2–7. Each set includes some, but not all, of the character-types of set 3.15.1. For each set one might infer that the missing character-types are true absences or, alternatively, that the absence of certain character-types is a mere artifact of sampling.

Two general types of inferences are possible: the absence of character-types in a particular set reflects: (1) the absence of such character-types in the real world or (2) sampling error, with the implication that the missing character-types might exist but are absent from a particular set because of chance alone. Each general type of inference leads to a different prediction: that further search for characters will produce (1) data sets identical to the first or (2) data sets that vary among themselves in the presence or absence of particular character-types. The predictions may be tested by sampling, which in general may be expected to show some variation in the presence and absence of some character-types from sample to sample. It is reasonable to expect that, with sampling sufficiently extensive, all possible character-types will be represented, that the combined samples will conform to set 3.15.1, and that the most parsimonious tree for any two taxa will be a bifurcating tree similar to tree 3.25.1.

That tree 3.25.1 seems to be the general solution to the problem of the most parsimonious tree for any two taxa again does not prove that tree 3.25.1 is true, and the other trees are wrong, as historical statements of ancestry and descent. Tree 3.25.1 might indeed be the most parsimonious tree for a certain set of character-types, or sets of character-types in general, even though another tree, say 3.25.2, might be historically correct. If so, historical truth must be judged by some criterion other than the most parsimonious tree.

Consider the possibility that further sampling shows no variability. Suppose an initial sample of information is summarized by set 3.15.2 and represented by tree 3.25.2, and that some number of additional samples of information all yield identical summaries and trees. At some point one might incline to the judgment that characters of type A do not, in fact, exist in the real world, i.e., that the absence is true. But there is a problem with this judgment, for it would be false if there existed a single character of type A. If that character were discovered, the discovery would automatically change the summary of all samples to set 3.15.1, and would change the most parsimonious tree to 3.25.1. If characters of type A actually exist in very low numbers, so as only rarely to occur in samples of information, their discovery would be improbable because of chance alone, but their occurrence would nevertheless be real. To conclude that characters of type A do not exist simply because none appears in a sample would seem unwarranted. In cases of extensive and exhaustive sampling, however, one might expect that "characters" of type A would normally be "discovered" through misidentification, which, because of chance alone, becomes ever more probable as sampling becomes more extensive.

A most parsimonious tree seems to offer no basis for inference beyond that of the corresponding cladogram for the reasons mentioned above. But one may suppose that a tree such as 3.25.1 is in some sense an aspect of the real world and, as such, deserves an explantion on its own terms. With omission of the "trees" with only one line, or no lines at all 3.25.4–8), there are four possible causal explanations in a historical sense, if trees can be regarded as such: trees 3.25.1–3 and some hybrid scheme of which there is an endless number. If tree 3.25.1 were true historically, the implication would be that evolution worked so as to produce a tree that is not only most parsimonious in a general sense, but also is historically true; if one of the other trees were true, e.g., tree 3.25.2, the implication would be that evolution worked so as to produce a tree that is most parsimonious in a general sense, but also is historically false. With these different possibilities, there seems no reason to equate the notions of most parsimonious tree and historical truth. Indeed, it may be best to divorce these two concepts and to inquire whether historical truth may be estimated by means other than the search for the most parsimonious tree.

The implication of the possible causal explanations, with the exception of reticulate trees, is evolutionary loss or evolutionary transformation of characters. If tree 3.25.1 is truly most parsimonious in a general sense, and if tree 3.25.2 is nevertheless true in a historical sense, characters of type A must have been lost during the evolution from lamprey to shark. Similarly, if tree 3.25.1 is truly most parsimonious, and if tree 3.25.3 is historically true, characters of type B must have been lost during the evolution from shark to lamprey. Searching without success for characters of type A in the shark, or characters of type B in the lamprey, one might judge that the characters are, in fact, truly absent. This judgment would mean also that tree 3.25.1 is truly most parsimonious. The possibilities always exist, however, that characters of type A are present in a transformed, but yet undiscovered, state in the shark; and that characters of type B are present in a transformed, but yet undiscovered, state in the lamprey. Discovery of transformed characters of types A and B would establish their presence in species previously thought to lack them. The discovery would also convert them into characters of type C. If all characters of type A were discovered in a transformed state in the shark, the combined set of all known character-types would change from set 3.15.1 to set 3.15.2, for the transformed characters would be type B; and the most parsimonious tree would change from tree 3.25.1 to tree 3.25.2 (table 3.16:set 1). Similarly, if all characters of type B were discovered in a transformed state in the lamprey, the combined set of all known character-types would change from set 3.15.1 to set 3.15.3, for the transformed characters would be of type A; and the most parsimonious tree would change from tree 3.25.1 to tree 3.25.3 (table 3.16:set 2). Interestingly, if all characters of type A were discovered in a transformed state in the shark, and all characters of type B were discovered in a transformed state in the lamprey, nothing would change, for the transformed characters would constitute a new set of characters of both types A and B (table 3.16:set 2A).

Thus the most parsimonious tree can be false in a historical sense if characters of types A or B have been lost or, which amounts to the same thing, if all the characters of one type have been so greatly changed when transformed into characters of the other type that the relation (transformation) between them is undiscoverable.

Thus the possibility exists that a character of type A is, in reality, a transformed character apparently of type B; and, conversely, that a

Table 3.16. Five Sets of Characters (of types A, B, C) for Two Taxa (lamprey, shark) with Possible Change ($\longrightarrow$) of a Character (+) into a Transformed State (t), and the Resulting Most Parsimonious Tree (MPT) (cf. figure 3.25) and Set of Character Types (cf. table 3.15)

Set	Species	Character-Types				MPT (Figure 3.25)	Resulting Character-Types (Table 3.15)		
		C	A	B	C			A	B	C
1	Lamprey	−	+	−	A	3.25.2	3.15.2	−	−	+
			$\downarrow$							
	Shark	−	t = +		A(t)			−	+	+
2	Lamprey	−	+ = t		B(t)	3.25.3	3.15.3	+	−	+
			$\uparrow$							
	Shark	−	−	+	B			−	−	+
2A	Lamprey	−	+ = t		B(t)	3.25.1	3.15.1	+	−	+
			$\downarrow\ \uparrow$							
	Shark	−	t = +		A(t)			−	+	+
3	Lamprey	X $\rightarrow$ t		−	X(t)	3.25.1	3.15.1	+	−	+
	Shark	X(t)	−	t $\leftarrow$ X				−	+	+
4	Lamprey	−	+ = ?		?	3.25.1	3.15.1	+	−	+
			$\downarrow\ \uparrow$							
	Shark	−	? = +		?			−	+	+

character of type B is, in reality, a transformed character apparently of type A. To assume that such might be true in a particular case is to employ the notion of homology in its conventional, and evolutionary, sense. A lamprey, for example, has a single nostril (table 3.2:character 2), and a shark has two nostrils (table 3.2:character 5). To assume that the nostril of the lamprey and the nostrils of the shark are structures homologous in an evolutionary sense is to suppose that both kinds of nostrils have a common origin in some ancestral structure. Possibilities for a common origin include the following: that the ancestral structure was (1) a single nostril (A) like that of the lamprey, which was transformed somehow into the two nostrils of the shark (figure 3.26.1:B); (2) two nostrils (B) like those of the shark, which were transformed somehow into the single nostril of the lamprey (figure 3.26.2:A); (3) a structure (X) other than (1) and (2), which was transformed somehow into the single nostril (A) of the lamprey, on the one hand, and which was transformed into the two nostrils (B) of the

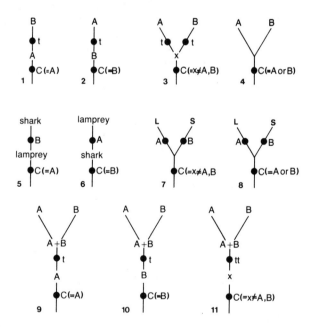

Figure 3.26. Trees implying transformations (t) of characters (A, B, X) for two species, lamprey (L) and shark (S). 1–4: Character trees. 5–8: Phyletic trees. 9–11: Character trees.

shark, on the other. Possibility (1) means that a single nostril is not a character of type A, but rather a character of type C, for it is present (in a transformed state) in the shark (figure 3.26.1); possibility (2) means that two nostrils are not a character of type B, but rather a character of type C, for they are present (in a transformed state) in the lamprey (figure 3.26.2); possibility (3) means that a single nostril is a character of type A, and that two nostrils are a character of type B, for neither character is present in a transformed state in another species, but rather both characters are transformations of a third character (X) of type C, present (in a transformed state) both in the lamprey and in the shark (figure 3.26.3).

According to possibility (1) in other words, the character of a single nostril is primitive or ancestral (plesiomorphic) relative to the character of two nostrils, which is advanced or derived (apomorphic; figure 3.26.1). According to possibility (2), the character of two nostrils is primitive or ancestral (plesiomorphic) relative to the character of one nostril, which is advanced or derived (apomorphic; figure 3.26.2). According to possibility (3), the character of a single nostril and the character of two nostrils are both advanced or derived (apomorphic) relative to yet another character (X) that is primitive or ancestral (plesiomorphic; figure 3.26.3).

A homology statement merely invokes the notion of a character in common or, in this example, a character of type C (nostril or nostrils present). A statement of evolutionary homology adds the possibility, but not the necessity, that a character previously thought to be of type A, or of type B, is in reality, a character of type C. In an evolutionary sense, therefore, a homology statement may be understood generally to imply the possibility of a particular character transformation: that, of two characters, one is primitive relative to the other, which is advanced.

One may compare the three possibilities, each in the form of a character tree (figures 3.26.1–3). In each tree the line leading to both characters represents the homologous relation (a character of type C) supposed to exist between them; and the line leading from one character to another represents an implied transformation (t). In terms of evolutionary gains and losses of characters and their transformations, two trees (3.26.1–2) are more parsimonious than the third (3.26.3), but the two trees are equally parsimonious relative to each other (table 3.17).

Table 3.17. Evolutionary Gains and Losses of Characters (A, B, X) and Character Transformations (A ⟶ B, B ⟶ A, X ⟶ A, X ⟶ B) Implied by Different Character Trees (cf. figure 3.26)

| | | | | *Gains of Characters and Transformations* | | | | | *Total Gains* |
Trees	*A*	*B*	*X*	*A ⟶B*	*B ⟶A*	*X ⟶A*	*X ⟶B*	*Losses*	*and Losses*
3.26.1	1	0	0	1	0	0	0	0	2
3.26.2	0	1	0	0	1	0	0	0	2
3.26.3	0	0	1	0	0	1	1	0	3

A parsimonious representation in a single character tree would, therefore, have to represent, and allow for, two alternative and conflicting possibilities (figures 3.26.1–2). The possibilities are allowed by a bifurcating character tree (figure 3.26.4) wherein there are no lines extending from one character to another. In phyletic trees the characters would be represented by lines interconnecting species (figures 3.26.5–8).

There is correspondence between the character trees and the phyletic trees, such that each character tree is included within, and implied by, the corresponding phyletic tree (3.26.1, 3.26.5; 3.26.2, 3.26.6; 3.26.3, 3.26.7; 3.26.4, 3.26.8). The phyletic trees may, therefore, be considered generalized character trees, i.e., trees that represent all character trees combined. The set of characters represented in each pair of corresponding trees is given in table 3.16. The character trees (figures 3.26.1–4) represent only the characters, not the distribution of the characters among the species. The phyletic trees (figures 3.26.5–8) include both types of information, the characters and their distribution among the species. Again, tree 3.26.8 (cf. tree 3.25.1) seems to be the general solution to the problem of the most parsimonious tree for any two taxa, even in cases wherein character transformations occur. In its form and implications it is the same as the cladogram accepted on intuitive grounds (figure 3.14.1).

Tree 3.26.8 does not in itself preclude, or imply the preclusion of, the historical possibility of independent transformation of characters A and B from character X (figure 3.26.7), for this possibility is merely an example of nonparsimonious representation of characters in trees. Nevertheless the most parsimonious tree for this possibility (figure 3.26.7) is the same as the generally most parsimonious tree (figure 3.26.8). Nor does tree 3.26.8 preclude, or imply the preclusion of, the historical possibility that characters of types A and B were merely segregated from an ancestral species in which both types were present (figures 3.26.9–11). Segregation is merely another example of non-parsimonious representation of characters in trees, but in this example there exist the same possibilities for character transformations: $A \longrightarrow B$ (figure 3.26.9); $B \longrightarrow A$ (figure 3.26.10); $X \longrightarrow A$ and $X \longrightarrow B$ (figure 3.26.11). Nevertheless, the most parsimonious tree for these possibilities is still the generally most parsimonious tree (figure 3.26.8). Again, that there exists a generally most parsimonious tree does not prove that the tree is correct as a historical statement of ancestry and descent.

For three taxa there are 54 different possible sets of character-types, each of which specifies a different "tree" as its most parsimonious representation. Most of the "trees" are fragmentary in lacking enough lines to form an interconnected structure. In the sense of an interconnected structure in which each of the taxa is separated by at least one line from the other two taxa, there are only 22 trees. Of these, 18 trees are primary trees (figure 3.27) and four trees are tertiary trees (figure 3.28).

The 22 trees may be compared in their representation of certain information about the lancelet, lamprey, and shark (tables 3.3, 3.18). Without consideration of possible character transformations, one tree (3.27.1) is most parsimonious in its implications about evolutionary gains and losses. Interestingly, tree 3.27.1 corresponds to the cladogram that best summarizes the same set of character-types (figure 3.15.1).

Each of the 22 trees is a parsimonious representation of a different set of character-types (table 3.19). The 18 primary trees form three groups, and each group is defined by a different character of type D. In group 1 (table 3.19:left, figure 3.27:left) characters of type D occur in the lamprey and shark; in group 2 (table 3.19:center, figure 3.27:center) characters of type D occur in the lancelet and shark; in group 3 (table 3.19:right, figure 3.27:right) characters of type D occur in the lancelet and lamprey.

Among the six trees of each group, one tree has all possible character-types (e.g., tree 3.27.1), and of the other five trees, each has some but not all possible character-types (e.g., trees 3.27.4–5, 3.27.10, 3.27.13, 3.27.16). For the same reasons given in the discussion of trees for two taxa, the tree in which all character-types are represented is the generally most parsimonious tree of its group, and in that sense corresponds to a cladogram. Consequently the problem of the generally most parsimonious tree for three taxa involves only three primary trees (figures

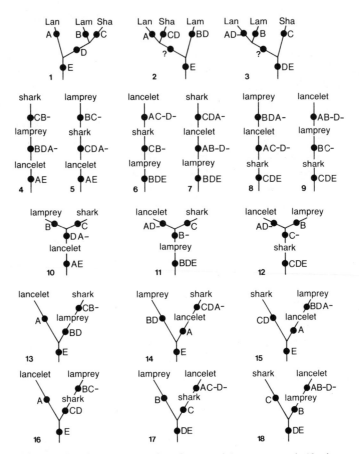

Figure 3.27. Parsimonious representation of one set of character-types in 18 primary trees for three species (cf. tables 3.3, 3.18).

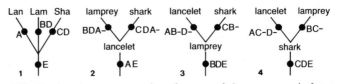

Figure 3.28. Parsimonious representation of one set of character-types in four tertiary trees for three species (cf. tables 3.3, 3.18).

Table 3.18. Evolutionary Gains and Losses Implied by Parsimonious
Representation of Characters in Phyletic Trees of Three Species
(cf. figures 3.27–3.28 and table 3.3)

Primary Trees	Evolutionary Gains						Evolutionary Losses						Total Gains and Losses
	A	B	C	D	E	Total	A	B	C	D	E	Total	
3.27.1	1	1	1	1	1	5	0	0	0	0	0	0	5
3.27.2	1	1	1	2	1	6	0	0	0	0	0	0	6
3.27.3	1	1	1	1	1	5	0	0	0	1	0	1	6
3.27.4	1	1	1	1	1	5	1	1	0	0	0	2	7
3.27.5	1	1	1	1	1	5	1	0	1	0	0	2	7
3.27.6	1	1	1	1	1	5	0	1	1	1	0	3	8
3.27.7	1	1	1	2	1	6	1	1	0	1	0	3	9
3.27.8	1	1	1	2	1	6	1	0	1	1	0	3	9
3.27.9	1	1	1	1	1	5	0	1	1	1	0	3	8
3.27.10	1	1	1	1	1	5	1	0	0	0	0	1	6
3.27.11	1	1	1	1	1	5	0	1	0	1	0	2	7
3.27.12	1	1	1	1	1	5	0	0	1	1	0	2	7
3.27.13	1	1	1	1	1	5	0	1	0	0	0	1	6
3.27.14	1	1	1	2	1	6	1	0	0	0	0	1	7
3.27.15	1	1	1	2	1	6	1	0	0	0	0	1	7
3.27.16	1	1	1	1	1	5	0	0	1	0	0	1	6
3.27.17	1	1	1	1	1	5	0	0	1	1	0	2	7
3.27.18	1	1	1	1	1	5	0	1	0	1	0	2	7
Tertiary Trees	A	B	C	D	E	Total	A	B	C	D	E	Total	Total Gains and Losses
3.28.1	1	1	1	2	1	6	0	0	0	0	0	0	6
3.28.2	1	1	1	2	1	6	2	0	0	0	0	2	8
3.28.3	1	1	1	1	1	5	0	2	0	1	0	3	8
3.28.4	1	1	1	1	1	5	0	0	2	1	0	3	8
table 3.3	?	?	?	?	?	?	?	?	?	?	?	?	?

3.27.1–3), and in that sense is equivalent to the problem of choosing the one cladogram, of three possible cladograms, that best summarizes the information. For choosing the cladogram that best summarizes the information, what is relevant are only positive occurrences of characters of type D. For choosing the most parsimonious tree, what is relevant are the implied total evolutionary gains and losses. But in choosing among the three generally most parsimonious trees, what is relevant are the gains and losses of characters of type D. Hence cladograms and generally most parsimonious trees turn out, at least in cases of three taxa, to be determined by the same factors. Cladograms and generally most parsimonious trees, therefore, seem merely to imply alternative

Table 3.19. Twenty-Two Sets of Character-Types for Three Species (lancelet, lamprey, shark) and the Corresponding Most Parsimonious Trees, Primary (MPPT) and Tertiary (MPTT) (cf. figures 3.27–3.28)

Group 1

MPPT	Species	A	B	C	D	E
3.27.1	Lan	+	–	–	–	+
	Lam	–	+	–	+	+
	Sha	–	–	+	+	+
3.27.4	Lan	–	–	–	–	+
	Lam	–	–	–	+	+
	Sha	–	–	+	+	+
3.27.5	Lan	–	+	–	–	+
	Lam	–	–	–	+	+
	Sha	–	–	–	+	+
3.27.10	Lan	–	–	–	–	+
	Lam	–	+	–	+	+
	Sha	–	–	+	+	+
3.27.13	Lan	+	–	–	–	+
	Lam	–	–	–	+	+
	Sha	–	–	+	+	+
3.27.16	Lan	+	–	–	–	+
	Lam	–	+	–	+	+
	Sha	–	–	–	+	+
MPTT	Species	A	B	C	D	E
3.28.1	Lan	+	–	–	–	+
	Lam	–	+	–	–	+
	Sha	–	–	+	–	+
3.28.4	Lan	+	–	–	–	+
	Lam	–	+	–	–	+
	Sha	–	–	+	–	+

Group 2

MPPT	Species	A	B	C	D	E
3.27.2	Lan	+	–	–	–	+
	Lam	–	+	–	+	+
	Sha	–	–	+	+	+
3.27.6	Lan	+	–	–	–	+
	Lam	–	–	–	+	+
	Sha	–	–	–	+	+
3.27.7	Lan	–	–	–	+	+
	Lam	–	–	–	+	+
	Sha	–	–	+	+	+
3.27.11	Lan	+	–	–	–	+
	Lam	–	–	–	+	+
	Sha	–	–	+	+	+
3.27.14	Lan	–	–	–	+	+
	Lam	–	+	–	+	+
	Sha	–	–	+	+	+
3.27.17	Lan	+	–	–	–	+
	Lam	–	+	–	+	+
	Sha	–	–	–	+	+
MPTT	Species	A	B	C	D	E
3.28.2	Lan	–	–	–	–	+
	Lam	–	–	+	+	+
	Sha	–	+	–	–	+

Group 3

MPPT	Species	A	B	C	D	E
3.27.3	Lan	+	–	–	+	+
	Lam	–	+	–	+	+
	Sha	–	–	+	–	+
3.27.8	Lan	–	–	–	+	+
	Lam	–	+	–	+	+
	Sha	–	–	–	–	+
3.27.9	Lan	+	–	–	+	+
	Lam	–	–	–	+	+
	Sha	–	–	–	–	+
3.27.12	Lan	+	–	–	+	+
	Lam	–	–	–	+	+
	Sha	–	–	+	–	+
3.27.15	Lan	–	–	–	+	+
	Lam	–	+	–	+	+
	Sha	–	–	+	–	+
3.27.18	Lan	+	–	–	+	+
	Lam	–	+	–	+	+
	Sha	–	–	+	–	+
MPTT	Species	A	B	C	D	E
3.28.3	Lan	+	–	–	–	+
	Lam	–	–	+	–	+
	Sha	–	–	+	–	+

strategies for arriving at the same result—namely a generalization about the distribution of character-types among species in the real world. Among other of its attributes, the generalization restricts the range of possible historical explanations (if phyletic trees may be considered to be such) to a relatively small group of trees that are alike in their agreement with the generalization about the distribution of character-types. In the above example of three species, both the cladogram and the generally most parsimonious tree specify that characters of type D occur in the lamprey and shark, and suggest as a restricted group of possible historical explanations the phyletic trees of group 1, all of which agree, in the above respect (characters of type D), with the cladogram, the generally most parsimonious tree, and each other.

It seems superfluous to analyze the sets of character-types of tables 3.6–3.8 (figures 3.17–3.19), except in a cursory fashion (table 3.20). Of

Table 3.20. Evolutionary Gains and Losses Implied by Phyletic Trees, and the Generally Most Parsimonious Tree (GMPT) for Three Sets of Character-Types (cf. tables 3.6–3.8, figures 3.17–3.19)

Tree	Gains	Losses	Total	GMPT
3.17.1 } 3.17.3	{ 6 7	1 0	7 7	3.17.4
3.17.2 } 3.17.4	{ 6 7 8	2 1 0	8 8 8	
Table 3.6	?	?	?	
3.18.1 3.18.2 3.18.3	{ 7 8 9	2 1 0	9 9 9	3.18.4
3.18.4	{ 7 8 9 10	3 2 1 0	10 10 10 10	
Table 3.7	?	?	?	
3.19.1	{ 8 9 10	2 1 0	10 10 10	3.19.1
3.19.2 } 3.19.3	{ 8 9 10 11	3 2 1 0	11 11 11 11	
3.19.4	{ 8 9 10 11 12	4 3 2 1 0	12 12 12 12 12	
Table 3.8	?	?	?	

Table 3.21. Numbers of Characters of Different Types in Three Species
(lancelet, lamprey, and shark) (cf. figure 3.29)

Species	A	B	C	D1	D2	D3	E
			Character-Types				
Lancelet	3	–	–	–	4	2	7
Lamprey	–	2	–	8	4	–	7
Shark	–	–	9	8	–	2	7

the diagrams in figure 3.17 (viewed as trees in relation to the character-types of table 3.6), two trees (3.17.1 and 3.17.3) are more parsimonious than the others, but equally parsimonious relative to each other. For each of the two trees (3.17.1 and 3.17.3), alternative representations of the character-types, equally parsimonious, are possible: according to one representation, there are six gains and one loss; according to the other, there are seven gains and no losses. Because there are two primary trees of equal (and maximum) parsimony, the generally most parsimonious tree is the tertiary tree 3.17.4. Of the diagrams in figure 3.18 (viewed as trees in relation to the character-types of table 3.7), all three primary trees (3.18.1–3) are equally parsimonious, and there are three alternative representations of the character-types; the generally most parsimonious tree is the tertiary tree 3.18.4. Of the diagrams in figure 3.19 (viewed as trees in relation to the character-types of table 3.8), one tree (3.19.1) is the most parsimonious, and there are three alternative representations of the character-types. In all three of the above examples, the generally most parsimonious tree duplicates the cladogram previously accepted as the best summary.

If extensive sampling is apt to result in some number of characters of each possible character-type, a realistic problem must concern the relative numbers of characters. Consider the hypothetical set of characters of table 3.21, wherein characters of type D occur in all possible combinations in unequal numbers: eight characters of type D1 (in lamprey and shark); four characters of type D2 (in lancelet and lamprey); two characters of type D3 (in lancelet and shark). Because the characters occur in different numbers, the numbers may be used as a basis for comparing the efficiency of different cladograms, and the parsimony of different trees. For three taxa, counting positive occurrences and counting evolutionary gains and losses give the same result (table 3.22), for no pattern of loss can give a total more parsimonious

Table 3.22. Positive Occurrences (= evolutionary gains) of Characters of Type D in Cladograms (= generally most parsimonious trees) and Some Reticulate Trees for Three Species (lancelet, lamprey, and shark) (cf. table 3.21 and figure 3.29)

Trees	D1	D2	D3	Total
3.29.1	8	8	4	20
3.29.2	16	8	2	26
3.29.3	16	4	4	24
3.29.4	8	4	4	16
3.29.5	16	4	2	22
3.29.6	8	8	2	18
3.29.7	8	4	4	16
3.29.8	8	2	4	14
3.29.9	8	2	4	14

than a pattern of multiple gains. Thus branching diagram 3.29.1 may be considered (as a cladogram) as the most efficient summary, and (as a tree) as the generally most parsimonious tree.

If hybridization is considered, as exemplified by reticulate trees (e.g., figures 3.29.4–6), the trees will be found yet more "parsimonious" in number of evolutionary gains (table 3.22). Such trees, however, are

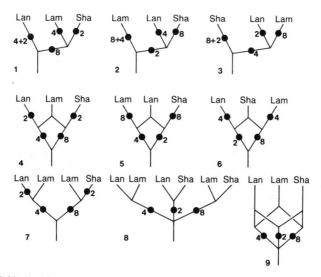

Figure 3.29. Positive occurrences (= evolutionary gains) of characters of type D (D1, D2, D3) in cladograms (= generally most parsimonious trees) and in some reticulate trees for three species, lancelet, lamprey, and shark (cf. tables 3.21–3.22).

equivalent to branching diagrams in which one or more species appears twice. Compare, for example, diagrams 3.29.4 and 3.29.7, which represent the same characters by the same lines. A tree with more complex reticulation is comparable to a branching diagram wherein all taxa appear more than once (3.29.8), and can be represented by a reticulate tree that is best viewed as three-dimensional (figure 3.29.9). Complex reticulation allows all combinations of characters of type D (D1, D2, D3) to be represented by single lines. For this reason, reticulate trees are always most "parsimonious" in their implied numbers of evolutionary gains and losses, even though they include more lines than nonreticulate diagrams. Hence, reticulate trees are not appropriate as possible candidates for the best summary (cladogram) or generally most parsimonious tree. In other words, reticulate trees do not integrate information; they merely reproduce it, and the reticulate tree that reproduces the information most "parsimoniously" (completely) is that tree that allows all possibilities to be represented by single lines (e.g., tree 3.29.9).

For three taxa, there are three possible kinds of characters of type D (D1, D2, D3). The cladogram that is the most efficient summary merely represents the greatest number of characters of type D by the single line available for such representation. Diagrams with more than one line for representing characters of type D for three taxa are equivalent to reticulate trees.

Reticulate trees with two lines available for representing characters of type D may be compared among themselves. Of trees 3.29.4–6, for example, tree 3.29.4 is more parsimonious than trees 3.29.5–6, for tree

Table 3.23. Relations Between Numbers of Characters of Type D (D1, D2, and D3) and (1) Positive Occurrences (total evolutionary gains and losses) and (2) Positive-Negative Occurrences (overall similarity) for Three Species (cf. tables 3.21, 3.24, and figure 3.29)

Tree	Character-Types Represented by Single Lines	Number of Characters Represented	Total Positive Occurrences (D1 + D2 + D3)	Total Positive-Negative Occurrences (Overall Similarity)		
				+	−	Total
3.29.1	D1	8	20	20	25	45
3.29.3	D2	4	24	24	19	43
3.29.2	D3	2	26	26	26	52

3.29.4 does represent more information with its two lines than do trees 3.29.5–6.

The relation between numbers of characters of type D (D1, D2, D3) and total positive occurrences (total evolutionary gains and losses) is demonstrated in table 3.23: as characters decrease, positive occurrences increase by a like amount. Such being the case, the notion of a most efficient cladogram, or generally most parsimonious tree, may be grasped immediately, at least in cases of three taxa, as indicated by the character-type (D1, D2, or D3) represented by the most characters (e.g., D1).

A POSTSCRIPT ON PARSIMONY

If parsimony and historical truth are best divorced, what is the significance of parsimony? One significance is its role as a criterion of representing characters in a branching diagram. One branching diagram of the many possible diagrams may be judged the most parsimonious for a particular sample of characters. If there is more than one sample, each sample will specify one diagram as its most parsimonious representation. Two or more such diagrams permit comparisons to be made among them, and ultimately permit a judgment as to whether they agree for reasons other than chance alone.

If there is agreement that is non-random, what is the cause? There are only two possibilities: the agreement is either (1) an artifact imposed upon the samples by the methods of the investigator, or (2) a reflection of some real factor that is independent both of method and investigator. Discovery of factors in category (1) constitutes increased knowledge of particular methods and their limitations; discovery of factors in category (2) constitutes increased knowledge of the world.

If our knowledge of the world ultimately stems from comparison of results that need not agree except by chance, but in fact do agree for some reason other than chance, we are well advised to understand, as clearly as possible, the nature of the comparisons that we make. How, then, are different samples of characters to be compared, if each sample, as is the case, may be represented by many different branching diagrams? One procedure is simply to limit comparisons to those few diagrams, one for each sample, that are most parsimonious representations of the samples. The limitation in itself can neither cause nor imply agreement among the diagrams for reasons other than chance alone.

Does parsimonious representation presuppose or imply that evolution (or some other causal factor) is parsimonious? To some persons, such seems to be the case. But let us try to be exact in our argument. If we observe nonrandom agreement, and that agreement is between diagrams most parsimonious for different samples, we conclude that a causal factor is at work. But what do we know of the causal factor? Only that it is the cause of the agreement among the diagrams, not that it is the cause either of parsimony or of the diagrams. If we liken parsimonious representation to a magnifying lens through which we look in order to see what otherwise would be invisible to our naked eye, we may better understand parsimony as a procedure, not a discovery. No one who observes, say, a dividing amoeba would assert that mitosis magnifies rather than multiplies amoebas.

That evolution may seem parsimonious, rather than merely orderly, is easy to understand. If we perceive nonrandom agreement (order) through comparison only of most parsimonious diagrams, our perception might be said to depend upon, or presuppose, parsimony (as our perception of a dividing amoeba might be said to presuppose a magnifying lens). That parsimony is presupposed may be misconstrued to imply that evolution is, or is presupposed to be, parsimonious (as mitosis would be misconstrued to cause magnification rather than multiplication of amoebas).

Does parsimony have a greater significance? If it does the significance lies in the relation between perceived order (as represented, for example, in a cladogram) and some set of historical explanations (as represented, for example, in a set of trees), restricted in their number by the use of a parsimony criterion (as represented, for example, by a set of trees denoted by one cladogram). Thus, the parsimony criterion may be used to specify a restricted set of historical explanations (trees). If one member of the set (of trees) were in fact true, then the notion that evolution is parsimonious may be defined and understood to mean exactly that (namely, that one tree of the set is true; and that an approximation to the truth is achieved first by specifying the set). Here the difficulties that attend the study of historical explanation (trees and their significance) become apparent. The cladogram may be considered to have a truth of its own even if the true historical explanation (the true tree) happens not to be a member of the set denoted by the cladogram (if evolution is truly not parsimonious in a given case). Thus, the notion that evolution is or is not parsimonious, in a given case, always implies comparison between a (true) cladogram and a (true) tree: if the (true) tree is a member of the set denoted by the (true) cladogram, evolution is (truly) parsimonious; if the (true) tree is not a member of the set denoted by the (true) cladogram, evolution is (truly) not parsimonious.

How might one form a judgment that evolution is or is not parsimonious (in the above sense) in a given case or in general? Ideally, some notion of the truth should be in hand, so that the use of a parsimony criterion could be judged according to how well its results match the truth. Yet the truth is not available to us in any particular case or in general. Without a notion of truth, what else is possible? We imagine one possibility, namely that the cladograms (restricted sets of historical explanations) for different groups of organisms might agree in the geographical dimension (biogeographically through area-clado-grams, as detailed in Chapters 6–8 below). If there is geographical agreement among cladograms for different groups of organisms, then there is a reason to infer a common causal factor (historical explanation). In such a case there would be reason to infer that evolution was indeed parsimonious (that the true historical explanations lie within the restricted sets of trees).

Suppose that there is no geographical agreement among cladograms for different groups of organisms? Does the lack of agreement mean that evolution is not parsimonious? Failure to find agreement is not decisive unless it is supposed that the reason for failure is that the agreement does not exist to be found. But even under this supposition the true lack of agreement offers no basis for judgment as to which (sets of) trees are true. Failure to find agreement, then, is not evidence that evolution is not parsimonious, only that evolution possibly is not parsimonious.

To show that evolution is nonparsimonious requires that a certain tree and a certain cladogram are true, and that the tree is not a member of the set denoted by the cladogram. Is it possible that evidence could show such in a given case?

If there is geographical agreement among some cladograms, and we infer that there is a common historical explanation, what of a group distributed in a similar way but whose area-relationships conflict with the agreement? One possibility is that the conflict arises from nonparsimonious evolution: namely that there is one (true) historical explanation for all groups, one of which has a (true but) incongruent cladogram.

Thus, the use of a parsimony criterion does not presuppose anything in particular about the nature of evolution. Rather, a parsimony criterion makes possible certain comparisons, according to which evolution may be judged parsimonious, or possibly not parsimonious, or nonparsimonious, as the case may be. This kind of judgment may have little relevance in itself; it arises only as a by-product, as it were, of comparison between cladogram and tree, both of which are assumed true in a given case. Of course, *any* cladogram or tree may be assumed to be true. Judgment of truth is a matter independent of such assumption. We consider that a cladogram may be judged true on the basis of agreement among samples of characters (that agree for reasons other than chance alone or methodological artifact); and that a tree (or set of trees) may be judged historically true on the basis of agreement among cladograms of different groups as considered in the geographical dimension.

Biogeography is often asserted to constitute some of the strongest evidence in favor of evolution. The sense of the assertion is easy to perceive if it stems from agreement of cladograms in the geographical dimension. If agreement is perceived it permits judgment about the historical truth of (a set of) trees—a judgment that is perhaps unattainable through any other considerations independent of parsimony. Such agreement is direct evidence not of evolution, but rather of historical process that is parsimonious, i.e., of historical process that binds cladogram and tree as one.

To the extent that cladograms for different groups agree in the geographical dimension, what is perceived as order (a cladogram) and what is inferred as its historical cause (a tree) are bound together as one and the same notion. To a person with a materialistic philosophy, the notion easily transforms into evolution—as the by-product of mutation and recombination of genes. To a person with an idealistic philosophy, the notion easily transforms into evolution—as the by-product of movement of form or idea. For persons who contemplate final causes, the notion transforms into creation (if final cause is considered supernatural) or to historical process that is inexplicable (if final cause is considered natural but either unknown or unknowable).

Evolution thus seems to depend upon the notion that cladogram and tree are in fact bound as one, as revealed by actual evidence in hand. Our considerations suggest that relevant evidence consists only of agreement of cladograms in the geographical dimension—agreement that is by no means abundantly available at present. In fact, this evidence is notable by its very scarcity. Historically speaking, this evidence seems to have been taken for granted—which is understandable enough if only because cladograms and trees were only recently distinguished as different concepts. We conclude, therefore, that biogeography (or geographical distribution of organisms) has not been shown to be evidence for or against evolution in any sense. The significance of biogeography has been merely that biogeography has raised the possibility of agreement between cladograms as considered in the geographic dimension—a possibility that has been little studied, but one worthy of further investigation.

INFORMATION, PHYLETICS, AND PHENETICS

That there are more characters of type D1 than types D2 or D3 (D1 > D2, D3) in a particular sample of characters need not be considered significant in itself, for what is true of one sample need not be true of other samples. If significant, the relation D1 > D2, D3 should be general, in the sense that the relation is true of samples generally—a matter that is open to empirical investigation through further sampling.

A generalization about the distribution of characters among species (e.g., D1 > D2, D3) is interpretable in an evolutionary sense to indicate relative recency of common ancestry. In other words,

(1) If D1 > D2, D3 is generally true, then
(2) the lamprey and shark had an ancestor in common that was not an ancestor of the lancelet (as exemplified in figure 3.29.1).

According to this mode of interpretation, characters of type D1 are true homologies (synapomorphies, or marks of common ancestry) that evolved prior to the last occurrence of a common ancestor, and were inherited from that ancestor without subsequent loss in any of the descendant species. Also, characters of types D2–3 are interpretable in one of two ways:

(3) as true homologies inherited from a common ancestor with subsequent loss in one or more of the descendant species (if loss is an interpretation more parsimonious than multiple gains); or

(4) as characters not inherited from a common ancestor, but rather as characters developed independently two or more times subsequent to the last occurrence of the common ancestor (if multiple gain is an interpretation more parsimonious than loss).

For three taxa, the two interpretations are equally parsimonious in any given case.

This mode of interpretation, which has been employed commonly enough within systematic biology to be termed "traditional," leads to a seeming difficulty in its implications. The mode may be stated in an abbreviated form:

(5) If $D1 > D2$, D3 is generally true, then

(6) characters of type D1 were present in a common ancestral species, and characters of types D2–3 either were present in an ancestor and were lost in some descendant species, or they were not present in an ancestor at all.

The difficulty arises if one asks, how might one determine what characters were present in a common ancestor? If the only answer to this question is that

(7) characters present in a common ancestor are those of type D1, if

(8) $D1 > D2$, D3 is generally true,

then the difficulty is not seeming but real. The implication is circular, leads nowhere, and suggests that the notion of common ancestry, as inferred from character distributions, is meaningless. The difficulty is seemingly multiplied by a reconsideration of an earlier inference:

(1) If $D1 > D2$, D3 is generally true, then

(2) the lamprey and shark had an ancestor in common that was not an ancestor of the lancelet.

The additional difficulty arises if one asks, how might one determine whether the lamprey and shark had an ancestor in common that was not an ancestor of the lancelet? If the only answer to this question is that

(3) the lamprey and shark had an ancestor in common that was not an ancestor of the lancelet, if

(4) D1 > D2, D3 is generally true,

then the difficulty is not seemingly, but really, multiplied.

The difficulties may be obviated, at least temporarily, if common ancestry is regarded not as real knowledge, in addition to generalizations about character distributions, but merely as a mode of interpretation of generalizations about character distributions. Indeed, common ancestry, or simply "evolution" or "coming into being," may be regarded as a causal explanation, dictated by other factors, of whatever character distributions seem really to exist in a general sense. If, for example, D1 > D2, D3 is generally true, i.e., if the distribution really exists, one may assert simply as an article of faith that the character distribution exists because it (or, more exactly, the species for which it exists) evolved, or came into being.

In this connection, it may be noted that the nature of character distributions is sometimes differently construed. In the above example, only positive occurrences of characters of type D are considered relevant. Sometimes, however, and most commonly in phenetic studies, positive and negative occurrences are combined. Consider, for example, the characters of table 3.21. Three characters are known only for the lancelet, but eight characters (of type D1) are unknown only for the lancelet. In this sense there are 3 + 8 = 11 "characters" known for the lancelet. In table 3.24, the positive and negative occurrences are combined for all character-types of table 3.21.

The relative numbers of "characters" of type D (positive and negative occurrences) may be considered a measure of "overall similarity" among the species (table 3.23). For three species, the effect is to add negative "characters" of type D (D1, D2, D3), in numbers equal to the characters of types A, B, and C. Adding negative "characters" derived in this way can have no effect on the relative numbers of characters of type D (D1,

Table 3.24. Numbers of "Characters" (positive and negative occurrences) of Different Types in Three Species (based on table 3.21; cf. table 3.23)

				Character-Types			
Species	*A*	*B*	*C*	*D1*	*D2*	*D3*	*E*
Lancelet	11	–	–	–	13	4	35
Lamprey	–	4	–	11	13	–	35
Shark	–	–	13	11	–	4	35

D2, D3) unless characters of types A, B, and C occur in different numbers. Adding such negative "characters" implies that, in some sense, they are real information, i.e., that there are real differences in the true relative abundance of characters of types A, B, and C, as reflected in their relative numbers in a particular sample. Thus in table 3.21, there are three characters known for the lancelet, two for the lamprey, and nine for the shark; in table 3.24 each of these values is added to the appropriate character of type D (3+D1 = 11, 2+D3 = 4, 9+D2 = 13).

Considering positive occurrences alone is one approach to systematics, which is sometimes termed phyletic ("cladistic"); considering positive and negative occurrences together is another approach to systematics, which is sometimes termed phenetic. Both approaches assume that, if characters of type D occur in different relative numbers within a sample of information, the differences may be informative. In a phyletic sense, the information is construed to mean that some characters are true homologies and others are parallelisms or convergences, i.e., that some characters are, in a general way, true and others are false. In a phenetic sense, the information is construed to mean that some character-types truly occur more abundantly than others, but that all occurrences of characters are true, and none is false. In a phyletic sense the information relevant for estimating common ancestry is positive occurrences of characters of type D. In a phenetic sense, the information relevant for estimating "overall similarity" is positive occurrences of characters of type D and, in addition, negative occurrences of characters of types A, B, and C.

To the extent that the relative numbers of characters of type D vary randomly in samples, both approaches are liable to mistake random differences for real ones. To the extent that relative numbers of characters of types A, B, and C vary randomly in samples, the phenetic approach is liable to mistake random differences for real ones. Both approaches face one hazard (random variation in numbers of characters of type D); the phenetic approach faces another (random variation in numbers of characters of types A, B, and C).

Interestingly, both approaches would assume that all characters of types A, B, and C, no matter what their relative numbers, are, or might be, real. The approaches differ with respect to characters of type D: in the phyletic sense, some are real and some are not, and the problem is to find out which are real and which are not; in the phenetic sense, all are real, and the problem is to estimate their true abundances.

The "overall similarities" of table 3.23 may be represented by a "most parsimonious" tree (3.29.3; lancelet and lamprey grouped together as most similar), different from that tree most parsimonious for positive occurrences only (3.29.1; lamprey and shark grouped together). Again, that there are more "characters" of type D2 than of types D1 and D3 (D2± > D1, D3) in a particular sample of characters need not be considered significant. If significant, the relation D2± > D1, D3 should be general, in the sense that the relation is true of samples generally—a matter that also is open to empirical investigation through further sampling.

A generalization about "overall similarity" (e.g., D2± > D1, D3) also is interpretable in an evolutionary sense to indicate relative recency of common ancestry. In other words,

(1) if D2± > D1, D3 is generally true, then
(2) the lancelet and lamprey had an ancestor in common that was not an ancestor of the shark (as exemplified in figure 3.29.3).

This mode of interpretation has been employed to some extent within systematic biology, by persons who assume that degree of "overall similarity" reflects relative recency of common ancestry. The mode, not surprisingly, leads to difficulties the same as those discussed above. One may ask, for example, how might one determine whether the lancelet and lamprey had an ancestor in common that was not an ancestor of the shark? If the only answer to this question is that

(3) the lancelet and lamprey had an ancestor in common that was not an ancestor of the shark, if
(4) D2± > D1, D3 is generally true,

then the same circularity is repeated. But there is an additional difficulty about a fundamental principle: how information about "characters" is to be represented by the lines of a phyletic tree.

Consider a simple case of two species (lamprey and shark) and two characters (A and B), each observed to occur in one species (table 3.25:left). Parsimonious representation is achieved in a "tree" having two lines (figure 3.30.1); a third line might be added (figure 3.30.2), but

Table 3.25. Relation Between Number of Characters of Different Types in Two Species (lamprey and shark) and Positive-Negative Occurrences, or "Overall Similarity" (cf. figure 3.30)

	Character-Types			"Overall Similarity"		
Species	*A*	*B*	*C*	*A*	*B*	*C*
Lamprey	1	0	0	1+	1−	2
Shark	0	1	0	1−	1+	2

what would it represent if no homology between A and B is implied? Dividing each of the two characters into positive and negative occurrences would produce two "characters" of type C, each present in both species (table 3.25:right). What would constitute parsimonious representation of the "characters"? Two lines, connected or not, would be "unparsimonious" (figure 3.30.3), inasmuch as no line would represent the "characters" of type C. A third line may be added for this purpose (figure 3.30.4). The line would represent both "characters" of type C, but would leave in doubt the nature of such "characters."

Analysis of "character" A may be considered in relation to character trees. If the third line indicates "homology" between A+ and A−, what can be said of the ancestral or primitive (plesiomorphic) condition (figure 3.31.1)? There are three possibilities: A+, A−, and some third condition Ax (figures 3.31.2–6). Parsimonious representation of each possibility is shown in figures 3.31.3, 3.31.5, and 3.31.6, respectively. One ancestral possibility (A−) results in a representation more parsimonious than the other two, in its implied numbers of gains and transformations (table 3.26).

Analysis of "characters" A and B together may be considered in relation to bifurcating "character" trees. What can be said of the ancestral condition? There are nine possibilities (figures 3.32.1–9), two of which (figures 3.32.2–3) are parsimoniously represented by non-

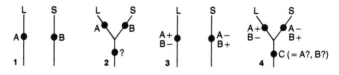

Figure 3.30. Parsimonious representation of character-types (A, B, C) and positive and negative occurrences of character-types (A+, A-, B+, B-) in phyletic trees for two species, lamprey and shark (cf. table 3.25).

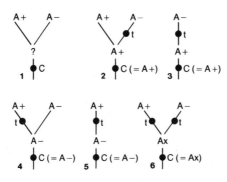

Figure 3.31. Trees implying transformations (t) of characters (A+, A-, Ax; cf. table 3.26).

bifurcating trees, because one line in each tree represents nothing, and could just as well be omitted. One ancestral condition (A–B–; figure 3.32.4) is more parsimonious than the other eight, in its implied gains and transformations (table 3.27). Thus, one step toward parsimonious representation may be achieved by specifying the ancestral condition in a phyletic tree (A–B–; figure 3.33.1). Another step may be achieved by eliminating redundancy of character representation (figure 3.33.2; eliminating B– from the line leading to the lamprey, and A– from the line leading to the shark). The resulting tree (figure 3.33.2) is a contradiction; its third line represents a "character" shared by both the lamprey and the shark, but there is no such "character": neither "character" A– nor B– is found in both taxa. The tree represents no more than a "tree" with two lines (figure 3.33.3), which is the final step toward parsimonious representation.

Examination of the possible "character" trees (figure 3.32) shows that

Table 3.26. Evolutionary Gains of Characters (A, Ax) and Character Transformations (A– —· →A+, A+ ——→ A–, Ax——→ A+, Ax ——→ A–) Implied by Different Character Trees (cf. figure 3.31)

Trees	Gains of Characters		Gains of Character Transformations				Total
	A	Ax	A– —→A+	A+ ——→A–	Ax ——→A+	Ax ——→A–	
3.31.2	1	0	0	1	0	0	2
3.31.3	1	0	0	1	0	0	2
3.31.4	0	0	1	0	0	0	1
3.31.5	0	0	1	0	0	0	1
3.31.6	0	1	0	0	1	1	3

Form

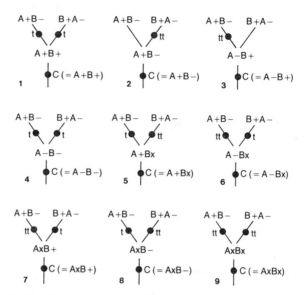

Figure 3.32. Trees implying transformations of characters (A+, A-, Ax, B+, B-, Bx; cf. table 3.27).

a bifurcating form is required only when ancestral conditions include positive occurrences (A+, Ax, B+, Bx). Hence, representation of both positive and negative occurrences in a bifurcating tree (e.g., figure 3.30.4), or trees generally, can be parsimonious only if positive occurrences are ancestral, i.e., if all negative occurrences are transformed positive occurrences (e.g., A+ $\longrightarrow$ A-, Ax $\longrightarrow$ A-). The conclusion seems unavoidable that under the assumption that relative "overall similarity" reflects relative recency of common ancestry, positive and negative occurrences may be represented in a phyletic tree only through a criterion other than parsimonious representation.

Is such a criterion possible? If not, then there is no way that positive and negative occurrences considered together can specify one phyletic tree rather than another. If such a criterion is possible, then two kinds of generalizations about character distributions are possible, as exemplified by those discussed above:

(1) Dl > D2, D3
(2) D2± > Dl, D3

And it is possible that, in a given case, both might be generally true for the same three species. If so, each would lead to a different and conflicting statement about the relative recency of common ancestry. Such conflict would represent no more than a methodological artifact due to different criteria for representation of characters in trees.

But it is possible that, in a given case, both kinds of generalizations might agree. If, for example,

(1) D1 > D2, D3 is generally true, and if
(2) D1± > D2, D3 is generally true,

then what might explain the agreement? If the agreement in a given case is due to chance alone, then there is no need for further explanation. Hence, at this point an important question is: how may one judge whether the agreement is random (due to chance alone) or nonrandom (due to a cause other than chance)?

Purely random agreement would be expected to occur in 33 percent of pairs of generalizations about character distributions in three taxa. This percentage is too high to allow confidence in a judgment that a particular agreement is nonrandom. One way to decrease the probability is to consider samples of characters rather than pairs of generalizations. If, for example,

(1) D1 > D2, D3 is true for sample 1, and
(2) D1 > D2, D3 is true for sample 2, and
(3) D1± > D2, D3 is true for sample 3, and
(4) D1± > D2, D3 is true for sample 4,

then what is the probability that the agreement among the samples is due to chance alone, if there is no reason to expect agreement due to other factors? Given (1), the probability that (2) would agree is 33 percent; given (1) and (2), the probability that (3) would agree is 33 percent (with a combined probability of 11 percent); given (1), (2), and (3), the probability that (4) would agree is 33 percent (with a combined probability of 4 percent). A probability of 4 percent would indicate that the agreement is probably not due to chance alone.

One might ask how many characters are sufficient to serve as a sample. If the samples are truly independent, the probability that four characters would agree by chance alone is also 4 percent. Thus, four characters are sufficient to indicate a nonrandom distribution of characters in a case with three taxa, under the assumption of independence.

Table 3.27. Evolutionary Gains of Characters (A, Ax, B, Bx) and Character Transformations (A+ ⟶ A−, A− ⟶ A+, Ax ⟶ A+, Ax ⟶ A+, Ax ⟶ A−, B+ ⟶ B−, B− ⟶ B+, Bx ⟶ B+, Bx ⟶ B−) Implied by Different Character Trees (cf. figure 3.32)

Trees	Characters				Gains of Character Transformations								Total
	A	Ax	B	Bx	A+⟶A−	A−⟶A+	Ax⟶A+	Ax⟶A−	B+⟶B−	B−⟶B+	Bx⟶B+	Bx⟶B−	
3.32.1	1	0	0	0	1	0	0	0	1	0	0	0	4
3.32.2	1	0	1	0	1	0	0	0	0	0	0	0	3
3.32.3	0	0	0	0	0	1	0	0	1	0	0	0	3
3.32.4	0	0	0	0	0	1	0	0	0	1	1	0	2
3.32.5	1	0	0	1	1	0	0	0	0	0	0	1	5
3.32.6	0	0	1	1	0	1	1	0	1	0	1	1	4
3.32.7	0	1	0	0	0	0	1	1	0	0	0	0	5
3.32.8	0	1	0	0	0	0	1	1	0	1	1	0	4
3.32.9	0	1	0	1	0	0	1	1	0	0	1	1	6

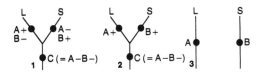

Figure 3.33. Steps toward parsimonious representation of positive and negative occurrences of character-types in phyletic trees for two species, lamprey and shark.

Suppose there is, in a given case, agreement between both factors (positive occurrences alone, and positive and negative occurrences combined). Is the inference of common ancestry more meaningful?

(1) If the agreement is not due to chance alone, then
(2) the lamprey and shark had an ancestor in common that was not an ancestor of the lancelet.

The inference (2) is open to criticism, and may actually be unwarranted in this hypothetical case, for the agreement may itself be an artifact due to nonindependence of the two factors that agree. But the inference illustrates that relative recency of common ancestry may be considered a causal principle not merely of a particular generalization about character distributions, but of an agreement between different kinds of generalizations that need not agree, except by chance, in a given case. If chance agreement can be ruled out, then some other causal principle may be sought. "Evolution" in the sense of "coming into being" by itself is no causal principle of nonrandom agreement, for "evolution" or "coming into being" implies nothing about nonrandom agreements or the lack of them. Relative recency of common ancestry, however, does imply that, if the lamprey and shark had an ancestor in common that was not an ancestor of the lancelet, then the lamprey and shark might be expected to exhibit the marks of that ancestry unique to themselves.

It is interesting to note that generalizations about character distributions relate to current controversies about biological classification. Three points of view may be distinguished, which are sometimes termed phenetic, phyletic ("cladistic"), and gradistic ("evolutionary"). In terms of the above example of three species (lancelet, lamprey, and shark), these points of view may be illustrated as follows:

Phenetic: (1) If $D1\pm > D2, D3$ is generally true, then

    (2) the lamprey and shark should be classed together in a group that does not include the lancelet, even if

    (3) $D2 > D1$, $D3$ or $D3 > D1$, $D2$ is also generally true.

Phyletic: (1) If $D1 > D2$, $D3$ is generally true, then

    (2) the lamprey and shark should be classed together in a group that does not include the lancelet, even if

    (3) $D2\pm > D1$, $D3$ or $D3\pm > D1$, $D2$ is also generally true.

Gradistic: (1) If $D1$ defines a biologically meaningful group (grade), then

    (2) the lamprey and shark should be classed together in a group that does not include the lancelet,

    (3) no matter what character distribution or distributions might generally be true.

At present, gradistic theory is not precisely formulated, and there is some doubt whether it can be, but the notion of a biologically meaningful group (grade) is sometimes said to be essentially the same as a phenetic group, and, if so, gradistic classification would be illustrated by the phenetic formulation above.

Because notions of "overall similarity" seem to be the root of current dispute about systematic philosophies, some additional remarks are appropriate. Consider tables 3.21 and 3.24. The former records information in terms of positive occurrences only; the latter records information in terms of positive and negative occurrences combined. Of interest in the former (table 3.21) are relative numbers of characters of type D (D1, D2, D3), which may serve as a basis of inference about the abundance of characters shared by two of the three species. Of interest in the latter also are relative numbers of characters of type D, which may serve as a basis of inference about the abundance of characters shared by two of the three species, but also of characters unique to each of the three species. Thus in the latter, positive occurrences of characters of type A increase the number of "characters" of type D1; characters of type B increase the number of "characters" of type D3; characters of type C increase the number of "characters" of type D2. In table 3.21 characters of type D1 are more numerous than characters of types D2–3; in table 3.24 "characters" of type D2 are more numerous than "characters" of types D1, D3. The reasons for the discrepancy are that in table 3.21 there are more characters of type C (9) than of types A (3) and B (2), and that when characters of type C (9) are added to those of type D2 (4), the total (13) is larger than the totals of A + D1 = 11 or of B + D3 = 4.

What are the implications of adding characters of types A–C to characters of type D? Consider table 3.21: 3 characters of type A, 2 characters of type B, 9 characters of type C. What might one infer to be true on the basis of these numbers? The only possibility would seem to be that the character-types really differ in their abundance. If the real differences in abundance are reflected in the sample, then

(1) C > A > B is true.

This sort of inference is open to empirical investigation through additional sampling; in short, one may predict that additional samples would show the same relative abundance. But there is a philosophical difficulty with this sort of inference:

(1) if C > A > B is true, then
(2) A, B, and C are unequal quantities and at least A and B (and possibly all three quantities) are therefore finite.

Are characters of a certain type finite in number? And, if so, how might that fact be learned? What seems beyond dispute is that characters are defined, or recognized, by systematists, who regularly count them, or compile lists of them, and then interpret the counts, or lists, in one fashion or another. Are such counts, or lists, theoretically finite in length? Or, alternatively, are such counts and lists always incomplete in the sense that they may always be increased? If so, there is some reason to believe that

(3) A = B = C = ∞ is true.

If so, then the different numbers of characters of types A, B, and C in a particular sample are purely the result of chance (sampling error). And adding their numbers to those of characters of type D (D1, D2, D3) adds only random variation to whatever real information characters of type D might contain.

But what of characters of type D? What might one infer to be true on the basis of the numbers of them in table 3.21? One possibility is that

(4) D1 > D2 > D3 is generally true.

But this is the same sort of inference, because

(4) if D1 > D2 > D3 is true, then

(5) at least D2 and D3 are finite and unequal quantities.

Doubts the same as those mentioned above apply to characters of type D. Thus, there is the possibility that

(6) D1 = D2 = D3 = ∞ is true.

But there is an alternative inference, namely that

(7) D1 = ∞ and D2 = D3 = 0.

Or, in other words,

(8) that the lamprey and shark are members of a group that does not include the lancelet is true, and

(9) that the lancelet and shark are members of a group that does not include the lamprey is false, and

(10) that the lancelet and lamprey are members of a group that does not include the shark is false.

The sense of inference (7) is reflected in the notion of evolutionary homology, in the sense that, for example, characters of type D1 may be interpreted as "true homologies" and characters of types D2 and D3 may be interpreted as "convergences" or "parallelisms." Interpreted as "convergences," they would be equivalent to "false homologies," i.e., "characters" that are not really characters at all, such that

(11) x "convergences" = 0 characters of type D.

Interpreted as "parallelisms," they are equivalent either to "false homologies," such that

(12) x "parallelisms" = 0 characters of type D;

or to characters of type E, present in some sense even in species from which they are absent, such that they are really plesiomorphic:

(13) x "parallelisms" = x characters of type E = 0 characters of type D.

Traditional notions of "convergence" and "parallelism" thus seem to exemplify inference (7), and seem to be at odds with inferences (5), (6), and, by implication, (2). Particular interpretations of "convergence" or "parallelism" may be based on generalizations about character distributions. But,

(14) if D1 > D2, D3 is true, and

(15) characters of types D2 and D3 are, therefore, convergences or parallelisms,

then one may always ask, how might one determine what characters are "convergent" or "parallel"? If the only answer is that

(16) characters of types D2, D3 are "convergent" or "parallel," if

(17) D1 > D2, D3 is true,

then circularity again emerges. What is needed is another kind of answer, so that there exists the possibility of nonrandom agreement.

With the above analysis certain philosophical differences underlying phenetic and phyletic ("cladistic") approaches seem clarified. Certain general conclusions may be reached. One possibility is that,

(18) if $A = B = C = \infty$ is true, and

(19) if $D1 = D2 = D3 = \infty$ is true, then

(20) all differences in relative numbers of characters of different types in a particular sample are random and due purely to sampling error, and

(21) relative numbers of characters furnish no basis for inference.

If so, then it is futile to use characters as a basis for inference, for no real parameter would be estimated either by a phenetic or by a phyletic ("cladistic") approach. There are, of course, other possibilities:

(22) if $A = B = C = \infty$ is true, and

(23) if $D1 = \infty$ and $D2 = D3 = 0$ is true, then

(24) D1 > D2, D3 is true and may be reflected in the relative numbers of characters of type D in a particular sample, so that

(25) relative numbers of characters of type D furnish a basis for inference.

If so, then it is futile to combine characters of types A, B, and C with characters of type D in the hope that a better basis for inference would be obtained. All that could possibly be obtained is greater sampling error (random variation) with no change in the single parameter estimated by both approaches. Still another possibility is that

(26) if A, B, and C are finite and unequal quantities, and

(27) if D1, D2, and D3 are finite and unequal quantities, then

(28) D1 > D2, D3 and D2$\pm$> D1, D3 might both be true in a given case.

If so, then it is futile to compare the results of phenetic and phyletic approaches because each would estimate a different, and equally real, parameter.

How may these different possibilities be evaluated? Items (18)–(21) imply nothing but random variation, which is belied by the history of systematics. Items (22)–(25) and (26)–(28) allow a basis for inference to both phenetic and phyletic approaches. In the former case, (22)–(25), there is a single parameter estimated by both approaches; in the latter case, (26)–(28), there are two parameters, one for each approach. But the two parameters are not independent, for they contain some common elements (D1, D2, D3). If there really are two parameters, however, then

(29) of A, B, and C, some are finite and unequal quantities, such that
(30) characters of each type would not randomly vary in relative frequency among samples,

and the question would seem amenable to empirical investigation through sampling.

INFORMATION AND COMPONENTS

The 3-taxon problem plays a special role in systematics. Firstly, three taxa are the minimum that allows for a choice among cladograms; for two taxa there is only one possible cladogram, but for three taxa there are three possible primary cladograms. Secondly, the number (three) of possible cladograms for three taxa is not so large that a complete analysis (e.g., table 3.21, figures 3.29.1–3) is either prohibitively complex or beyond intuitive appraisal; complete analysis of four taxa (15 possible primary cladograms) is significantly, and prohibitively, more complex— perhaps beyond intuitive appraisal, and complete analysis of five taxa (105 primary cladograms), unaided by a computer, is out of the question. Thirdly, any complex problem, involving four or more taxa, may be reduced to a series of 3-taxon problems. No one would dispute that some systematic problems have been posed and subsequently solved without aid of computers. It is seldom realized, however, that virtually all such problems are, or imply, 3-taxon problems.

Consider a fourth species, the salmon. Suppose that study of the salmon shows that, of the three characters previously known only for the lancelet (table 3.21:A), one character occurs in the salmon (table

Table 3.28. Numbers of Characters of Different Types in Four Species (lancelet, lamprey, shark, and salmon) (derived from table 3.21, except for the 4 characters of type D)

Species	A	B	C	D	Character-Types E1	E2	E3	E4	E5	E6	F1	F2	F3	F4	G
Lancelet	2	–	–	–	3	1	1	1	–	–	–	–	1	–	6
Lamprey	–	1	–	–	3	–	–	2	1	–	–	1	–	6	6
Shark	–	–	6	–	–	1	–	2	–	3	1	–	1	6	6
Salmon	–	–	–	4	–	–	1	–	1	3	–	1	1	6	6

3.28:E3); of the two characters previously known in the lamprey (table 3.21:B), one character occurs in the salmon (table 3.28:E5); of the nine characters previously known in the shark (table 3.21:C), three characters occur in the salmon (table 3.28:E6); of the eight characters previously known in the lamprey and shark (table 3.21:D1), six characters occur in the salmon (table 3.28:F4); of the four characters previously known in the lancelet and lamprey (table 3.21:D2), one character occurs in the salmon (table 3.28:F2); of the two characters previously known in the lancelet and shark (table 3.21:D3), one character occurs in the salmon (table 3.28:F3); of the seven characters previously known in the lancelet, lamprey, and shark (table 3.21:E), six characters occur in the salmon (table 3.28:G). Suppose also that study of the salmon reveals four characters unique to that species (table 3.28:D). The total information known for the four species would be shown in table 3.28, which is a sample "realistic" in the sense that all possible character-types are represented by some characters. Character-types relevant for the most efficient cladogram (generally most parsimonious tree) are increased from three (for three taxa: D1, D2, D3) to ten (table 3.28:E1–6, F1–4). Also the kinds of relevant character-types are increased from one (for three taxa: characters of type D that occur in two of three taxa) to two (for four taxa: characters of type E that occur in two of four taxa, and characters of type F that occur in three of four taxa). Unlike the assembled data for three taxa (table 3.21), for which the most efficient cladogram may be determined at a glance, the data for four taxa seem immune to quick assessment.

An alternative to a complete analysis of the 15 primary cladograms for four taxa is reduction of the four taxa to three. On the basis of previous considerations (table 3.21), one may infer that the lamprey and shark form a group that does not include the lancelet; hence the lamprey

Table 3.29. Numbers of Characters of Different Types in Four Species (lancelet, lamprey, shark, and salmon), Two of Which (lamprey and shark) Are Combined into a Single Taxon (derived from table 3.28)

Species	Character-Types						
	A	B	C	D1	D2	D3	E
Lancelet	2	–	–	–	5	1	8
Lam + Sha	–	9	–	10	5	–	8
Salmon	–	–	4	10	–	1	8

and shark may be combined in one taxon, so as to reduce the number of character-types (table 3.29, derived from table 3.28):

Table 3.28 (lamprey + shark)	Table 3.29
A	A
B + C + E4	B
D	C
E5 + E6 + F4	D1
E1 + E2 + F1	D2
E3	D3
F2 + F3 + G	E

The problem may be viewed as beginning with four taxa: lancelet, salmon, lamprey, shark (figure 3.34.1), three of which (lancelet, lamprey, shark) are selected as an initial 3-taxon problem. If the lamprey and shark are grouped together, the result is cladogram 3.34.2. The next step concerns the placement of the salmon, for which there are seven possibilities (figures 3.34.2:1–7), three of which (1, 3, 5) define another 3-taxon problem (figure 3.34.3), for each possibility specifies a different cladogram (figures 3.34.4–6). These three possibilities for placement of the salmon (1, 3, 5) are those subsumed by the secondary cladogram (figure 3.34.2).

A glance at table 3.29 suffices to show that the salmon has its relationships with the lamprey-shark taxon, i.e., that the most efficient cladogram is the secondary cladogram of figure 3.34.6. If true, the secondary cladogram (3.34.6) implies that two other 3-taxon clado-grams are also true (figures 3.34.7–8), and the characters relevant to these two cladograms (tables 3.30 and 3.31) may also be derived from table 3.28:

Table 3.28 (- shark)	Table 3.30	Table 3.28 (- lamprey)	Table 3.31
A + E2	A	A + E1	A
B + E4	B	C + E4	B
D + E6	C	D + E5	C
E5 + F4	D1	E6 + F4	D1
E1 + F1	D2	E2 + F1	D2
E3 + F3	D3	E3 + F2	D3
F2 + G	E	F3 + G	E

A glance at tables 3.30 and 3.31 suffices to show that each implied

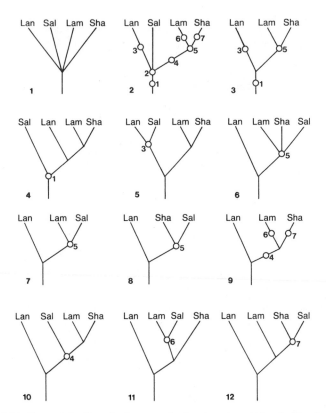

Figure 3.34. Four species (lancelet, lamprey, shark, salmon), considered with reference to three 3-taxon problems. Numbers 1–7 are possibilities for placement of a fourth species, the salmon (cf. tables 3.28–3.32).

Table 3.30. Numbers of Characters of Different Types in Three Species (lancelet, lamprey, and salmon) (derived from table 3.28)

Species	A	B	C	D1	D2	D3	E
			Character-Types				
Lancelet	3	–	–	–	4	2	7
Lamprey	–	3	–	7	4	–	7
Salmon	–	–	7	7	–	2	7

Table 3.31. Numbers of Characters of Different Types in Three Species (lancelet, shark, and salmon) (derived from table 3.28)

Species	A	B	C	D1	D2	D3	E
			Character-Types				
Lancelet	5	–	–	–	2	2	7
Shark	–	8	–	9	2	–	7
Salmon	–	–	5	9	–	2	7

3-taxon cladogram (figures 3.34.7–8) is the most efficient for the relevant data. The only problem remaining concerns the interrelationships of the lamprey, shark, and salmon. This problem may also be viewed as the placement of a fourth taxon (salmon) in a cladogram previously resolved for three taxa, with the possibilities reduced to three (figure 3.34.9) from the original seven (figure 3.34.2), or in other words to three 3-taxon cladograms (figures 3.34.10–12). These three possibilities for placement (4, 6, 7) are those subsumed by the secondary cladogram (figure 3.34.6). Together with the two possibilities previously considered (figures 3.34.4 and 3.34.5:1,3), these three (4, 6, 7) exhaust the five possibilities for dichotomous resolution allowed under the assumption that the lamprey and shark together form a group that does not include the lancelet.

The relevant data (table 3.32) may also be derived from table 3.28:

Table 3.28 (– lancelet)	Table 3.32
B + E1	A
C + E2	B
D + E3	C
E6 + F3	D1
E4 + F1	D2
E5 + F2	D3
F4 + G	E

A glance suffices to show that the salmon has its relationships with the shark, not the lamprey (figure 3.34.12).

In the above account, all possible placements of the fourth taxon (salmon) have been considered. That placement 2 subsumes possibilities 1, 3, and 5, and that placement 5 subsumes possibilities 4, 6, and 7, mean that placement 2 subsumes all (1–7) possibilities for dichotomous

Table 3.32. Numbers of Characters of Different Types in Three Species (lamprey, shark, and salmon) (derived from table 3.28)

Species	A	B	C	D1	D2	D3	E
Lamprey	4	–	–	–	3	2	12
Shark	–	7	–	4	3	–	12
Salmon	–	–	5	4	–	2	12

resolution. This conclusion has an important implication. Given the cladogram of figure 3.34.2, for example, one may assert that, if the cladogram is true, the lamprey and shark together form a group; but it is not clear whether both the lancelet and salmon are excluded from the group, or merely the lancelet or the salmon is excluded from the group; and if one and not the other is excluded, it is not clear which is which.

Given information about characters in four taxa (e.g., table 3.28), one may select any three taxa for an initial 3-taxon problem: for example, lancelet, lamprey, shark (figure 3.34.3); lancelet, lamprey, salmon (figure 3.34.7); lancelet, shark, salmon (figure 3.34.8); lamprey, shark, salmon (figure 3.34.12). Whatever the result, there is at least one additional 3-taxon problem, but there is a maximum of two additional 3-taxon problems, in the placement of the fourth species in a dichotomous cladogram. If the initial placement of the fourth species results in a primary (dichotomous) cladogram of the four taxa (e.g., figure 3.34.12), there would have been only one additional 3-taxon problem (for a total of two). If the initial placement of the fourth species results in a secondary (trichotomous) cladogram of the four taxa (e.g., figure 3.34.6), there would have been one, but there would be yet another, additional 3-taxon problem (for a total of three).

Similarly, for five taxa (105 possible dichotomous cladograms), there is a minimum of three 3-taxon problems, and a maximum of six, required for dichotomous resolution (table 3.33). For six taxa (945 possible dichotomous cladograms), there is a minimum of four 3-taxon problems, and a maximum of ten, required for dichotomous resolution. For ten taxa (34,459,425 possible dichotomous cladograms), there is a minimum of eight 3-taxon problems, and a maximum of 36, required for dichotomous resolution. It is apparent that, for a group of many species, no cladogram could possibly be achieved by complete analysis of all possibilities without either the aid of a computer, or the reduction of the

Table 3.33. Number of Taxa in Relation to Number of Possible 3-Taxon Problems (for completely dichotomous resolution), Number of Implied 3-Taxon Cladograms, Number of Character-Types Shared by Two or More Taxa, and Number of Possible Dichotomous Cladograms

	3-Taxon Problems		Implied 3-Taxon Cladograms		Shared	Dichotomous
Taxa	Minimum	Maximum	Maximum	Minimum	Character-Types	Cladograms
2	0	0	0	0	1	1
3	1	1	0	0	4	3
4	2	3	3	2	11	15
5	3	6	9	7	26	105
6	4	10	19	16	57	945
7	5	15	34	30	120	10,395
8	6	21	55	50	247	135,135
9	7	28	83	77	502	2,027,025
10	8	36	119	112	1013	34,459,425

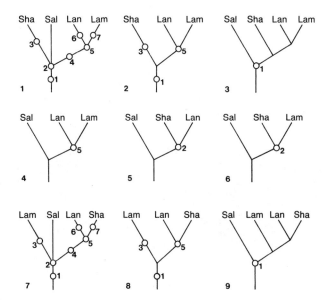

Figure 3.35. Four species (lancelet, lamprey, shark, salmon), considered with reference to two 3-taxon problems. Numbers 1–7 are possibilities for placement of a fourth species, the salmon (cf. tables 3.34–3.35).

many species to a series of 3-taxon problems. That cladograms for large groups have been achieved without the aid of a computer (e.g., Hennig 1969) suggests that the 3-taxon approach was either the method of choice or the method intuitively applied.

Although the above examples include taxa represented by species (lancelet, lamprey, etc.), the 3-taxon approach may be applied generally to taxa of any number of species. Thus, instead of the lancelet, lamprey, shark, and salmon, one may inquire into the relationships of groups more inclusive, such as the Echinodermata, Vertebrata, Mollusca, and Annelida. One may, of course, simply accept such groups as given, in some current classification, without concern for their composition. But if so, problems relating to their composition might possibly be uncovered and go unrecognized. An example of such a problem may be illustrated with reference to the lancelet, lamprey, shark, and salmon, under the assumption that the lancelet and lamprey are members of a group that does not include the shark (figure 3.35.1). Information relevant to this example (table 3.34) may be derived from table 3.28:

Table 3.28	Table 3.34
(lancelet + lamprey)	
A + B + E1	A
C	B
D	C
E6	D1
E2 + E4 + F1	D2
E3 + E5 + F2	D3
F3 + F4 + G	E

As before, there are seven possibilities for placement of the salmon, and possibility 2 subsumes all others (figure 3.35.1). The three possibilities of the next 3-taxon problem are those of figure 3.35.2. A

Table 3.34. Numbers of Characters of Different Types in Four Species (lancelet, lamprey, shark, and salmon), Two of Which (lancelet and lamprey) Are Combined in a Single Taxon (derived from table 3.28)

Species	*A*	*B*	*C*	*D1*	*D2*	*D3*	*E*
Lan + Lam	6	–	–	–	4	3	13
Shark	–	6	–	3	4	–	13
Salmon	–	–	4	3	–	3	13

glance at table 3.34 shows that possibility 1 specifies the most efficient cladogram (figure 3.35.3), which is a primary (dichotomous) cladogram. Accordingly, there is no additional 3-taxon problem. If cladogram 3.35.3 is true, there are, nevertheless, three implied 3-taxon cladograms that also must be true (figures 3.35.4–6), the data relevant to which have already been assembled for another purpose (tables 3.30–3.32). A glance at tables 3.30–3.32 shows that each implied cladogram is not the most efficient summary of the relevant data; hence the implied cladograms would seem false. That the cladograms seem false suggests in turn that the initial grouping (lancelet + lamprey) is also false. Such is apparent only because the implied cladograms, and the data relevant to them, are at hand.

Study of inclusive taxa, such as Echinodermata, Vertebrata, etc., might similarly lead to an erroneous conclusion for the same reason— namely that one of the groups assumed to exist does not exist. Such would be apparent only if data relevant to the implied cladograms were at hand, so that the implied cladograms, or some sufficient number of them, could be checked.

Another example of the same sort arises if one groups together the lancelet and shark, which would likewise allow for seven possible placements of the salmon (figure 3.35.7). The three possibilities of the next 3-taxon problem are 1, 3, and 5 (figure 3.35.8), and the relevant data (table 3.35) may be derived from table 3.28:

Table 3.28 (lancelet + shark)	Table 3.35
A + C + E2	A
B	B
D	C
E5	D1
E1 + E4 + F1	D2
E3 + E6 + F3	D3
F2 + F4 + G	E

A glance at table 3.35 shows that possibility 1 is the most efficient cladogram (figure 3.35.9), which is a primary (dichotomous) cladogram. Accordingly, there is no additional 3-taxon problem. If cladogram 3.35.9 is true, there are, nevertheless, three implied 3-taxon cladograms that also must be true, and they are the same as those of the previous

Table 3.35. Numbers of Characters of Different Types in Four Species (lancelet, lamprey, shark, salmon), Two of Which (lancelet and shark) Are Combined in a Single Taxon (derived from table 3.28)

Species	A	B	C	D1	D2	D3	E
			Character-Types				
Lan + Sha	9	–	–	–	6	5	13
Lamprey	–	1	–	1	6	–	13
Salmon	–	–	4	1	–	5	13

example (figures 3.35.4–6), all of which again seem false and suggest in turn that the initial grouping (lancelet + shark) is also false.

Groupings such as the lancelet + lamprey, which seem false in relation to available information, are sometimes termed "nonmonophyletic." Interpreted within the context of a phyletic tree, such a grouping would include some, but not all, of the descendant species of a common ancestral species. Interpreted in relation to character distributions,

(1) a grouping is false ("nonmonophyletic") if it conforms to D2 or D3, when
(2) D1 > D2, D3 is true.

False groupings, to the extent that they occur in classification, are apt to mislead any investigator who accepts them as true, unless he can examine the cladograms implied by his results in relation to relevant data sufficient to expose the initial grouping(s) as false. Discovery of false groupings is one of the general purposes of systematic research, but it is a goal the value of which is sometimes disputed. False groupings, even when recognized as such, are sometimes valued if they can be defined by "characters" easily perceived. In table 3.28, for example, there are three "characters" shared by the lamprey and lancelet. If these "characters" are conspicuous, or otherwise easily perceived, they might work well in an identification key, by separating the lancelet and lamprey as a "group" distinct from other species. Hence it is helpful to distinguish groups that seem really to exist from "groups" that may be defined. That a "group" may be defined by one or more "characters" does not mean, therefore, that the group has any existence in the real world.

That all implied cladograms are false, as in the above examples

Table 3.36. Character-Types (A, B, C, D, E1, F1, G) and Their Occurrence in Four Taxa (1–4)

Taxa	A	B	C	D	E1	F1	G
				Character-Types			
1	+	–	–	–	+	+	+
2	–	+	–	–	+	+	+
3	–	–	+	–	–	+	+
4	–	–	–	+	–	–	+

(figures 3.35.4–6), suggests that the "solutions" of all the 3-taxon problems are also false (figures 3.35.3 and 3.35.9). Such need not always be the case. Consider the data of table 3.36, which is "unrealistic" in the senses that (1) not all possible character-types for the four taxa (1–4) are represented by some characters, and (2) the character-types represented cannot possibly conflict no matter how the taxa might be combined. The four taxa may be reduced to three by combining any two of them, e.g., taxa 3 and 4 (table 3.37; figure 3.36.1). The problem may be visualized as the placement of taxon 2, with three placements possible (1, 3, 5). A glance at table 3.37 shows that taxon 2 is best placed with taxon 1 (figure 3.36.2: placement 3), with the implications that three 3-taxon clado-grams must also be true (figures 3.36.3–4 and 3.36.6). Of the three implied cladograms, two cladograms are true (figures 3.36.3–4), and one cladogram is false (figure 3.36.6). The results are not totally false, for they allow taxa 1 and 2 to be grouped (figure 3.36.7), and combined for a new 3-taxon problem (figure 3.36.8). Data relevant to the new problem (table 3.38) show that taxon 3 may be grouped with taxa 1–2 (figure 3.36.9). Interrelationships of taxa 1–3 are another 3-taxon problem, the data relevant to which (table 3.39) show that taxa 1 and 2 may be grouped together relative to taxon 3 (figure 3.36.10).

Table 3.37. Character-Types (A, B, C, D2, E) and Their Occurrence in Four Taxa (1–4), Two of Which (3 + 4) Have Been Combined (derived from table 3.36)

Taxa	A	B	C	D2	E
			Character-Types		
1	+	–	–	+	++
2	–	+	–	+	++
3 + 4	–	–	++	–	+++

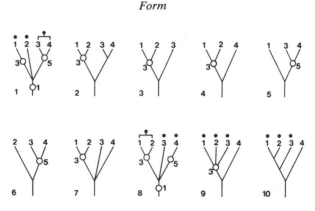

Figure 3.36. Four taxa (1–4), considered with reference to various 3-taxon problems. Circles 1, 3, and 5 are possibilities for placement of a fourth taxon. Black dots specify the three taxa of each 3-taxon problem to be solved (cladograms 1, 8, 9) or the 3-taxon problem solved (cladogram 10).

The minimum and maximum numbers of 3-taxon problems (table 3.33) characterize what may be termed minimum and maximum "modes" of resolving dichotomous cladograms. For six taxa (1–6), an example of resolution in the minimum mode is shown in figure 3.37. The first 3-taxon problem concerns taxa 4, 5, and 6 (shown as solved in figure 3.37.1); the second 3-taxon problem, taxa 3, 4, and 5+6 (solved in figure 3.37.2); the third 3-taxon problem, taxa 2, 3, and 4+5+6 (solved in figure 3.37.3); the fourth 3-taxon problem, taxa 1, 2, and 3+4+5+6 (solved in figure 3.37.4). In this instance, the minimum mode is a unique and stepwise resolution, for it consists of particular 3-taxon problems solved in a particular order. If the final resolution (figure 3.37.4) is correct for the information available, it cannot be reached in four steps other than those of figure 3.37; nor can it be reached in four steps except

Table 3.38. Character-Types (A, B, C, D2, E) and Their Occurrence in Four Taxa (1–4), Two of Which (1 + 2) Have Been Combined (derived from table 3.36)

Taxa	Character-Types				
	A	B	C	D2	E
1 + 2	++ ++	–	–	++	++
3	–	+	–	+	+
4	–	–	+	–	+

Table 3.39. Character-Types (A, B, C, D2, E) and Their Occurrence in Three Species (1–3; derived from table 3.36)

Taxa	Character-Types				
	A	*B*	*C*	*D2*	*E*
1	+	−	−	+	++
2	−	+	−	+	++
3	−	−	+	−	++

as in the order shown. An example of greater-than-minimum resolution is shown in figure 3.38, wherein the first two steps (figures 3.38.1–2) are the same as those of figure 3.37 (3.37.1–2). The third step (figure 3.38.3) differs in concerning taxon 1 rather than taxon 2 (cf. figure 3.37.3). The result is that taxon 2 requires two additional problems (figures 3.38.4–5) for final resolution; and the entire resolution requires five steps instead of four (five 3-taxon problems instead of four).

An example of resolution in the maximum mode is shown in figure 3.39, wherein ten 3-taxon problems are required for resolution of the same dichotomous cladogram (3.39.10; cf. figures 3.37.4 and 3.38.5). The maximum mode is not a unique and stepwise resolution, as is the minimum mode, for the steps toward resolution need not follow in the same sequence. In figure 3.40, for example, the same ten steps occur in a different order.

To determine whether the minimum or maximum mode characterizes a given resolution requires retrospective comparison between the fully resolved cladogram and the various steps toward it. A solved first 3-taxon problem does not in itself indicate either the minimum or maximum mode. A solved second 3-taxon problem will give an indication of the minimum mode (e.g., figures 3.37.2 and 3.38.2) or a greater-than-minimum mode (e.g., figures 3.39.2 and 3.40.2), if the taxa of the first problem are subsumed in the second problem. Taxa 4–6, for

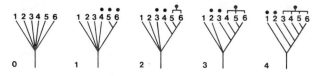

Figure 3.37. A dichotomous cladogram (4) resolved for six taxa (1–6) in the minimum mode, through the solution of four successive 3-taxon problems (shown as solved in cladograms 1–4).

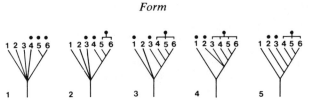

Figure 3.38. The same dichotomous cladogram (5) resolved for six taxa (1–6) in a greater-than-minimum mode, through the solution of five successive 3-taxon problems (shown as solved in cladograms 1–5; cf. figure 3.37).

example, constitute the first problem of figures 3.37 and 3.38, and are subsumed in the second problem of figures 3.37 and 3.38. Similarly, taxa 1–3 constitute the first problem of figures 3.39 and 3.40, and are subsumed in the second problem of figures 3.39 and 3.40.

Resolution of a cladogram in the minimum mode, if such could be done without advance knowledge of the final resolution, would require intuition equivalent in its effect to such knowledge. The role of intuition in systematics, and in science generally, is difficult to assess, mainly because the thrust of scientific investigation is toward nonintuitive analysis, solution, and synthesis of particular research problems. In some general sense, however, intuition might be imagined to play a role in the selection of particular problems to be analyzed, solved, or synthesized. In this sense, resolution in the minimum mode, to the extent that it actually might occur, can be imagined to result from an intuitive sense of the appropriate 3-taxon problem to be solved at the appropriate time within a suite of such problems. With a starting point for a minimum-mode resolution (e.g., figure 3.37.1), what is next

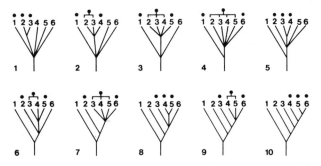

Figure 3.39. The same dichotomous cladogram (10) resolved for six taxa (1–6) in the maximum mode, through the solution of ten successive 3-taxon problems (shown as solved in cladograms 1–10; cf. figures 3.37–3.38).

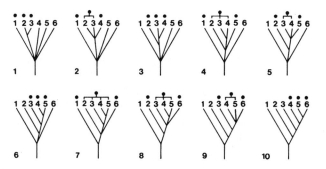

Figure 3.40. The same dichotomous cladogram (10) resolved for six taxa (1–6) in the maximum mode, through the solution of ten successive 3-taxon problems in a different order (shown as solved in cladograms 1–10; cf. figure 3.39).

required is the taxon to be added so as to frame the next 3-taxon problem (e.g., figure 3.37.2: taxon 3), and so on, until the suite of 3-taxon problems gives a final resolution (e.g., figure 3.37.4). In the case of branching of a single taxon at a time (e.g., figure 3.37), the taxon to be added is always single. In a case of more complex branching (e.g., figure 3.41), the taxon to be added will sometimes be two taxa that together form a group of the final resolution (e.g., figure 3.41.3: taxa 1 and 2). Minimum-mode resolution is sufficiently complex so that it probably can never be consistently achieved in practice.

Minimum-mode resolution seems always specifiable retrospectively. For example, consider the final resolution of figure 3.37.4, which is a dichotomous cladogram defined by its five groups, or components: 5,6; 4,5,6; 3,4,5,6; 2,3,4,5,6; 1,2,3,4,5,6. From the components one may derive the suite of four 3-taxon problems for which the informative components are the solutions (figures 3.37.1–4; table 3.40): 4,5,6, with the solution 4(5,6); 3,4(5,6), with the solution 3(4,5,6); 2,3(4,5,6), with the solution 2(3,4,5,6); 1,2(3,4,5,6), with the solution 1(2,3,4,5,6). Consider also the final resolution of figure 3.41.4, which is a dichotomous cladogram defined by five components: 5,6; 4,5,6; 1,2; 1,2,3; 1,2,3,4,5,6. In this case, there are two possible suites of four 3-taxon problems in the minimum mode. One suite (figures 3.41.1–4) is: 4,5,6, with the solution 4(5,6); 3,4(5,6), with the solution 3(4,5,6); 1,2,3, with the solution (1,2)3; (1,2)3(4,5,6), with the solution (1,2)(3)(4,5,6). The second suite (figures 3.41.5–8) is: 1,2,3, with the solution (1,2)3; (1,2)3,4,

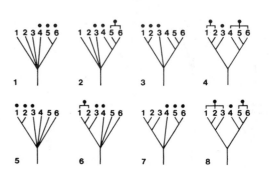

Figure 3.41. Cladograms (4, 8) resolved for six taxa (1–6) in the minimum mode, through the solution of suites of four successive 3-taxon problems (shown as solved in cladograms 1–4 and 5–8).

with the solution (1,2,3)4; 4,5,6, with the solution 4(5,6); (1,2,3)4(5,6), with the solution (1,2,3)(4,5,6).

The minimum mode, then, is a suite of 3-taxon problems that, once solved, result in the informative components of the cladogram that, with respect to a certain sample of information, is the true and final resolution. Thus a cladogram is definable in two different, but related,

Table 3.40. Comparison of Some Cladograms, 3-Taxon Problems and Solutions (minimum mode), and Components of Cladograms

Cladograms	Problems	Solutions	Components
3.37.1	4,5,6	4(5,6)	5,6
3.37.2	3,4(5,6)	3(4,5,6)	4,5,6
3.37.3	2,3(4,5,6)	2(3,4,5,6)	3,4,5,6
3.37.4	1,2(3,4,5,6)	1(2,3,4,5,6)	2,3,4,5,6
3.37.1–4	–	–	1–6
3.41.1	4,5,6	4(5,6)	5,6
3.41.2	3,4(5,6)	3(4,5,6)	4,5,6
3.41.3	1,2,3	(1,2)3	1,2
3.41.4	(1,2)3(4,5,6)	(1,2,3)(4,5,6)	1,2,3
3.41.5	1,2,3	(1,2)3	1,2
3.41.6	(1,2)3,4	(1,2,3)4	1,2,3
3.41.7	4,5,6	4(5,6)	5,6
3.41.8	(1,2,3)4(5,6)	(1,2,3)(4,5,6)	4,5,6
3.41.1–8	–	–	1–6
3.42.1	?	1(2,3,4)	2,3,4
3.42.2	?	1–2(3,4)	3,4
3.42.1–2	–	–	1–4

senses: (1) as a suite of components; (2) as a suite of 3-taxon problems for which the solutions are the suite of components.

For dichotomous cladograms, there is no difficulty of definition in either sense. For nondichotomous cladograms, there are difficulties for the 3-taxon sense of definition. For example, consider the cladogram of figure 3.42.1, defined by two components: 2,3,4 and 1,2,3,4. What is the 3-taxon problem solved? None seems specifiable without reference to a final resolution. In terms of one possible resolution (figure 3.42.3), the problem solved is 1,2(3,4), with the solution 1(2,3,4); in terms of another possible resolution (figure 3.42.4), the problem solved is 1,3(2,4), with the solution 1(2,3,4); in terms of a third possible resolution (figure 3.42.5), the problem solved is 1,4(2,3), with the solution 1(2,3,4). Such being the possibilities, no one of which is specified by figure 3.42.1, the problem solved is unspecifiable except in terms of its solution—1(2,3,4).

Consider the cladogram of figure 3.42.2, defined by two components: 3,4 and 1,2,3,4. What is the 3-taxon problem solved? In this case there are only two possibilities: 1,3,4, with the solution 1(3,4); and 2,3,4, with the solution 2(3,4). Such being the possibilities, neither of which is specified by figure 3.42.2, the problem solved is again unspecifiable except in terms of two possibilities, 1–2,3,4, or their solutions, 1–2(3,4). Thus it would seem that definition of cladograms is not generally possible in terms of 3-taxon problems, but is generally possible in terms of the solutions to such problems. If the solutions and components are compared, it is easy to see that the components are merely abbreviated forms of the solutions, with the addition of one component including all taxa (table 3.40).

Thus, a component may be understood as the solution to a particular 3-taxon problem in the minimum mode; and a cladogram may be understood as the combined solutions to a suite of 3-taxon problems. The information of a cladogram is consequently reflected in the quantity of components that correspond to solved 3-taxon problems. Consider

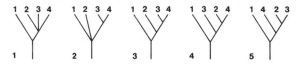

Figure 3.42. Cladograms (1–2) that represent solutions to unspecifiable 3-taxon problems in the minimum mode, with (1) or without (2) reference to final resolutions (3–5).

Table 3.41. Total Components, Informative Components, and Component
Information of Some Cladograms (cf. figure 3.41)

Cladograms	Total Components	Informative Components	Component Information
3.41.1	2:1–6; 5,6	1:5,6	1
3.41.2	3:1–6; 4–6; 5,6	2:4–6; 5,6	2
3.41.3	4:1–6; 1,2; 4–6; 5,6	3:1,2; 4–6; 5,6	3
3.41.4	5:1–6; 1–3; 1,2; 4–6; 5,6	4:1–3; 1,2; 4–6; 5,6	4

the cladograms of figure 3.41, with respect to the quantity of their
components, and the quantity of components that correspond to solved
3-taxon problems—the component information (table 3.41). The
component information is always one less than the total number of
components.

The component information is only part of the total information of a
cladogram. Another part concerns the taxa, or terms, included in the
components. Consider the cladograms of figures 3.42.1–2. Each cladogram has two components (one of which is informative), but the
informative components differ in the quantity of taxa, or terms, that
they contain, and in the quantity of term information (table 3.42). For
reasons mentioned below, the term information is always one less than
the total terms of a component.

Whereas the components relate to the number of minimum-mode
problems, the terms of a component relate to the number of maximum-
mode problems. Cladogram 3.42.1 represents one problem solved in the
minimum mode and, at the same time, two problems solved in the
maximum mode. Similarly, cladogram 3.42.2 represents one problem
solved in the minimum mode, and one problem in the maximum.

Because there is a unit common to both modes (3-taxon problem), the
quantities of solved problems of both modes may be summed as a
measure of the total information of a cladogram:

Table 3.42. Total Components, Informative Components, Total Terms, and
Term Information of Some Cladograms (cf. figure 3.42)

Cladograms	Total Components	Informative Components	Total Terms	Term Information
3.42.1	2:1–4; 2–4	1:2–4	3:2,3,4	2
3.42.2	2:1–4; 3,4	1:3,4	2:3,4	1

$$\frac{\text{Component}}{\text{information}} + \frac{\text{Term}}{\text{information}} = \frac{\text{Total}}{\text{information}}$$

The total information is equal also to the sum of the terms of all informative components of a cladogram:

$$\frac{\text{Terms of}}{\text{component 1}} + \frac{\text{Terms of}}{\text{component 2}} + \cdots \frac{\text{Terms of}}{\text{component n}} = \frac{\text{Total}}{\text{information}}$$

Considered as some number of 3-taxon problems solved, the total information may be divided by 2, to give the average of minimum- and maximum-mode resolutions:

$$\frac{\text{Total inf}}{2} = \frac{\text{Average}}{\text{information}}$$

The component, term, total, and average information of all cladograms of figures 3.37–3.42 are listed in table 3.43.

INCOMPLETE INFORMATION, MULTIPLE BRANCHING, AND RETICULATION

Because secondary cladograms (such as 3.42.1 and 3.42.2) represent solutions to unspecifiable 3-taxon problems, the meaning of the multiple branchings they contain is open to question. An instance of multiple branching in a cladogram may reflect nothing more than ignorance of certain character distributions (those that would be represented by, and allow the resolution of, a primary cladogram). Yet there are situations in which cladograms should exhibit multiple branchings that in some sense reflect real phenomena: for example, character distributions caused by cases of "simultaneous" multiple speciation, instances of hybridization, or groups wherein one species is ancestral to two or more others (as in speciation by the sequential isolation of two or more peripheral populations without change in the central population of a "mother" species, or cases in which studied fossil species are actually the ancestors of other studied species). Multiple branching, therefore, plays an important role in cladistic theory, inasmuch as it is used to represent a variety of character distributions that cannot be represented by a primary (dichotomous) cladogram.

Critics of cladistics have frequently misconstrued it as a theory of

Table 3.43. Component, Term, Total, and Average Information of Some
Cladograms (cf. figures 3.37–3.42)

	Information			
Cladogram	*Component*	*Term*	*Total*	*Average*
3.37.1	1	1	2	1.0
3.37.2	2	3	5	2.5
3.37.3	3	6	9	4.5
3.37.4	4	10	14	7.0
3.38.1	1	1	2	1.0
3.38.2	2	3	5	2.5
3.38.3	3	6	9	4.5
3.38.4	3	7	10	5.0
3.38.5	4	10	14	7.0
3.39.1	1	1	2	1.0
3.39.2	1	2	3	1.5
3.39.3	1	3	4	2.0
3.39.4	1	4	5	2.5
3.39.5	2	5	7	3.5
3.39.6	2	6	8	4.0
3.39.7	2	7	9	4.5
3.39.8	3	8	11	5.5
3.39.9	3	9	12	6.0
3.39.10	4	10	14	7.0
3.40.1	1	1	2	1.0
3.40.2	1	2	3	1.5
3.40.3	2	3	5	2.5
3.40.4	2	4	6	3.0
3.40.5	2	5	7	3.5
3.40.6	3	6	9	4.5
3.40.7	3	7	10	5.0
3.40.8	3	8	11	5.5
3.40.9	3	9	12	6.0
3.40.10	4	10	14	7.0
3.41.1	1	1	2	1.0
3.41.2	2	3	5	2.5
3.41.3	3	4	7	3.5
3.41.4	4	6	10	5.0
3.41.5	1	1	2	1.0
3.41.6	2	3	5	2.5
3.41.7	3	4	7	3.5
3.41.8	4	6	10	5.0
3.42.1	1	2	3	1.5
3.42.2	1	1	2	1.0
3.42.3	2	3	5	2.5
3.42.4	2	3	5	2.5
3.42.5	2	3	5	2.5

evolution, rather than a method of systematics. The critics have seized upon the variety of evolutionary events which should produce character distributions that cannot be represented by primary cladograms, and have claimed that these events either falsify cladistics as a theory, or render it useless, for all practical purposes, as a method. In contrast to the critics, we view multiple branching as an essential feature of cladistics—one that is put to a variety of uses.

It is easy to understand how multiple branching of a cladogram can represent character distributions caused by real multiple speciation; all that is necessary is to view the cladogram as a tree (and the lines as evolving lineages). It is almost as easy to understand how multiple branching of a cladogram can represent character distributions caused by an ancestral species that gives rise to two descendant species; all that is necessary is to ask the question (and to realize that it is answerable only in the negative): of the three species, are two of them more closely related to each other than either is to the third? It is less easy to understand how multiple branching can represent character distributions caused by hybridization. As a representation of hybridization, a reticulate pattern (of a tree) seems intuitively meaningful, whereas multiple branching (of a cladogram) seems counterintuitive. Ignorance, finally, is no problem at all if it is complete, but complete ignorance is an ideal seldom achieved in practice. What if ignorance is only partial? If all attempts at generalization about character distribution (by way of a cladogram) reflect partial ignorance, then ignorance is a factor that is perennially present.

Consider cladogram 3.42.1; cladistic interpretation allows for three different primary cladograms (figures 3.42.3–5), each with a component 234. Under the assumption that cladogram 3.42.1 is correct (that component 234 is real), further resolution would seem to be limited to only one of the three primary cladograms.

Consider also cladogram 3.43.1; cladistic interpretation again allows for three different primary cladograms (figures 3.43.2–4), each with a component 34. Under the assumption that cladogram 3.43.1 is correct (that component 34 is real), further resolution would seem to be limited to only one of the three primary cladograms.

The above analysis of possible resolutions, however, is based on the notion (hereafter referred to as interpretation 1) that the information

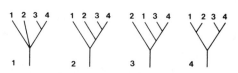

Figure 3.43. Resolution of a secondary cladogram with a basal trichotomy under interpretation 1.

contained in cladogram 3.43.1, for example, is that taxa 3 and 4 are more closely related to each other than either of them is to taxa 1 *and*2. This would require that two 3-taxon problems had been solved: 1,3,4, with the solution 1(3,4); and 2,3,4, with the solution 2(3,4). Cladogram 3.43.1, however, does not specify that both of these 3-taxon problems have been solved.

There is, therefore, an alternative notion (hereafter referred to as interpretation 2): that the information contained in cladogram 3.43.1, for example, is that taxa 3 and 4 are more closely related to each other than either of them is to taxon 1 *and/or* taxon 2. Under interpretation 2, there are, as before, two possible 3-taxon solutions, only one of which need be true. Each 3-taxon solution (figures 3.44.1 and 3.44.7) allows a fourth taxon to be added at any of five different positions; hence each solution allows a suite of five possible primary cladograms (figures 3.44.2–6 and 3.44.8–12). Three cladograms are common to both suites (figures 3.44.4–6 and 3.44.10–12). If both 3-taxon solutions are true in a given case, they would jointly allow only three primary cladograms (the three cladograms common to both suites, which are the same as those allowed under interpretation 1: figures 3.43.2–4). If only one 3-taxon solution is true, and if it is specified, one (figures 3.44.2–6) or the other

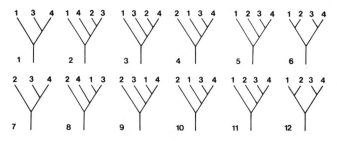

Figure 3.44. Resolution of a secondary cladogram with a basal trichotomy under interpretation 2.

(figures 3.44.8–12) suite of five primary cladograms is allowed. If only one 3-taxon solution is true, and if it is unspecified, a suite of seven primary cladograms is allowed (the three cladograms common to both suites of five, and four other cladograms unique to one suite).

Inasmuch as cladograms may be judged true or false, as the case may be, only on the basis of characters at hand, we may distinguish real and implied information content. Real information in the above sense comprises those 3-taxon solutions that are true on the basis of characters at hand. Implied information comprises those 3-taxon solutions that are derived logically from real information; implied information cannot be determined to be true, except by logical implication, on the basis of characters at hand.

Consider cladogram 3.42.1 and its information in the form of three problems and solutions: 1,2,3, with the solution 1(2,3); 1,2,4, with the solution 1(2,4); 1,3,4, with the solution 1(3,4). Consider the solutions in relation to certain characters (table 3.44): with reference to character A (considered by itself), all three solutions are true (and real information); with reference to characters B–D (considered by themselves), all three solutions are true (and real information); with reference to characters E–F (considered by themselves), only two solutions are true (and real); the third solution is only implied.

Cladogram 3.42.1 might serve as a preferred summary of each of the samples of characters (characters A; B,C,D; E,F), but the real information of the cladogram is lower for one sample (characters E,F). In the case of that sample, cladogram 3.42.1 could serve as a summary under interpretation 1, but only an appeal to logical necessity (rather than characters at hand) could justify the truth of one solution: 1(2,4).

Consider cladogram 3.43.1 and its information in the form of two problems and solutions: 1,3,4, with the solution 1(3,4); 2,3,4, with the solution 2(3,4). Consider the solutions in relation to certain characters

Table 3.44. Occurrence of Some Characters (A–I) in Four Taxa (1–4)

Taxa	Characters								
	A;	B	C	D;	E	F;	G;	H;	I;
1	−;	−	−	−;	−	−;	−;	−;	?;
2	+;	−	+	+;	−	+;	−;	?;	−;
3	+;	+	+	−;	+	+;	+;	+;	+;
4	+;	+	−	+;	+	−;	+;	+;	+;

{table 3.44): with reference to character G (considered by itself), both solutions are true (and real information); with reference to character H (considered by itself), only one solution is true (the other is not even implied); character I gives the same result as character H.

Cladogram 3.43.1 might serve as a preferred summary for each sample (characters G; H; I), but the information of the cladogram is lower for two samples (H; I). For those samples, cladogram 3.43.1 can serve as a summary only under interpretation 2, for there is no basis for an appeal to logical necessity as a justification for the truth of one solution in each case: 2(3,4) for character H; 1(3,4) for character I. Secondary cladograms incorporating basal trichotomies (or multiple branchings) thus differ from those incorporating terminal trichotomies (or multiple branchings) in that they are subject to two different interpretations reflecting varying degrees of completeness in the information they summarize. Basal trichotomies (or multiple branchings), under interpretation 2, can accommodate taxa for which available characters are inadequate to allow their placement on the cladogram under interpretation 1.

The two interpretations can be compared with reference to the primary cladograms that they allow, or, alternatively, that they prohibit. Interpretation 1 (figure 3.43) permits three, and therefore prohibits 12, of the 15 possible dichotomous cladograms for four taxa. Interpretation 2 (figure 3.44) permits seven, and therefore prohibits eight, of the 15 possible dichotomous cladograms for four taxa. Thus, interpretation 2 is less prohibitive, or less restrictive, than interpretation 1. Both interpretations allow the group 34 to be real, but the interpretations differ in their specifications of the limits of the group's reality. Under interpretation 1, neither 1 nor 2 can be a member of the group 34; under interpretation 2, either 1, or 2 (but not both) can be a member of the group 34.

Consider a hypothetical group of three species (1,3,4) whose relationships have been established, for example in the pattern specified by cladogram 3.44.1: solution 1(3,4). Suppose that a fourth species (2) is discovered, that species 2 is determined to be a member of the group already including species 1, 3, and 4, and that the precise relationships of species 2 can be determined with no further degree of accuracy. How may species 2 be added to the cladogram (3.44.1)?

There are two possibilities: under interpretation 1, the only possible placement is the tertiary (unresolved) cladogram (figure 3.9.1); under interpretation 2, the only possible placement is shown in figure 3.43.1. The latter placement (figure 3.43.1) would not be possible under interpretation 1, because that interpretation prohibits dichotomous cladograms (figures 3.44.2–3) that cannot reasonably be prohibited in the absence of evidence pertaining to the precise relationships of species 2 within the group. To arrive at cladogram 3.43.1 under interpretation 1, we would need to know not only that species 2 is a member of the group already including species 1, 3, and 4, but also that species 3 and 4 are more closely related to each other than either is to species 2 (i.e., that cladograms 3.44.2 and 3.44.3 are false). In any particular case, such information may or may not be available.

Species, or taxa generally, which can be placed in a higher taxon, but whose relationships are otherwise obscure, are commonplace. Their occurrence accounts for some of the trichotomies (and multiple branchings) found in the cladogram of any large group. If systematic practice operated exclusively under interpretation 1, any such species (or taxon generally) would effectively collapse the internal cladistic structure of the group to which the species (or taxon) was assigned. An extreme example would be a species that could be recognized as living but whose relationships were otherwise unspecifiable. Under interpretation 1, all cladistic structure would collapse into a basal branching as numerous as there are species.

Interpretation 1 is not universally adopted in systematics (except perhaps in previous theoretical discussion) simply because it is sometimes unworkable. Interpretation 2 seems sometimes to be adopted in systematics; at least it workably merges with, or is implicit within, routine taxonomic practice. If so, then the problem of multiple branching can be considered in a somewhat different light. In a given case one may ask: under which interpretation (1 or 2) is a trichotomy (or multiple branching) to be understood? If under interpretation 2, as occurs sometimes with recent species and perhaps more commonly with fossils, then the trichotomy (or multiple branching) is not a final solution, but rather a problem that, until solved, injects considerable ambiguity into the cladogram. Progress in the taxonomy of a given group may thus involve a gradual shift from interpretation 2 to

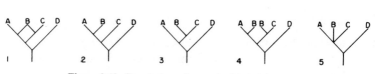

Figure 3.45. Resolution of a terminal hybridization.

interpretation 1 (and finally to a primary cladogram) as additional 3-taxon problems are solved.

Hybridization is sometimes viewed as a problem for cladistics, but hybridization can also be considered in the light of interpretations 1 and 2. A simple case of hybridization involves four species (A, B, C, D), two of which (A and C) hybridize and produce a third species (B; figure 3.45.1). Study of the cladistic relationships of the four species should reveal two conflicting patterns (figures 3.45.2–3), combinable in one branching diagram with reticulation (figure 3.45.1) or in a branching diagram wherein species B (the hybrid) appears twice (figure 3.45.4). The only possible cladogram (without reticulation or repetition of taxa) is one with a terminal trichotomy (figure 3.45.5).

Another simple instance of hybridization involves four species (A, B, C, D), two of which (B and D) hybridize and produce a third species (C; figure 3.46.1). Study of the cladistic relationships of the four species should reveal two conflicting patterns (figures 3.46.2–3), combinable in one branching diagram with reticulation (figure 3.46.1) or in a branching diagram wherein species C (the hybrid) appears twice (figure 3.46.4). Under interpretation 2, the only possible cladogram (without reticulation or repetition of taxa) is one with a basal trichotomy (figure 3.46.5). Under interpretation 1, the cladistic structure collapses (figure 3.9.1) into a tertiary cladogram. In this context, interpretation 2 operates to produce the "consensus tree" (Adams 1973) representing only that information contained in both of the two conflicting patterns (figures 3.46.2–3).

Hybrid species, or taxa generally, presumably occur. Their occurrence may account for some of the trichotomies (and multiple branchings) found in the cladogram of any large group. If systematic practice

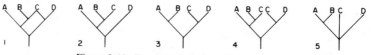

Figure 3.46. Resolution of a basal hybridization.

operated exclusively under interpretation 1, any such hybrid species (or hybrid taxon generally) would effectively collapse the internal cladistic structure of the groups involved in the hybridization. An extreme example would be a species produced by hybridization of the first two primordial species. Under interpretation 1, all cladistic structure would effectively collapse into a basal branching as numerous as there are species.

Multiple speciation, hybridization, and groups including actual ancestors seem cladistically indistinguishable from each other, and from simple ignorance of interrelationships, in that multiple speciation, hybridization, actual ancestors, and ignorance are all represented in cladograms in the same way: by trichotomies (or multiple branchings, be they terminal or basal). In addition, basal trichotomies (or multiple branchings), under interpretation 2, also represent partial ignorance, of whatever degree, of species interrelationships. That the limits of cladistics (Hull 1980) are thereby defined is advantageous in the sense that the discipline of cladistics is rendered intelligible. But what of multiple speciation, hybridization, and actual ancestors? They seem to belong to the suite of problems that arise from a consideration of trees rather than cladograms. If cladistics is that part of systematics concerned with cladograms, then perhaps it is time to speak of "arboristics" as that part concerned with trees and, specifically, modes of speciation in general, as well as particular histories of speciation. One might conceive of an "arboristic analysis" which attempts to determine what tree is the cause of a particular instance of trichotomy or multiple branching, and operates by investigating the particular character distributions found in a given instance and their relative compatibility with various evolutionary scenarios. The results of such an analysis, however, would be of questionable significance for classification if hierarchical classifications convey only the information contained in cladograms, and not the additional information contained in trees.

4

SYSTEMATIC RESULTS: CLASSIFICATIONS

CLASSIFICATION: SOME CONCRETE CONSIDERATIONS

This chapter on classification is devoted to analysis of some concrete examples. For any particular classification, there is a corresponding cladogram that specifies the information content of the classification. Consider, for example, the following classification:

Tetrapoda
  Amphibia (A)
  Amniota
    Reptilia (B)
    Aves (C)
    Mammalia (D)

The corresponding cladogram is secondary cladogram 3.7.1, specified by the tertiary component ABCD, and the secondary component BCD. Each of the components is a statement of general synapomorphy, which means that generalizations (homology-synapomorphy statements) are possible (and predicted) for ABCD and BCD. The classification does not specify in what context the components should be interpreted (phyletic, phenetic, gradistic, etc.). Nor does the classification specify that both components should be interpreted in the same context. At most, the classification is a prediction that future research will reveal the same pattern (ABCD, BCD) and no other, such as ACDE, ACD. If the prediction is corroborated, the classification will doubtlessly prove useful; if the prediction is falsified, the classification will doubtlessly be changed to reflect a more useful pattern.

With respect to this, or any other, classification, the information content may be specified (by specifying the general components of the corresponding cladogram)—without considering any of the specific homology statements (individual synapomorphies) that suggested the existence of the pattern (ABCD, BCD). Indeed, the information content of the classification consists only of the pattern—not the specific homology statements. The component BCD, for example, may be stated thus: Reptilia, Aves, and Mammalia are predicted to be synapomorphic, relative to all other organisms. They may be known to be synapomorphic in numerous ways, in only one way, or in no ways at all: the classification does not tell us, for the classification is merely a hypothesis of general synapomorphy (in exactly the same way as the cladogram). How particular synapomorphies might be conceived, as, for example, anatomical, behavioral, or physiological characters, or in what context the general synapomorphy might be interpreted—in, for example, phyletic, phenetic, or gradistic contexts—is of secondary relevance. Such matters may become relevant, but only if they suggest the existence of some other pattern. Suggestions to that effect, of course, frequently arise, creating the problem of incongruence. But much incongruence is more apparent than real (see above), and is apt to dissipate with further analysis. Persisting incongruence is either noise (random error), or a real indication of a different pattern. But only future research will show which is apt to be the case.

A PHYLETIC EXAMPLE

Simpson's (1945) classification of mammals includes the following major groups († indicates a group known only from fossils):

> Class Mammalia
>   Subclass Prototheria
> †Subclass Allotheria
>   Subclass Theria
>   †Infraclass Pantotheria
>   Infraclass Metatheria
>   Infraclass Eutheria

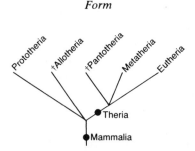

Figure 4.1. A cladogram derived from Simpson's (1945) classification of the major groups of mammals.

For this classification, a cladogram (figure 4.1) may be constructed to portray the information content (two components) of the classification. The two components correspond to the two inclusive taxa (Mammalia and Theria). This cladogram may be compared with the concept upon which Simpson based his classification, and about which he stated:

> The underlying considerations and concrete evidence have been presented here only in the barest possible outline, but they seem to support the following [the above classification] as the most convenient and most natural primary arrangement of mammals. (1945:165)

Simpson did not provide a diagram of his concept, but one may be constructed from his commentary (figure 4.2). With respect to the Theria, he states:

> Although they are too poorly known for certainty, there is good evidence that some of the very primitive and ancient Jurassic mammals, the † Pantotheria, are an offshoot of, and nearly represent, the common placental-marsupial ancestry before it had split up into the Metatheria and Eutheria properly definable. . . . If this is true, these mammals were also Theria but were not Metatheria or Eutheria. These are the conceptions formalized by recognizing a Subclass

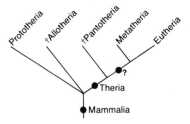

Figure 4.2. A cladogram (phylogram) summarizing the concept upon which Simpson based his classification of the major groups of mammals (cf. figure 4.1).

Theria with three secondary divisions, † Pantotheria, Metatheria, and Eutheria. (1945:165)

With respect to the Mammalia, he states:

There is one great division of fossil mammals that will not enter into either the Prototheria or the Eutheria, although these animals, the †multituberculates, have been referred to both on evidence now unacceptable. Everything now points to their having been distinct from all other mammals since the very beginnings of the Mammalia, and if this is true they can hardly be classified except as another subclass, a conclusion that I have supported and documented elsewhere. . . . The name †Allotheria is available for this subclass. . . . As far as present knowledge shows, the three subclasses have been separate since the beginning of the Class Mammalia, and there is no clear evidence of special relationship between any two of these subclasses exclusive of the third. (1945:165)

A comparison of Simpson's concept (figure 4.2) with the information (two components) actually contained in his classification (figure 4.1) shows that his concept includes three components (1, Mammalia; 2, Theria; and 3, Metatheria-Eutheria), and his classification only two (Mammalia and Theria). The Metatheria-Eutheria component is omitted from his classification.

A recent revision of mammalian classification includes the following major groups (McKenna 1975:40):

Class Mammalia
   Subclass Prototheria
   Subclass Theria
     †Superlegion Kuehneotheria (new)
     Superlegion Trechnotheria (new)
       †Legion Symmetrodonta (new rank)
       Legion Cladotheria (new)
         †Sublegion Dryolestoidea (new rank)
         Sublegion Zatheria (new)
           †Infraclass Peramura (new)
           Infraclass Tribosphenida (new)
             Supercohort Marsupialia (new rank)
             Supercohort Eutheria (new rank)

A cladogram may be constructed from this classification (figure 4.3), and it proves to be identical with the concept upon which the classification is based (figure 4.4). It includes six components, which

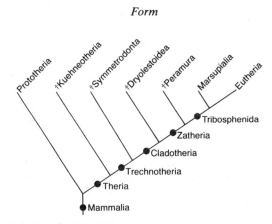

Figure 4.3. A cladogram derived from McKenna's (1975) classification of the major groups of mammals.

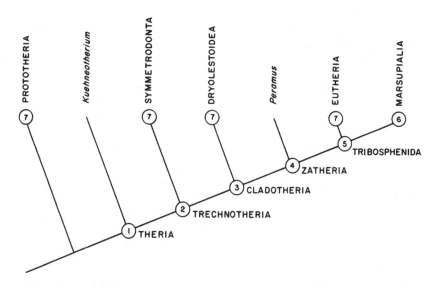

Figure 4.4. A cladogram (phylogram) summarizing the concept upon which McKenna based his classification of the major groups of mammals (cf. figure 4.3). After McKenna (1975), figure 1, p. 25.

correspond to the six inclusive groups (Mammalia, Theria, Trechnotheria, Cladotheria, Zatheria, Tribosphenida).

Comparison of this classification with Simpson's shows many differences: the disappearance of the Allotheria (included by McKenna in the Prototheria); the inclusion of a new taxon (unknown to Simpson), the Kuehneotheria; and the splitting of Simpson's Pantotheria into three taxa, Symmetrodonta, Dryolestoidea, and Peramura (the "Marsupialia" is merely another name for Simpson's Metatheria, and represents no change). Because there are four additional components in the cladogram, four new inclusive taxa are named (Trechnotheria, Cladotheria, Zatheria, Tribosphenida) and new ranks are used to accommodate them (superlegion, legion, sublegion, supercohort). Like that of Simpson, the classification is understandable as cladistic classification in a phyletic context. Unlike that of Simpson, however, the context is explicitly stated:

I propose here a cladistic reclassification of the therian groups Marsupialia, Eutheria, and Simpson's infraclass Pantotheria, emphasizing kinship and recency of common ancestry. (McKenna 1975:26)

And unlike that of Simpson, the classification includes all of the components of the concept upon which the classification is based. McKenna states:

To the critics who would ask, "Where will all this proliferation of names and ranks end?" I suggest that if the terms are not found useful to convey exact genealogical meanings dictated by phylogeny, then they can be ignored by those who so choose. They are, however, based upon cladistic principles, not on "art" or caprice. (1975:27)

McKenna's statements are unarguable on the principles of classification developed in this book (but, then, so are Simpson's). Noteworthy is that both of the components of Simpson's classification (Mammalia and Theria) are preserved in McKenna's classification. McKenna also names a component recognized but unnamed by Simpson (Tribosphenida). Thus, it would seem that Simpson's three components, considered as bases for prediction, have been found usefully predictive by McKenna. The components added by McKenna (Trechnotheria, Cladotheria, and Zatheria) may also be considered bases for prediction. Only future research will determine if they are usefully predictive.

McKenna's revision of mammalian classification may be considered in relation to the stability problem dealt with above, for it might seem that classification of this sort poses a new and serious threat to stability, a threat that might dwarf that posed by the arbitrariness of traditional taxonomy.

The main differences between the concepts and classifications of Simpson and McKenna concern (1) the inclusion of a new group, Kuehneotheria, and (2) different interpretation of Simpson's Pantotheria. Following McKenna's principles, any further reinterpretation, or any discovery of new groups, might cause further revision of the classification, involving new inclusive groups and new ranks.

Devices have been suggested to accommodate such reinterpretation and discovery without resorting to "this proliferation of names and ranks." Most involve discriminating between fossil and recent groups (Brundin 1966; Hennig 1966; Crowson 1970). According to one suggestion, new inclusive taxa are not needed to accommodate purely fossil groups, which may be listed in the order of their branching sequence—a suggestion in agreement with Simpson's omission from his classification of a metatherian-eutherian component distinct from the therian component. According to this suggestion, McKenna's cladogram would result in the following classification with two components (the same two components of Simpson's classification—Mammalia and Theria):

Class Mammalia
   Subclass Prototheria
   Subclass Theria
     †Infraclass Kuehneotheria
     †Infraclass Symmetrodonta
     †Infraclass Dryolestoidea
     †Infraclass Peramura
     Infraclass Eutheria
     Infraclass Marsupialia

Comparison of this classification with that of Simpson shows (1) a new group, Kuehneotheria, (2) in place of Simpson's Pantotheria, the three groups into which the Pantotheria were split, and (3) the same components (Mammalia, Theria).

This suggestion has the advantage that, for reinterpretation of the relationships of fossil groups, no change would result in the basic structure (the components) of the classification. Consider a reinterpretation (hypothetical) along the lines of figure 4.5. According to the above suggestion—that fossil groups be listed in the order of their branching sequence—fossil groups are, so to speak, free to wander about in the classification. A classification based on this purely hypothetical reinterpretation (figure 4.5) would change only in the rank and position of the fossil groups whose relationships were reinterpreted (Peramura, Symmetrodonta, Kuehneotheria):

Class Mammalia
  †Subclass Peramura
  †Subclass Symmetrodonta
  Subclass Prototheria
  Subclass Theria
    †Infraclass Dryolestoidea
    †Infraclass Kuehneotheria
    Infraclass Eutheria
    Infraclass Marsupialia

As noted above for the Kuehneotheria, this suggestion has the additional advantage that newly discovered fossil groups could be added to the classification in their appropriate position and rank, without change of the components of the classification. This suggestion, in effect, is that components involving one or more fossil groups and only one recent group be omitted from classification. Such omitted

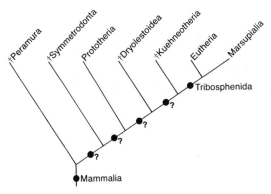

Figure 4.5. A cladogram summarizing a hypothetical reinterpretation of McKenna's concept of the interrelationships of the major groups of mammals (cf. figures 4.3, 4.4).

components constitute no great loss. Even if they were included, they could have only a very restricted generality—restricted to those few characters recoverable from fossils. With a restricted generality, they could never prove very useful, either as summaries of information already acquired or as predictions about information yet to be discovered, for the reason that the amount of information, even in its most complete form, would be meager. And their usefulness, such as it is, would extend only to paleontology. If the sign (†) for fossil groups also indicates use of this convention, the complete cladogram can be reconstructed from the classification with no loss of information content.

It is not the purpose here to argue the pros and cons of various devices that might enhance stability of classification. It is enough to mention that such devices exist (Patterson and Rosen 1977; Wiley 1979). But to mention them does not solve the stability problem. That problem is inherent in usage and custom, which are always at the mercy of future events.

A PHENETIC EXAMPLE

There are very few phenetic classifications in the form of a completely resolved hierarchical arrangement of taxa, for the reason that the phenogram is usually considered a classification in itself. With respect to component analysis, it is difficult to begin with an example of a phenogram for an actual group of species. Although many phenetic studies have been performed, they have usually involved groups of numerous species. Discussion of the many components of even one such phenogram would be tedious. Therefore, an example of a hypothetical group, with a small number of species, will be considered. This group comprises six species, A, B, C, D, E, F, and the results of the study are portrayed as a phenogram (figure 4.6). The components of the phenogram are

> ABCDEF
> BD
> ACEF
> ACE
> AC

One may inquire what these components represent in the phenetic

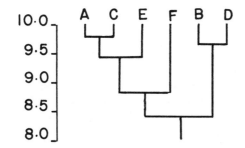

Figure 4.6. A phenogram for six species (A, B, C, D, E, F). After Sokal and Sneath (1963), figure A-2, p. 311. Copyright © 1963, W. H. Freeman.

context. Do the components represent generalizations (statements of general synapomorphy)? If they do, the general synapomorphy would be general phenetic synapomorphy: for example, with respect to "overall similarity" species A and C are more similar to each other than to any other species. Similarly, species A, C, and E are more similar to each other than to any other species; species A, C, E, and F are more similar to each other than to any other species; species B and D are more similar to each other than to any other species; and, finally, species A, B, C, D, E, and F are more similar to each other than to any other species.

Within the phenetic context, the sense of "similarity" is total or "overall" similarity—a concept of similarity that is estimated by sampling numerous characters of each species and combining positive and negative occurrences of them such that a numerical value of "overall similarity" can be computed for comparisons between species. A characteristic of the phenetic approach is that the similarity statements (components) can be associated with numerical levels of similarity. For example, species A and C are similar at a level of 9.8. The level of similarity may, therefore, be considered a generalization for each component:

$$
\begin{aligned}
\text{ABCDEF:} &\quad 8.4 \\
\text{BD:} &\quad 9.6 \\
\text{ACEF:} &\quad 8.8 \\
\text{ACE:} &\quad 9.4 \\
\text{AC:} &\quad 9.8
\end{aligned}
$$

These components allow two types of predictions: that independent samples of information will reveal (1) the same components at (2) the same levels. To date, neither type of prediction has proved useful, because in actual examples the components and the levels depend upon the characters sampled and the particular numerical methods used to compare and group the species (Farris 1977, 1979, 1980; Mickevich 1978). Different methods yield different phenograms with different components, or the same components at different levels (figures 4.7, 4.8).

In expositions of phenetic theory, "prediction" is considered important:

> Gilmour's dictum—that a system of classification is the more natural the more propositions can be made regarding its constituent classes—admits of objective measurement and testing, in contradistinction to Simpson's natural system. Furthermore, Gilmour's system has powerful predictive properties; it is therefore the one we recommend. (Sneath and Sokal 1973:27, also Sokal and Sneath 1963:19)

The nature of this "prediction" has been further clarified in expositions of phenetic theory, but it seems to have no relation with the two types of phenetic prediction considered above:

> We can neither list nor remember all the characteristics of various organisms and higher taxa, and we therefore need a system of grouping them into a manageable number of groups whose characters are preponderantly constant. Because of high constancy and mutual intercorrelations of characters, such a grouping will carry a high predictive value. Thus, if we read of a new aphid species we can immediately predict a number of characteristics that this species is expected to possess. An aphid will with almost complete certainty be a plant feeder, possess a particular type of wing venation, be parthenogenetic in part of its life cycle, produce males by nondisjunction of the sex chromosomes, produce honeydew, secrete wax from cornicles or other glands, and so on. Since an aphid is a homopteran, we can forecast with some accuracy the general construction of its mouth parts, the texture of its wings, and other homopteran characteristics. This type of argument can, of course, be extended to the hexapod and arthropod levels of classification and even higher. It is obviously much easier for us to remember this of the group Aphididae than of each individual aphid or species of aphid. Furthermore, it is impossible to remember or appreciate the innumerable relations between the various OTU's [species] to be classified, but this is easier when they are grouped into fewer inclusive taxa.
>
> It is clear that such considerations lead to a second, closely related purpose of taxonomy, namely predictive power. The more natural a taxonomy is, the more predictive it will be about characters known from part of the group that have not yet been investigated in another part. (Sneath and Sokal 1973:188–189)

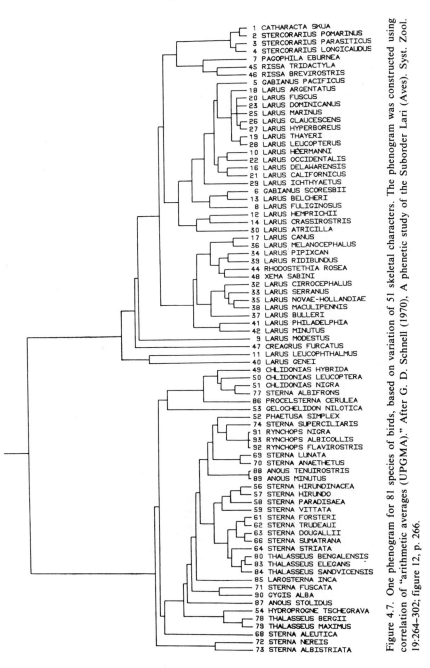

1 CATHARACTA SKUA
2 STERCORARIUS POMARINUS
3 STERCORARIUS PARASITICUS
4 STERCORARIUS LONGICAUDUS
7 PAGOPHILA EBURNEA
45 RISSA TRIDACTYLA
46 RISSA BREVIROSTRIS
5 GABIANUS PACIFICUS
18 LARUS ARGENTATUS
20 LARUS FUSCUS
23 LARUS DOMINICANUS
25 LARUS MARINUS
26 LARUS GLAUCESCENS
27 LARUS HYPERBOREUS
19 LARUS THAYERI
28 LARUS LEUCOPTERUS
10 LARUS HEERMANNI
22 LARUS OCCIDENTALIS
16 LARUS DELAWARENSIS
21 LARUS CALIFORNICUS
29 LARUS ICHTHYAETUS
6 GABIANUS SCORESBII
13 LARUS BELCHERI
8 LARUS FULIGINOSUS
12 LARUS HEMPRICHII
14 LARUS CRASSIROSTRIS
30 LARUS ATRICILLA
17 LARUS CANUS
36 LARUS MELANOCEPHALUS
34 LARUS PIPIXCAN
39 LARUS RIDIBUNDUS
44 RHODOSTETHIA ROSEA
48 XEMA SABINI
32 LARUS CIRROCEPHALUS
33 LARUS SERRANUS
35 LARUS NOVAE-HOLLANDIAE
38 LARUS MACULIPENNIS
37 LARUS BULLERI
41 LARUS PHILADELPHIA
42 LARUS MINUTUS
9 LARUS MODESTUS
47 CREAGRUS FURCATUS
11 LARUS LEUCOPHTHALMUS
40 LARUS GENEI
49 CHLIDONIAS HYBRIDA
50 CHLIDONIAS LEUCOPTERA
51 CHLIDONIAS NIGRA
77 STERNA ALBIFRONS
86 PROCELSTERNA CERULEA
53 GELOCHELIDON NILOTICA
52 PHAETUSA SIMPLEX
74 STERNA SUPERCILIARIS
91 RYNCHOPS NIGRA
93 RYNCHOPS ALBICOLLIS
92 RYNCHOPS FLAVIROSTRIS
69 STERNA LUNATA
70 STERNA ANAETHETUS
88 ANOUS TENUIROSTRIS
89 ANOUS MINUTUS
56 STERNA HIRUNDINACEA
57 STERNA HIRUNDO
58 STERNA PARADISAEA
59 STERNA VITTATA
61 STERNA FORSTERI
62 STERNA TRUDEAUI
63 STERNA DOUGALLII
66 STERNA SUMATRANA
64 STERNA STRIATA
80 THALASSEUS BENGALENSIS
83 THALASSEUS ELEGANS
84 THALASSEUS SANDVICENSIS
85 LAROSTERNA INCA
71 STERNA FUSCATA
90 GYGIS ALBA
87 ANOUS STOLIDUS
54 HYDROPROGNE TSCHEGRAVA
78 THALASSEUS BERGII
79 THALASSEUS MAXIMUS
68 STERNA ALEUTICA
72 STERNA NEREIS
73 STERNA ALBISTRIATA

Figure 4.7. One phenogram for 81 species of birds, based on variation of 51 skeletal characters. The phenogram was constructed using correlation of "arithmetic averages (UPGMA)." After G. D. Schnell (1970), A phenetic study of the Suborder Lari (Aves). Syst. Zool. 19:264–302; figure 12, p. 266.

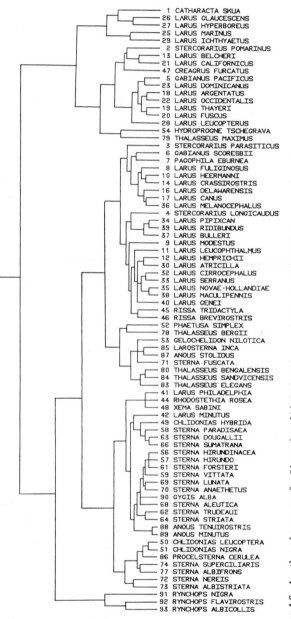

Figure 4.8. Another phenogram for 81 species of birds, based on variation of 51 skeletal characters. The phenogram was constructed using distance in relation to the "UPGMA method." After Schnell (1970; see caption for figure 4.7), figure 13, p. 267.

The key element in the above quotation is: " . . . if we read of a new aphid species we can immediately predict a number of characteristics that this species is expected to possess." This statement contains no reference to overall similarity, nor to levels of overall similarity. The statement does not, therefore, refer to prediction in the phenetic context. It is merely a statement that there are already known some generalizations (synapomorphies) for a group called Aphididae, generalizations that might apply to a presently unknown member (a species) of the group. To be recognized as a member of the Aphididae, the "new aphid species" would, of course, have to display at least some of the characters considered synapomorphic for the Aphididae (the element of prediction applies only to those aphid synapomorphies not already detected in the new species). In any event, the generalizations that pertain to the Aphididae are merely synapomorphies in the sense that that term has been used throughout this book. Considered as a basis for prediction, the generalizations would lead to the hypothesis of congruence: that a certain group of species, observed to be synapomorphic in one respect, with further study will be observed to be synapomorphic in other respects not yet discovered.

It seems, therefore, that in expositions of phenetic theory, the hypothesis of congruence has been confused with phenetic prediction. And although phenetic prediction of two types is possible, phenetic predictions are seldom if ever realized in practice. Predictions of overall similarity and the levels of overall similarity appear to be artifacts of the particular numerical methods used on particular sets of data.

These comments constitute something of a critique of phenetic taxonomy, but their main purpose is to relate component analysis to phenograms. The example shows that a phenogram has cladistic components. And the significance of the components, considered as a basis either of generalization or prediction, is the same significance as that considered earlier in relation to the hypothesis of congruence—a claim that the reader might like to explore independently, by reading the literature of phenetic taxonomy (an excellent bibliography is contained in Sneath and Sokal 1973).

A GRADISTIC EXAMPLE

As stated above, the concept of grade is not very well defined and, for all practical purposes, may be undefinable. This difficulty aside, some

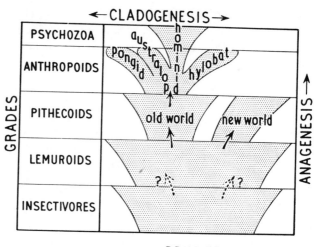

Figure 4.9. A diagram illustrating grades, published by Huxley in 1958. His legend reads: "Grades and clades in primate evolution. The Lemuroids are possibly and the Pithecoids certainly grades but not monophyletic clades. Cladogenetically, Man constitutes the family Hominidae, but anagenetically the major grade Psychozoa." After Huxley (1958), figure 12, p. 37.

attempts at gradistic taxonomy may be considered. A good place to begin is with Huxley's (1958:36) concept of grade: "I consider that *grade* is the best general term to denote readily delimitable or definable anagenetic units or assemblages." Huxley provided several diagrams of grades, one of which may be used here to illustrate component analysis of gradograms. Huxley's figure (figure 4.9) can be simplified to a gradistic tree (figure 4.10), and the general components of the tree may be summarized as a gradogram (figure 4.11):

Figure 4.10. A gradistic tree derived from Huxley's diagram (figure 4.9).

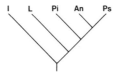

Figure 4.11. A gradogram derived from the gradistic tree of figure 4.10.

ILPiAnPs
LPiAnPs
PiAnPs
AnPs

These components do not, however, reflect Huxley's concept about these particular grades. With reference to the grade Psychozoa (including only *Homo*), he states: "Cladogenetically man constitutes only the single family Hominidae: but anagenetically he constitutes a grade equivalent in evolutionary importance to all other organisms taken together" (1957:455). The gradistic tree expressing this concept is that of figure 4.12; and the corresponding gradogram is that of figure 4.13.

It is unnecessary here to belabor the inconsistency between Huxley's diagrams and his concepts. But the nature of gradistic prediction may be briefly considered with reference to figure 4.13. The unique properties of humankind (Huxley's Psychozoa), such as intelligence, culture, language, etc., are a basis for the hypothesis of congruence respecting the psychozoans (e.g., the various kinds of humans): psychozoans are synapomorphic generally, relative to all other forms of life; therefore, they will prove synapomorphic in other ways not yet discovered. All other forms of life (here termed the "Apsychota"), however, are a group lacking any basis for generalization, except the negative generalization that "Apsychota" lack the unique properties of Psychozoa. Indeed, for the "Apsychota" there are known no generalizations that would not

Figure 4.12. A gradistic tree exemplifying Huxley's concept of the grade Psychozoa (including *Homo*) and the grade including all other organisms (here termed the "Apsychota").

Figure 4.13. A gradogram derived from the gradistic tree of figure 4.12.

apply also to Psychozoa. For the "Apsychota," therefore, the only possible "prediction" is that future research will demonstrate that newly discovered traits unique to Psychozoa will be absent among "Apsychota"—a prediction that in reality is a tautology. It is not surprising that Huxley's comments about the "Apsychota" have not been productive; indeed, they have been ignored, presumably because the "Apsychota" are a group for which useful prediction is not possible. Huxley's concept of grade, however, has been used, with whatever degree of success, despite the tautological nature of gradistic "prediction," which seems an ever-present problem and a serious defect of the gradistic approach.

For example, Simpson, in a branching diagram, has elaborated a more detailed gradistic scheme for humankind and its near relatives (figure 4.14), in which grades are considered "adaptive-structural-functional zones." He states (1963) that the diagram is a

combination of a dendrogram and an adaptive grid. . . . Interpretation of probable closeness of genetic connection is indicated by depth of branching, although it is to be emphasized that such a diagram is not a phylogenetic tree and has no time dimension. (p. 25)

Whatever Simpson meant by "genetic connection," the gradistic elements may be portrayed in a gradogram (figure 4.15) with the following components:

> HoHyPoPa
>   HyPoPa: Pongid Zone
>   PoPa: *Pongo-Pan* Subzone

The components correspond to Simpson's zones, subzones, and subsubzones (all of which are grades), and the gradogram is consistent with his concepts, in contradistinction to Huxley's diagrams and concepts, as may be appreciated by a complete listing (including implicit components):

ADAPTIVE AND STRUCTURAL-FUNCTIONAL ZONES

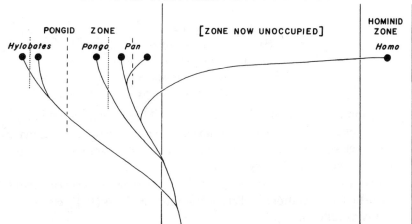

Figure 4.14. A diagram illustrating grades, published by Simpson in 1963. His legend reads: "Dendrogram of probable affinities of recent hominoids in relationship to their radiation into adaptive-structural-functional zones. The two major adaptive zones are bordered by solid lines. Pongid radiation into sub- and sub-sub-zones is schematically suggested by broken and dotted lines. A dendrogram of this sort has no time dimension and does not indicate lineages, but it is probable that divergences of lines showing affinities are topologically similar to the phylogenetic lineage pattern." After Simpson (1963), figure 5, p. 26. Reprinted from S. L. Washburn, ed., Classification and Human Evolution. Viking Fund Publications in Anthropology, no. 37. Copyright © 1963, Wenner-Gren Foundation for Anthropological Research, Inc., New York.

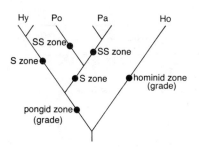

Figure 4.15. A gradogram derived from Simpson's diagram (figure 4.14).

HoHyPoPa
Ho: Hominid Zone
HyPoPa: Pongid Zone
Hy: *Hylobates* Subzone
PoPa: *Pongo-Pan* Subzone
Po: *Pongo* Subsubzone
Pa: *Pan* Subsubzone

Simpson has further developed his gradistic interpretation in another figure (figure 4.16), which has a time axis. The gradistic elements may, nevertheless, be isolated in a gradogram (figure 4.17), for which a listing of components seems superfluous.

The question may be asked, what is the relation between these gradistic interpretations and classification? Simpson's (1945) classification is as follows:

Order Primates
Suborder Prosimii
Suborder Anthropoidea
Superfamily Ceboidea
Superfamily Cercopithecoidea

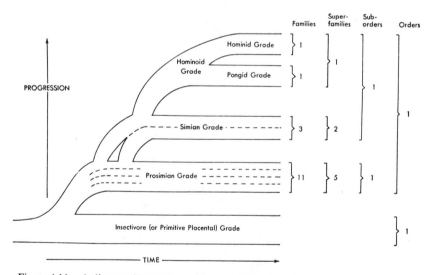

Figure 4.16. A diagram illustrating grades and their relation to taxa. After Simpson (1961), figure 29, p. 215.

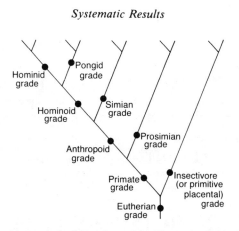

Figure 4.17. A gradogram derived from figure 4.16. The Anthropoid and Primate grades are represented in Simpson's diagram (figure 4.16) by the inclusive suborder and order (brackets). The Eutherian grade is not represented in Simpson's diagram.

Superfamily Hominoidea
Family Pongidae
Subfamily Hylobatinae
Subfamily Ponginae
Family Hominidae

The cladogram derived from this classification is that of figure 4.18. A comparison of this cladogram with the gradograms (figures 4.15, 4.17) shows many common components (the Simian grade, Ceboidea + Cercopithecoidea, is the only grade unrepresented by a taxon, and all taxa represent grades). The conclusion is that Simpson's groups are all

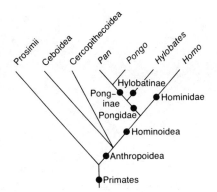

Figure 4.18. A cladogram derived from Simpson's classification of Primates.

grades, relative to his concepts. But Simpson had phyletic concepts as well (figures 4.14 and 4.16), and the phyletic concepts may be summarized in a phylogram (figure 4.19). Comparison of the phylogram (figure 4.19), gradograms (figures 4.15, 4.17), and the cladogram derived from his classification (figure 4.18) reveals the following:

Grades Unrepresented by Taxa	Monogradistic Taxa	Monophyletic Taxa	Polyphyletic Taxa
	Primates	Primates	
	Prosimii		
			Prosimii
	Anthropoidea	Anthropoidea	
	⎧ Ceboidea	Ceboidea	
Simian	⎨		
	⎩ Cercopithecoidea	Cercopithecoidea	
	Hominoidea	Hominoidea	
	Pongidae		Pongidae
	Ponginae		Ponginae
	Hylobatinae	Hylobatinae	
	Hominidae	Hominidae	

One hundred percent of these taxa are grades according to Simpson's concepts (figures 4.14–4.17), however clearly formulated they might be; 70 percent are also monophyletic according to Simpson's concept of

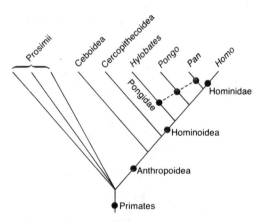

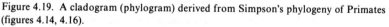

Figure 4.19. A cladogram (phylogram) derived from Simpson's phylogeny of Primates (figures 4.14, 4.16).

their phylogeny. The reader may independently investigate the reasons, if any can be discovered, for this mixture of taxa. Here it is enough to show that gradistic concepts may be analyzed into their components. Once isolated, the components may be considered in relation to classification, on the one hand, and phyletic components on the other. These considerations have importance for the problem of prediction.

Of interest in this regard are the developments in primate research during the twenty to thirty years since Simpson's publications. The summary of Goodman (1975) may be considered an example, in order to explore the predictive aspects of the various taxa recognized by Simpson. Goodman's figure of primate interrelationships (figure 4.20), based on studies of proteins and other molecules, may be simplified into

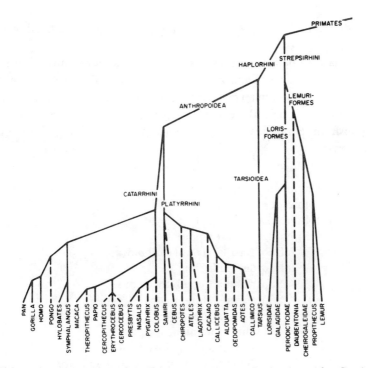

Figure 4.20. A cladogram of the Primates, based on molecular data. After Goodman (1975), part of figure 1, p. 226.

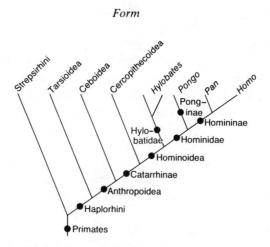

Figure 4.21. A simplified cladogram derived from Goodman's cladogram (figure 4.20), identical to one derived from Goodman's classification (see text).

a cladogram (figure 4.21) comparable in its level of detail with Simpson's figures.

Comparison between Simpson's concepts and Goodman's diagram shows the following: Simpson's Prosimii are divided into Strepsirhini and Tarsioidea; Goodman recognizes two new inclusive taxa, Haplorhini and Catarrhini; and there are two groups with a more inclusive composition, Hominidae and Homininae. Goodman's classification is as follows:

> Order Primates
>   Semiorder Strepsirhini
>   Semiorder Haplorhini
>   Suborder Tarsioidea
>   Suborder Anthropoidea
>   Infraorder Platyrrhini (Ceboidea)
>   Infraorder Catarrhini
>   Superfamily Cercopithecoidea
>   Superfamily Hominoidea
>   Family Hylobatidae
>   Family Hominidae
>   Subfamily Ponginae (*Pongo*)
>   Subfamily Homininae (*Pan* and *Homo*)

Comparison between Goodman's and Simpson's classification shows that Goodman's contains two ranks (semiorder and infraorder) absent in Simpson's, and certain names not used by Simpson. It is not the purpose here to decide which is the better classification. But we may ask which of Simpson's taxa have been retained, i.e., which taxa have proven predictive with respect to the new information about molecules, information unavailable to Simpson when he composed his classification? What is important for this question is not change of rank, or change in name, but change in composition (change in components).

Simpson's Taxa	Goodman's Taxa
Primates, unchanged	Primates
*Prosimii, split into	Strepsirhini Tarsioidea
Anthropoidea, unchanged	Anthropoidea
Ceboidea, unchanged	Platyrrhini (Ceboidea)
Cercopithecoidea, unchanged	Cercopithecoidea
Hominoidea, unchanged	Hominoidea
*Pongidae, split into	Hylobatidae Ponginae Homininae (in part: *Pan*)
*Ponginae, split into	Ponginae Homininae (in part: *Pan*)
Hylobatinae, unchanged	Hylobatidae
Hominidae, unchanged	*Homo*

*Taxa changed in composition (components).

The taxa that proved predictive were all of those visualized by Simpson as monophyletic. The only taxa that proved nonpredictive were those visualized by Simpson as polyphyletic (nonmonophyletic). Simpson, of course, viewed all of his taxa as grades, but one grade (Simian) visualized by Simpson, though not recognized as a taxon, proved nonpredictive.

It is not surprising that the only nonpredictive taxa are polyphyletic, i.e., groups that Simpson could not generally visualize as real (that is, groups that he thought were grades but not also monophyletic groups). It is not surprising because gradistic "prediction" seems to be tautological, rather than useful, prediction—a point that is not proved, however, by this discussion. Indeed, one might argue that, for one reason or another, Goodman's cladogram and classification do not constitute a useful test of gradistic prediction. Simpson, in fact, has already provided an argument of that kind. Referring to an earlier paper by Goodman and Moore (1971), he stated:

> Those authors report that . . . their results confirm the "established" taxonomic (classificatory) arrangement but show the "classical Pongidae" (i.e., the living apes) to be polyphyletic and require the reference of *Pongo*, subfamily Ponginae, to the Hominidae and of "*Pan*" . . . not only to the Hominidae but also to the Homininae.
>
> Those are subjective interpretations of data that are summed up in their Fig. 6, a "cladogram" or, as it should be designated not to prejudice interpretation, a dendrogram based on elaborately measured antigenic distances.
>
> The data of Goodman and Moore and the resulting dendrogram do not in fact show the classical Pongidae (or Pongidae plus Hylobatidae) to be polyphyletic. They are indicated as strictly monophyletic from the stem labeled "Hominoids" on the dendrogram. That dendrogram neither contradicts nor adds to the classical and still current consensus about relationships within the Catarrhini. It is interesting that as regards relationships of living catarrhines Goodman and Moore's dendrogram, although drawn in a different and unusual form, is topologically absolutely identical with that already presented by Haeckel (1866 . . .) more than a century ago. (1971:369)

With respect to Goodman's concept of primate interrelationships (figure 4.20), Simpson stated:

> The "classical" arrangement, with Hominoidea divided into Pongidae (all living apes) . . . and Hominidae (*Homo* only among living animals), is consistent with those relationships and was arrived at with them in mind. (*Ibid.*)

The relevant part of Haeckel's phyletic "tree" is reproduced here (figure 4.22) and summarized in a simplified cladogram (figure 4.23). The relevant part of Haeckel's classification is as follows (see figure 4.24 for a cladogram derived from this classification):

> Order Simiae
>   Suborder Arctopitheci
>   Suborder Platyrrhinae
>   Suborder Catarrhinae
>   Section Menocerca
>   Section Lipocerca
>   Family Tylogluta (*Hylobates*)
>   Family Lipotyla (*Pongo, Pan*)
>   Family Erecta (*Homo*)

Figure 4.22. Part of a phyletic tree, published by Haeckel in 1866, showing the interrelationships of some primates as he conceived them. After E. Haeckel (1866), Allgemeine Entwicklungsgeschichte der Organismen (Berlin: Reimer), part of plate 8.

Haeckel's taxa may be compared with those of Goodman:

Haeckel's Monophyletic Taxa	Haeckel's Polyphyletic Taxa	Equivalent Taxa in Goodman's Classification
Simiae, unchanged		Anthropoidea
Arctopitheci ⎫ Platyrrhinae ⎬ lumped as		Platyrrhini
Catarrhinae, unchanged		Catarrhini
Menocerca, unchanged		Cercopithecoidea
Lipocerca, unchanged		Hominoidea
Tylogluta, unchanged		Hylobatidae
	*Lipotyla, split into	⎧ Ponginae (*Pongo*) ⎨ ⎩ Homininae (in part: *Pan*)
Erecta, unchanged		*Homo*

*Taxon changed in composition (components)

Comparison between these classifications, concepts, and diagrams leads to the conclusion that Simpson is correct in asserting that a *Pongo-Pan* component was recognized by Haeckel in his classification, that the component was maintained by Simpson (who added a

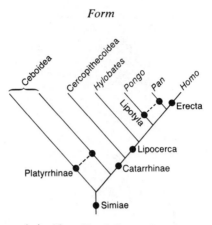

Figure 4.23. A cladogram derived from Haeckel's tree (figure 4.22). Along the top are the modern names of the groups recognized by Haeckel. Names of the inclusive taxa are the names in Haeckel's classification (see text).

Hylobates-Pongo-Pan component as well), and that the component was abandoned by Goodman (who abandoned Simpson's *Hylobates-Pongo-Pan* component as well).

Haeckel did not elaborate his reasons for grouping *Pongo* and *Pan*. Indeed, not much was known about the organisms in Haeckel's time. Simpson, of course, did elaborate his reasons to some extent, in relation to his concepts of grades (his zones, subzones, and subsubzones). These reasons and concepts are put into question by Goodman's results. It is not argued here that the *Pongo-Pan* controversy involving Simpson and

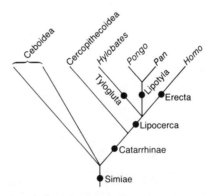

Figure 4.24. A cladogram derived from Haeckel's (1866, see caption for figure 4.22) classification of Primates.

Goodman has great intrinsic interest, but only that the controversy allows some general exploration of gradistic taxonomy.

Goodman's results, in effect, raise a question about the usefulness of the *Pongo-Pan* component: why was it not resolved by Goodman's study? And why did his study resolve instead a *Pan-Homo* component?

If gradistic prediction were possible, it would imply some generalization that would lead to the hypothesis of congruence. For example, a person might state that *Pongo* and *Pan* are gradistically synapomorphic, relative to all other organisms; and that independent samples of information will resolve a *Pongo-Pan* component rather than any other component involving those taxa. Such prediction presumes that this gradistic component has already been resolved at least once, or that it is potentially resolvable—a presumption that may have no basis in fact because of the apparently tautological nature of gradistic generalization and "prediction" (see above).

Simpson seems to have approached the problem of prediction. He stated:

It is abundantly established that anatomically, behaviorally, and in other ways controlled or influenced by total genetic makeup *Homo* is very much more distant from either "*Pan*" or [*Pongo*] . . . than they are from each other. That fact is not overbalanced by the failure of just one kind of data to reflect that distinction clearly or in equal degree. The distinction is real, and it still justifies the classical séparation of Pongidae and Hominidae in classification. (1971:370)

One problem with Simpson's generalization, and any "predictions" that might arise from it, is that it is a phenetic, rather than gradistic, prediction, and is subject to the difficulties of the phenetic approach. If it is not a phenetic "prediction," then it is a statement whose meaning will become clear when and if the "total genetic makeup" of these organisms becomes known. Simpson did not attempt to make a distinct gradistic prediction, in order to determine if it is realized or realizable, with particular reference to molecular information, but he did invoke a principle supposedly underlying all classification. With respect to Goodman's classification, Simpson stated:

The difference of opinion might thus be taken as purely formal and subjective.

Where does one prefer to draw arbitrary family lines? The consensus puts the division Pongidae-Hominidae at the dendrogram point of divergence between a line leading to "*Pan*" . . . and another leading to *Homo*. Goodman and Moore simply have a personal preference to draw it at the dendrogram point of divergence of a line leading to the gibbons from one leading to . . . [*Pan* and *Homo*] plus *Pongo*. (1971:369)

Here Simpson avoids the main problem (what groups to recognize) to focus on a trivial problem (what names should apply to the groups).

It is not the purpose here to discredit the gradistic approach, but rather to try to understand it. Simpson's words are quoted not to show that they are inadequate, but rather to inquire about their meaning and adequacy and, more precisely, to determine if they contain an element of useful gradistic prediction. They seem not to contain such, inasmuch as they lead into the area of phenetic "prediction" on the one hand, and arbitrariness of classification on the other.

An attempt to isolate gradistic synapomorphy for a group composed, say, of *Pongo* and *Pan* leads immediately to an appraisal of synapomorphy of organisms, such as *Homo*, distinct from the *Pongo-Pan* group. Interestingly, a statement of general synapomorphy for *Homo* is a statement that the various forms of *Homo* are synapomorphic relative not merely to *Pongo* and *Pan*, but to these as well as all other forms of life. In order to isolate *Pongo* and *Pan* from all other forms of life, and in effect to group them with *Homo*, another statement of general synapomorphy is required, or at least implied: that *Pongo, Pan,* and *Homo* are generally synapomorphic, relative to all other forms of life.

Gradistic "synapomorphy" (G_s), therefore, seems to be analyzable into two elements of general synapomorphy (S_1, S_2):

(1) S_1 (*Pongo, Pan, Homo*) – S_2(*Homo*) = G_s (*Pongo, Pan*)

Or, with reference to Simpson's classification:

(2) S_1 (Hominoidea) – S_2 (Hominidae) = G_{s1} ("Pongidae")

This conception of gradistic "synapomorphy" allows the possibility of numerous alternative grades (again, with reference to Simpson's classification):

(3) S_{1a} (Anthropoidea) – S_2 (Hominidae)
$$= G_{s2} \text{ (nonhuman anthropoids)}$$
(4) S_{1b} (Primates) – S_2 (Hominidae) = G_{s3} (nonhuman primates)
(5) S_{1c} (Eutheria) – S_2 (Hominidae) = G_{s4} (nonhuman eutherians)
(6) S_{1d} (Mammalia) – S_2 (Hominidae) = G_{s5} (nonhuman mammals)

(7) S_{1e} (Animalia) – S_2 (Hominidae) = G_{s6} (nonhuman animals)

(n) S_{1n} (All of life) – S_2 (Hominidae)

= G_{sn} (nonhuman organisms: "Apsychota")

If gradistic "synapomorphy" is as described above, analyzable into two elements of general synapomorphy, then there is a contradiction at a fundamental level in gradistic theory, at least as it is conceived and applied by Simpson: the groups Hominoidea, Anthropoidea, Primates, etc., if they include *Homo*, cannot be grades (figure 4.25). This conclusion suggests that a group characterized by gradistic "synapomorphy" (synapomorphy that is analyzable into two elements of general synapomorphy) cannot also be a monophyletic group. Neither can it be generally predictive. If so, there can be no generality of classification, based on congruence between grades and monophyletic groups, because such congruence is impossible. Indeed, such may be the case. Such would not be the case if some useful gradistic synapomorphy actually existed and could be resolved.

These considerations suggest various possibilities: (1) that, for example, the various taxa visualized by Simpson as both grades and monophyletic groups are either one or the other, but not both; (2) that the taxa he visualized as grades and polyphyletic groups may be both; (3) that there may be two kinds of grades, (*a*) those defined by gradistic synapomorphy (those that might also be monophyletic groups), and (*b*) those defined by gradistic "synapomorphy" (those that cannot also be monophyletic groups).

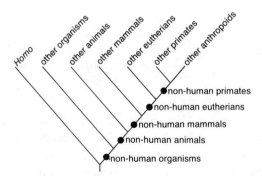

Figure 4.25. A gradogram of some grades (inclusive groups), defined with reference to two statements of general synapomorphy (see text).

Possibility (3) seems to reflect the actual situation, namely that some groups visualized as grades are predictive (monophyletic grades) and some are not (polyphyletic grades). Examples are furnished by Simpson's Hominidae (*Homo*) and Pongidae (*Pongo, Pan*), respectively. But if some grades are not usefully predictive, what then is their use? And why do taxonomists sometimes recognize them? In this case, Simpson has attempted to make clear his reasons. With reference to *Pongo, Pan,* and *Homo,* he stated:

Wilder boldly united all these forms in a single family, Hominidae, and Gregory and Hellman (e.g., 1939d) have adopted this arrangement. On the basis of usual diagnostic characters, such as the teeth, viewed with complete objectivity, this union seems warranted. I nevertheless reject it, for two reasons: (a) mentality is also a zoological character to be weighed in classification and evidently entitling man to some distinction, without leaning over backward to minimize our own importance, and (b) there is not the slightest chance that zoologists and teachers generally, however convinced of man's consanguinity with the apes, will agree on the didactic or practical use of one family embracing both. (1945:187–188)

It is doubtful that these reasons have much significance for gradistic taxonomy in general.

The previous examples, (1)–(n), involved comparisons of pairs of groups, both monophyletic, in order to resolve elements of gradistic "synapomorphy" (G_s). A more complex situation is discussed by Huxley (figure 4.26). His figure can be simplified to a gradistic tree (figure 4.27), and the general components of the tree can be summarized in a gradogram (figure 4.28):

$$AgPlOsAmReHo$$
$$PlOsAmReHo$$
$$OsAmReHo$$
$$AmReHo$$
$$ReHo$$
$$Ho$$

Of interest is the group "Homotherma," considered by Huxley to be polyphyletic ("diplyletic"). As subgroups it includes Aves and Mammalia. Because each of the two subgroups is monophyletic according to his conception, elements of gradistic "synapomorphy" can be resolved by comparing each of them with a more inclusive group, either individually:

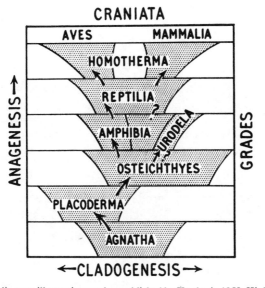

CRANIATA

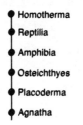

←CLADOGENESIS→

Figure 4.26. A diagram illustrating grades, published by Huxley in 1958. His legend reads: "Anagenesis and cladogenesis in Craniate evolution . . . . The grade Homotherma is diphyletic, consisting of the separate classes Aves and Mammalia." After Huxley (1958), figure 5, p. 28.

(1) S_1 (Amniota) − S_2 (Mammalia) = G_{s1} (nonmammalian amniotes)

(2) S_1 (Amniota) − S_3 (Aves) = G_{s2} (nonavian amniotes),

or jointly:

(3) S_1 (Amniota) − S_2 (Mammalia) − S_3 (Aves) = G_{s3} ("Reptilia").

Thus is it possible to create gradistic "synapomorphy" by subtracting from an inclusive group two or more of its included subgroups. Such

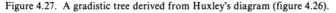

● Homotherma
● Reptilia
● Amphibia
● Osteichthyes
● Placoderma
● Agnatha

Figure 4.27. A gradistic tree derived from Huxley's diagram (figure 4.26).

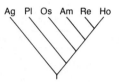

Figure 4.28. A gradogram derived from the gradistic tree of figure 4.27.

subgroups can also be combined in a "grade," which can be defined if the subgroups have characters in common:

(4) S_2 (Mammalia) + S_3 (Aves) = G_{s4} ("Homotherma").

Combining the subgroups produces another kind of gradistic "synapomorphy":

S_2 (Mammalia): a, b, c, d, e,
S_3 (Aves): d, e, f, g, h, i,
G_s ("Homotherma"): d, e

According to Huxley's concept in this case, the two subgroups together do not form a monophyletic group. The "Homotherma" are not, therefore, generally synapomorphic, and the defining characters of the group are parallelisms or convergences.

In another case, the combined subgroups might, however, form a monophyletic group, in which case the combined group will be generally synapomorphic. An example is furnished by Simpson's "Hominoid Grade," which is subdivided into the "Hominid Grade" and the "Pongid Grade" (figure 4.16):

(5) G_{s1} ("Hominoid Grade") = G_{s2}("Hominid Grade") +
G_{s3} ("Pongid Grade").

The gradistic "synapomorphy" of the "Hominid Grade" is merely the general synapomorphy of the Hominidae (*Homo*):

(6) G_{s2} ("Hominid Grade") = S_2 (Hominidae).

The gradistic "synapomorphy" of the "Pongid Grade" is derived by comparison of two groups:

(7) S_1 (Hominoidea) – S_2 (Hominidae) = G_{s3} ("Pongid Grade").

It follows, therefore, that

(8) G_{s3} ("Pongid Grade") + S_2 (Hominidae) = S_1 (Hominoidea)
$$= G_{s1} \text{ ("Hominoid Grade").}$$

Various types of gradistic "synapomorphy" may, therefore, be distinguished:

1. General: $G_s = S$
   Example: "Psychozoa" = *Homo*
2. Simple gradistic: $G_s = S_1 - S_2$
   Example: "Pongidae" = Hominoidea - Hominidae
3. Compound gradistic: $G_s = S_1 - S_2 - S_3$
   Example: "Reptilia" = Amniota - Mammalia - Aves
4. Complex gradistic:
   General: $G_{s1} = G_{s2} + G_{s3} = S$
   Example: "Hominoid Grade" = "Hominid Grade" + "Pongid Grade" = Hominoidea
   Convergent: $G_s = S_2 + S_3 \neq S$
   Example: "Homotherma" = Aves + Mammalia

One may compare the possibilities of prediction of these four types of gradistic "synapomorphy" (P_g) and the possibilities of prediction for general synapomorphy (P_s):

1. General: $P_g = P_s$
2. Simple gradistic: $P_g = P_{s1} - P_{s2}$
3. Compound gradistic: $P_g = P_{s1} - P_{s2} - P_{s3}$
4. Complex gradistic:
   General: $P_{g1} = P_{g2} + P_{g3} = P_s$
   Convergent: $P_g = P_{s2} + P_{s3} \neq P_s$

The only type of gradistic "prediction" that is conceivably different from that of general synapomorphy is that of convergent gradistic "synapomorphy." In this case, the "prediction" is limited to parallel or convergent characters.

The above analysis of gradistics is a preliminary attempt, not only to formulate gradistic principles, but to follow their implications. The anaylsis should not be considered definitive, but it does indicate severe weaknesses in gradistics, as exemplified in the writings of gradistic theorists. Either the weaknesses are misconstrued in the above analysis, or gradistic theory suffers from them.

In a recent discussion of grades, Schaefer (1976:4) asserts that "A higher group is the evolutionary result not merely of an accumulation of character differences (a clade), but also represents a new level of organization (a grade)." Paraphrased, Schaefer's assertion is that all grades are monophyletic. This assertion runs counter to Huxley's and Simpson's usages, which provide for grades that are explicitly nonmonophyletic. Schaefer's assertion thereby exemplifies the presently confused state of gradistic theory.

Inasmuch as Simpson's classification of mammals has served here in numerous discussions, both of phyletic and gradistic taxonomy, and his comments have been extensively quoted, it is appropriate to call attention to the overall nature of his classification. It cannot be argued that Simpson's classification is predominantly gradistic, even though Simpson may have tended to view his taxa as grades from time to time. Most of his taxa have precedents in older classifications of mammals. Moreover, most of his taxa are predictive, and have proven themselves so on numerous occasions through history, even if Simpson may on occasion have argued in ways that might suggest otherwise.

Simpson's classification can generally be interpreted as cladistic classification in a phyletic context, and his classification is, perhaps, best viewed as such—within the limitations imposed on any one person's ability fully to grasp or to explain taxonomic endeavors when applied on so broad a scale.

CLASSIFICATION AND GENERAL CHARACTERS

Having explored phyletic, phenetic, and gradistic classifications in relation to their components, we now turn to the obvious question: is there a way to resolve the differences? In particular, can the differences be resolved in a way that might be satisfactory to phyleticists, to pheneticists, and to gradists? One might ask if there is any common ground at all that is shared by workers of all three persuasions. Perhaps all systematists might agree that they wish to be able to predict from the components of their classifications the maximum possible number of unknown characters, in both studied and yet unstudied taxa. The views of Sneath and Sokal on such predictions (with regard to aphids) have been noted above. And Mayr (1969) says that one of the greatest assets of a sound classification is its predictive value. It permits extrapolation from known to previously unstudied characters. (p. 7)

If systematists can agree on this, the original question can be rephrased: is there a concept of "character" that might be satisfactory to all three schools? For if there is, the question of what kind of classification is best (i.e., is most successful at predicting unknown characters) becomes an empirical one, on which agreement should be attainable. This section will explore the possibility that the concept alternately called a general character, a synapomorphy, a homology-synapomorphy statement, or a defining character of a group, may provide such a resolution.

What, after all, is a taxonomic character? The conventional analysis indicates that a character consists of two or more different attributes (character states) found in two or more specimens that, despite their differences, can be considered alternate forms of the same thing (the character). A character is thus a theory, a theory that two attributes which appear different in some way are nonetheless the same (homologous). As such, a character is not empirically observable, and the hope of pheneticists to reduce taxonomy to mere empirical observation seems futile; to be able to consider the distance between two setae on the back of a mite a character, for example, one must hypothesize that given pairs of setae on two or more mite specimens, despite their different relative positions and interdistances, are nonetheless the *same* setae. But if alternate character states are in some sense the same, how can they be different? There seem to be only two possibilities: either one state is a modified form of the other, or both are modified forms of a third state. The "sameness" that constitutes the character (the homology) is thus the *unmodified* state, which all the organisms that show the character share, either in its original or in some modified form.

What, then, are the possibilities for prediction? Suppose that we have recognized a group (spiders) on the basis of seven character states believed unique to spiders, and that we find a new specimen about which we know only that it has one of these unique character states. Can we predict that the new specimen will have the other six character states as well? The new specimen might well have a different state of any or all of the six remaining characters, providing the different state represents a modified form of the character state found in all other spiders. Thus, given the information

Character A: State 1 (nonspiders)
                State 2 (spiders),

we can predict that all new spiders will have state 2, but they might have
instead a different character state:

Character A: State 1 (nonspiders)
                State 2 (most spiders)
                State 3 (some spiders),

if that different state is actually a further modification of state 2:

Character A: State 1 (nonspiders)
                State 2 (spiders):   Substate 1 (most spiders)
                                     Substate 2 (some modified spiders).

In this situation, of course, state 2 is functioning as a character (a unit of
"sameness"), not a character state; just like character A, it is divided into
an original and a modified form. If we agree that for the character (the
"sameness") to exist, the character states must be modifications, it is
apparent that predictions cannot really function at the level of character
states. Because of the ever-present possibility of further modifications,
predictions will hold only for characters (i.e., only for sets of original
plus modified character states).

In this sense, then, the concept of character states is misleading. To
view character A as being "composed" of three character states:

                            Character A
---State 1---   ---State 2---   ---State 3---

implies that the character states are alternatives, when they are actually
additions:

--- State 1 --
                    ---State 2 ------------------------------------
                                            ---State 3---

In this case, character A is actually equivalent to state 1 (i.e., it defines a
group, all the members of which have state 1, either in its original or
some modified form). States 2 and 3 are best regarded as new characters
(B and C), for which the same provisions hold. Further, state 1 itself is a
modification of some other character (state) and represents a restricted
subset of some other, more general, character. Thus, all characters can

be seen as modifications (or restrictions) of other characters, and the groupings of character states within characters can be seen as arbitrarily delimited clusters of separate characters that are increasingly more restricted in generality (i.e., that form nested sets of increasingly modified versions of other characters).

The implications of this for prediction can be readily seen in a simple example involving tetrapod and nontetrapod vertebrates. Systematists have long been in agreement that the limbs of tetrapods are homologous with the fins of nontetrapod vertebrates ("fishes"). If we regard fins and limbs as alternative states of a character (paired appendages), we might thereby sort out vertebrates into two groups:

---Vertebrata (paired appendages)---
---Pisces (fins)--------   ---Tetrapoda (limbs)---

and vertebrates were indeed classified in this way for many years. However, one of these groups (Pisces) proved not to be maximally predictive, in that many characters were found that are shared uniquely by tetrapods and some (but not all) fishes. If, however, the limbs of tetrapods are not only homologous with, but are also modifications of fins, the problem disappears:

---Vertebrata (paired appendages: fins) -------------------------------------
   ---Tetrapoda (modified paired appendages: limbs) --------------

We can now predict that any vertebrate will have fins, and that only some vertebrates (tetrapods) will have modified fins. The phyleticist should presumably be happy, because the groups that are recognized *are* believed to be monophyletic. The pheneticist should presumably be happy, because all of the phenetic information (that tetrapod vertebrates have limbs and nontetrapod vertebrates do not) *is* included. The gradist should presumably be happy, because the available phenetic and phyletic data *are* combined in the classification. If so, then all systematists can agree that what they are seeking are synapomorphies: hypotheses that particular groups, defined by particular characters, are synapomorphic generally, with respect to all other organisms, and will prove to be so, no matter what other characters we may eventually discover or consider.

One possible source of difficulty here is that some types of properties or attributes of organisms that have traditionally been used as characters by taxonomists may actually not be usable as defining characters of groups (even though they may still be useful for the practical task of identifying specimens). Size, for example, may fall into this category: we might say that a family of spiders is characterized by a total length of between 2 and 4 mm. But this is not a defining character of that family, for there are many other spiders, and many other organisms, with a total length of between 2 and 4 mm. Color might pose similar problems: we might say that a family of bugs is characterized by green wings, but many other insects have green wings. Of course, we may be able to show that bugs of this family have wings that are green because they contain a pigment not found in other organisms; if so, the pigment could be a defining character. Similarly, if a given pigment is found in two groups but we can show that the enzymatic pathways by which the pigment is synthesized are different in the two groups, then the two enzymatic pathways could each be defining characters. It is clear, at any rate, that the "best" groups, those that have proved maximally predictive through the years, have been explicitly based on general characters, like the spinnerets of spiders or mammary glands of mammals, which define groups (in the sense that any organism with spinnerets is a spider) that are generally synapomorphic with respect to all other organisms.

It would appear, then, that systematics in general consists of the search for defining characters of groups. Admittedly, the search seems to have been abandoned, on occasion, by persons who would search instead for overall phenetic similarity, or overall gradistic similarity. But what justification is there for abandoning the search for defining characters? Certainly we can make mistakes; we may fail to recognize that one character is a modified form of another, and thereby underestimate the generality of the latter one (plesiomorphy), or we may confuse two different characters as one (convergence), or we may mistakenly consider attributes (such as absences, or physical properties like total length and color) defining characters when they're not. But we do not need to fear such mistakes: should we make them, future research will reveal them quickly enough, through incongruences with other characters. Indeed, one might contend that the utility of cladograms is precisely that: to point up incongruences and allow us to predict that mistakes have been made, and that more intensive study will reveal them.

CLASSIFICATION AND GENERAL CLADOGRAMS

The preceding section explored the possibility of a synthesis of phyletics, phenetics, and gradistics involving the concept of a general character. In this section, we will explore another possibility for synthesis based on the concept of a general cladogram, using as an example a recent paper by Michener (1977) on the classification of bees.

Briefly summarized, Michener's paper includes five cladograms and three classifications of 14 genera of allodapine bees. The cladograms include what Michener terms a "cladogram . . . developed using the methods of Hennig" (his figure 1), and four diagrams, each of which he terms a "cladogram on which are shown certain distance coefficients" (his figures 7–10). These are the five cladograms, which include one phylogram (redrawn here as figure 4.29.1) and four phenograms (figures 4.29.2–5). Of each of Michener's figures 7–10, he states that "nested rings indicate subjective levels of similarity." The nested rings are the basis of the four phenograms of figures 4.29.2–5: one based on "54 characters of mature larvae" (his figure 7; cf. figure 4.29.2); a second, on "46 characters of pupae" (his figure 8; cf. figure 4.29.3); a third, on "144 external characters of adults" (his figure 9; cf. figure 4.29.4); and a fourth, on "25 genital and associated characters of adult males" (his figure 10; cf. figure 4.29.5). The five cladograms constitute a study of 14 genera coded as follows: A, *Halterapis*. B, *Compsomelissa*. C, *Allodape*. D, *Braunsapis*. E, *Nasutapis*. F, *Allodapulodes*. G, *Dalloapula*. H, *Allodapula*. I, *Eucondylops*. J, *Exoneurella*. K, *Brevineura*. L, *Inquilina*. M, *Exoneura*. N, *Macrogalea*. The classifications include what Michener terms "different lists of genera," of which there are three, specified by the cladograms of figure 4.30: one based on "all variables so far analyzed" (figure 4.30.1); a second, on "larval variables only (Fig. 7)" (figure 4.30.2); and a third, on "adult external variables only (Fig. 9)" (figure 4.30.3).

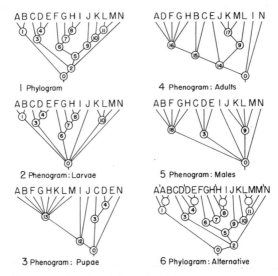

Figure 4.29. Cladograms based on information from Michener's (1977, figures 1, 7-10) study of allodapine bees (cladograms 1-5) and an alternative phylogram (cladogram 6).

There are 21 different components in the cladograms of figures 4.29 and 4.30. Twenty are listed in table 4.1, along with their distribution among the various cladograms. Component 0, defined as that branch point leading to all branch tips, has properties unlike components 1-20, for it appears in all cladograms of the 14 taxa. Components 1-20 need not appear in all cladograms, and, indeed, none of them does. For cladistic analysis, component 0 is irrelevant.

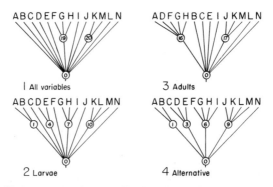

Figure 4.30. Cladograms based on classifications from Michener's (1977:51-52) study of allodapine bees (cladograms 1-3) and an alternative classification (cladogram 4).

The information of a component relates to the number of taxa in its definition. A branch point leading to two branch tips (terminal taxa, or terms) has an information content of one unit (one 3-taxon problem solved); a branch point leading to 12 branch tips, 11 units; and so on. The number of terms and the information of each of the 20 components are listed in table 4.1.

FUNDAMENTAL AND DERIVATIVE CLADOGRAMS

Two major types of cladograms may be distinguished: (1) fundamental cladograms, such as those of figure 4.29, which summarize data about the interrelationships of organisms; and (2) derivative cladograms (classifications), such as those of figure 4.30, which specify the cladistic aspect of hierarchical classifications. Systematic study of the interrelationships of the species of a given group generally results first in one or more fundamental cladogram, which is used as a basis for constructing one or more derivative cladogram. In a given study, the fundamental and derivative cladograms may agree or differ. If they agree, they agree because they include the same components. If they differ, they differ only because they include different components, or different sets of components. Of the cladograms in figures 4.29–4.30, each specifies a different set of components (table 4.1).

Figure 4.29 includes an additional fundamental cladogram (4.29.6), which is an alternative phylogram somewhat more detailed than Michener's (cf. figures 4.29.1, 4.29.6) obtained by splitting four of Michener's genera and creating four new taxa (A', D', H', M'). Figure 4.30 includes an additional derivative cladogram (4.30.4) derived from the alternative phylogram by omitting the new taxa mentioned above. Omitting them makes for easier comparison among the different classifications.

INFORMATION CONTENT OF A CLADOGRAM

The information of a cladogram has two parts: the component information and the term information. The component information, which relates to the number of components in the definition of a cladogram, is one unit less than the total (N–1, 3-taxon problems solved). A cladogram defined by 12 components has a component information of 11 units; a cladogram defined by 8 components, 7 units; and so on. The component information of the cladograms is listed in table 4.1.

Table 4.1. Analysis of Cladistic Components of Cladograms, and of Their Information and Efficiency[a]

| | Components | | | Fundamental Cladograms | | | | | Derivative Cladograms | | | |
| | | | | | Phenograms | | | | | | | |
Number	Definition	Terms/Information	Probability of Replication (%)	Phylogram Fig. 4.29.1	Larvae Fig. 4.29.2	Pupae Fig. 4.29.3	Adults Fig. 4.29.4	Males Fig. 4.29.5	All Variables Fig. 4.30.1	Larvae Fig. 4.30.2	Adults Fig. 4.30.3	Alternative Fig. 4.30.4
1	AB	2/1	1.1	+	+	−	−	−	−	+	−	+
2	C-N	12/11	1.1	+	−	−	−	−	−	−	−	−
3	C-E	3/2	0.27	+	+	+	−	+	−	−	−	+
4	DE	2/1	1.1	+	+	+	−	−	−	+	−	−
5	F-M	8/7	0.033	+	−	−	−	−	−	−	−	−
6	F-I	4/3	0.10	+	+	−	−	−	−	+	−	+
7	G-I	3/2	0.27	+	+	−	−	−	−	−	−	−
8	HI	2/1	1.1	+	+	−	−	−	−	+	−	−
9	J-M	4/3	0.10	+	−	−	+	+	−	−	−	+
10	K-M	3/2	0.27	+	+	−	−	−	−	−	−	−
11	LM	2/1	1.1	+	−	−	−	−	−	−	−	−
12	ABFG-M	10/9	0.27	−	−	+	−	−	−	−	−	−
13	ABF-HK-M	8/7	0.033	−	−	+	−	−	−	−	−	−
14	A-HJ-M	12/11	1.1	−	−	−	+	−	−	−	−	−
15	A-DF-H	7/6	0.029	−	−	−	+	−	−	−	−	−
16	ADF-H	5/4	0.050	−	−	−	+	−	−	−	+	−
17	JKM	3/2	0.27	−	−	−	+	−	−	−	+	−
18	ABF-H	5/4	0.050	−	−	−	−	+	−	−	−	−
19	FGH	3/2	−	−	−	−	−	−	+	−	−	−
20	KM	2/1	−	−	−	−	−	−	+	−	−	−
Information, Component:				11	7	4	5	3	2	4	2	4
Information, Term:				34	12	19	26	9	3	6	6	9
Information, Total:				45	19	23	31	12	5	10	8	13
Efficiency, Component (%):[b]				−	−	−	−	−	0	57	40	36
Efficiency, Term (%):[b]				−	−	−	−	−	0	50	23	26
Efficiency, Total (%):[b]				−	−	−	−	−	0	53	26	29

[a] Component present (+); component absent (−).

[b] Efficiency for cladogram 4.30.4 is computed by comparison with cladogram 4.29.1.

The term information is the sum of the information of all informative components of a cladogram. Consider cladogram 4.29.1. It has 11 informative components (components other than component 0). Component 1, with two terms, has an information content of 1 unit; component 2, with 12 terms, 11 units; component 3, with 3 terms, 2 units; etc. For cladogram 4.29.1 the total term information is the sum of 1 + 11 + 2 etc., = 34 units. The term information of the cladograms is listed in table 4.1.

The component information plus term information is the total information, which is equal to the sum of the terms of all informative components. Consider cladogram 4.29.1. Its component information is 11 units; its term information is 34 units; its total information is 45 units. The total equals the sum of terms of all informative components: 2 + 12 + 3 etc., = 45 units. The total information of the cladograms is listed in table 4.1.

DERIVATIVE CLADOGRAMS AND EFFICIENCY

The three derivative cladograms (4.30.1–3) may be compared with the fundamental cladograms (4.29.1–6); two of the three invite comparison with particular fundamental cladograms (4.30.2 with 4.29.2, and 4.30.3 with 4.29.4).

Comparison 1 (4.30.2 and 4.29.2): The classification of larvae contains components 1, 4, 7, 10. The phenogram for larvae contains components 1, 3, 4, 6, 7, 8, 10. The classification may be said to be derived by accepting components 1, 4, 7, 10 and rejecting components 3, 6, 8.

Comparison 2 (4.30.3 and 4.29.4): The classification of adults contains components 16 and 17. The phenogram for adults contains components 9, 14, 15, 16, 17. The classification may be said to be derived by accepting components 16 and 17 and rejecting components 9, 14, 15.

Comparison 3 (4.30.1 and all fundamental cladograms): The classification based on all variables contains two components (19, 20), which are absent from all fundamental cladograms (4.29.1–6). The classification may be said to be derived by accepting components 19 and 20 (from unknown sources) and rejecting components 1–18.

Other comparisons are of course possible; they would reveal various agreements or the lack thereof.

Inasmuch as derivative cladograms invite comparison with particular fundamental cladograms, such that one might be said to be derived from the other, a given derivative cladogram may be said to be more or less efficient in representing a given fundamental cladogram. Component efficiency may be expressed as the fraction of derived to fundamental component information. Consider derivative cladogram 4.30.2 in relation to fundamental cladogram 4.29.2: four of seven units of component information are represented; the derivative cladogram may be said to have a component efficiency of $4/7$, or 57 percent. Term efficiency may be expressed as the fraction of derived to fundamental term information. Consider again cladograms 4.30.2 and 4.29.2: 6 of 12 units of term information are represented; the derivative cladogram may be said to have a term efficiency of 50 percent. Total efficiency may be expressed as the fraction of derived to fundamental total information, 53 percent. Efficiency (component, term, and total) of the derivative cladograms is listed in table 4.1.

CHAOS VERSUS ORDER

Consider derivative cladogram 4.30.2 (the classification of larvae). All of its components are represented in fundamental cladogram 4.29.2 (the phenogram for larvae); all of its components are represented also in fundamental cladogram 4.29.1 (the phylogram). One may ask: what is the nature of the classification? Is it purely phenetic, purely phyletic, or both? It would seem to be both.

Consider derivative cladogram 4.30.3 (the classification of adults). Its components are represented only in fundamental cladogram 4.29.4 (the phenogram for adults). What is the nature of the classification? It would seem to be purely phenetic.

Consider derivative cladogram 4.30.1 (the classification based on all variables). Its components are represented in no fundamental cladogram. What is the nature of the classification? It would seem to have no nature that can be specified, except to say that it is derived from unknown sources.

That Michener presented three classifications, all differently derived, might be considered curious or unusual. One may ask, why three and three only? Or alternatively, why not just one? Confronted with five different fundamental cladograms, one might assume that each tells a different story of its own: that nature is chaotic; evolution, discordant; convergence, common; etc. This assumption might lead to the idea that the task of systematics is to impose order upon a chaotic nature (in which case many classifications would be justifiable). An alternative assumption is that each cladogram tells the same, or part of the same, story: that nature is ordered; that discordant evolution and convergence are mere artifacts of human perception. This assumption might lead to the idea that the task of systematics is to discover and record nature's order and to embody it in (a single) classification. Thus, different philosophies of systematics are possible, as exemplified in the discussion of artificial vs. natural systems—a discussion some 200 years old.

To some extent, one may try to avoid the pitfalls of metaphysics by a judicious choice of questions. For example, rather than assume that nature is chaotic (and opt for some artificial system), or that nature is ordered (and ponder the one and only natural system), one may ask how chaos or order might be discovered, granting the possibility that each might exist. For chaos, an immediate (but facile) answer is possible, with reference to the diversity of fundamental cladograms that might be at hand (figure 4.29). For order, in contrast, an immediate answer is problematical. The question, however, can be restated: given a diversity of fundamental cladograms, what single story might they all be telling?

The question could be answered if diverse fundamental cladograms can be combined to form a single general cladogram. There might seem to be various approaches to this objective. One might assume, for example, that the phylogram (figure 4.29.1), if it has been derived by combining various data according to some phyletic clustering technique, is *the* general cladogram, and that any departure from it is discordance, convergence, etc. Alternatively, one might assume that some phenogram (figures 4.29.2–5), if it has been derived by combining various data according to some phenetic clustering technique, is *the* general cladogram. One might question, however, whether such assumptions are satisfactory at a general level.

Again, one may turn toward a judicious choice of questions. Rather than assume that the general cladogram must be phyletic, or phenetic, or gradistic, or whatever, one may ask how a general cladogram might be discovered, given the possibility that it exists? What follows is an attempt to answer this question.

COMPONENT RELATIONS

One component is more or less like another, such that any two components share one of four possible relations:

Exclusion:
   components are combinable and exclusive if their definitions are different and nonoverlapping. Example: component 1, with the definition AB, and component 4, with the definition DE.

Inclusion:
   components are combinable and inclusive if their definitions are different and overlapping, such that one is included in the other. Example: component 3, with the definition CDE, and component 4, with the definition DE. For each inclusive relation one component is a whole (3), and the other (4) is a part.

Noncombinability:
   components are noncombinable if their definitions are different and overlapping, such that neither is completely included in the other. Example: component 11, with the definition LM, and component 17, with the definition JKM.

Replication:
   components are replicated if their definitions are the same, in which case there are not two components, but merely one that is replicated.
      These four relations may alternatively be viewed as two:

Combinability (Inclusion, Exclusion, and Replication):
   components are combinable if they can be parts of the same cladogram.

Noncombinability:
   components are noncombinable if they cannot be parts of the same cladogram.

Replicated components are significant because the probability of replication for any given component, due to chance alone, is small. In this case (Michener's 14 genera of bees) the probability is 1 percent or less (table 4.1). Of interest is the fact that the probability is not a function of the fundamental cladogram in which a component appears, but rather a function of the number of terms in the component's definition. Consider, for example, component 1, with two terms. It is one of 91 possible 2-term components that might result from clustering 14 terminal taxa. Hence its probability of replication, due to chance alone, is about 1 percent.

In the fundamental cladograms (table 4.2), there are eight 2-term

Table 4.2. Probabilities (P) of Replication of Components (C) in Fundamental Cladograms [a]

Number of Terms	Number of Possible Components	4.29.1 C	4.29.1 P:%	4.29.2 C	4.29.2 P:%	4.29.3 C	4.29.3 P:%	4.29.4 C	4.29.4 P:%	4.29.5 C	4.29.5 P:%	P for components of same number of terms, %	P cumulative %
2	91	1	–	1	4.40							3.90×10^{-4}	3.90×10^{-4}
		4	–	4	4.55	4	4.40						
		8	–	8	4.44								
		11	–										
3	364	3	–	3	3.75	3	1.29			3	1.10	8.37×10^{-7}	3.27×10^{-12}
		7	–	7	4.08								
		10	–	10	3.85			17	–				
4	1001	6	–	6	16.7			9	0.59	9	0.59	5.73×10^{-4}	1.87×10^{-17}
		9	–										
5	2002							16	–	18	–	–	10^{-17}
7	3432							15	–			–	10^{-17}
8	3003	5	–			13	–					–	10^{-17}
10	1001					12	–					–	10^{-17}
12	91	2	–					14	–			–	10^{-17}

[a] For a component that replicates a previously resolved component, the probability is that of replication due to chance alone, under the assumption that the components are resolved one at a time in the order listed for each cladogram. The probability fractions for replicates are: cladogram 4.29.2 (4/91, 3/66, 2/45, 3/80, 2/49, 1/26, 2/12); 4.29.3 (4/91, 3/232); 4.29.4 (2/341); 4.29.5 (4/364, 2/341). A component that is resolved for the first time has 0 probability of replicating a previously resolved component.

components, of which four are replicates ($P = 10^{-4}$ percent); there are nine 3-term components, of which five are replicates ($P = 10^{-7}$ percent); and there are five 4-term components, of which three are replicates ($P = 10^{-4}$ percent). One may doubt that these replicates are due to chance alone ($P = 10^{-17}$ percent). There are three different 2-term components that are replicated; three different 3-term components; and two different 4-term components. The replicated components may be combined in one cladogram (figures 4.31.1–2). One may doubt that this combination is due to chance alone ($P = 10^{-17}$ percent).

The cladogram specified by the replicates may be considered a first step toward a general cladogram. Inspection of the definitions of components (table 4.1) shows that several are noncombinable with the replicates (cladogram 4.31.2): components 13, 14, 15, 16, 17, 18. There is no reason to doubt that these components are due to chance alone. Of the remaining components (2, 5, 11, 12), each is combinable (either exclusively or inclusively) with all replicates, but components 2 and 12 are mutually noncombinable. There is no reason to doubt that these components are due to chance alone. Nevertheless, four categories of components may be recognized: (1) replicates (1, 3, 4, 6, 7, 8, 9, 10); (2) components noncombinable with replicates (13, 14, 15, 16, 17, 18); (3) components combinable with replicates and with each other (5, 11); (4) components individually combinable with replicates, but not with each other (2, 12). Only category (3) is combinable with the replicates in a single cladogram. Hence, categories (1) and (3) may be combined in a general cladogram (figure 4.32), which may be considered a best estimate of one, and the only apparent, cladistic parameter.

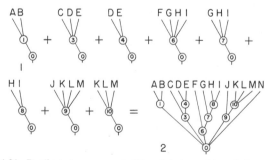

Figure 4.31. Replicated components (1) and their resulting cladogram (2).

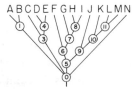

Figure 4.32. The general cladogram derived from the fundamental cladograms of figure 4.29.

Consider cladogram 4.31.2 (replicates) and cladogram 4.32 (general cladogram). Category (3) contributes two additional components (5 and 11)—25 percent of the replicates. What can be said on behalf of the added components? Not much beyond the statement that they are combinable with the replicates and with each other—a statement of less significance than, and a statement of which the significance depends upon, the phenomenon of improbable replication. Component 5 is one of 15 possible components that could be formed by component 6 (or component 9) in exclusive relation with components 1 and 3 and term N. Component 11 is one of three possible components in inclusive relation with component 10. Hence, there are $15 \times 3 = 45$ possible combinations. And there is only 1 chance in 45 that components 5 and 11 might accurately estimate the cladistic parameter, under the assumption that the replicates are an accurate estimation. That the probability is on the order of 2 percent, rather than 10^{-3} percent, is due wholly to the improbable pattern of replication, which determines that there are 45, rather than 400,000, possible combinations.

The general cladogram (figure 4.32) may be compared with the fundamental cladograms (figure 4.29.1-6). Its components are all included in the phylogram (figure 4.29.1), and it might, therefore, be termed purely "phyletic." Most of its components are included also in the phenogram for larvae (figure 4.29.2), and at least one of its components appears in each of the other phenograms (figures 4.29.3-5), and it might, therefore, also be termed somewhat "phenetic." In what follows, the components of the general cladogram are termed "true components"; those noncombinable with the general cladogram (category 2), "false components"; those mutually noncombinable (category 4), "ambiguous components."

Given a general cladogram with its "true components," it may be compared with fundamental and derivative cladograms (table 4.3), and their information content may be broken down into three categories: true, false, ambiguous (by "true" and "false" we mean only agreement or disagreement with the general cladogram). For fundamental cladogram 4.29.1, which includes 11 components, 10 components are true, and one is ambiguous (component 2). One-half of the ambiguity may be allocated to the true category, resulting in 10.5 true units. For fundamental cladogram 4.29.3, which includes four components, two components are true; one is false; and one, ambiguous. The false may be subtracted from the true, and the ambiguous allocated as above, resulting in 1.5 true units. General component information for all cladograms is listed in table 4.3. General term information and general total information may be treated in a similar way. They are listed in table 4.3.

Efficiency is calculable when there is a cladogram that can serve as a standard for comparison. General efficiency is calculable for both fundamental and derivative cladograms, with the general cladogram serving as a standard (10 component-units + 23 term-units = 33 units of information). Consider fundamental cladogram 4.29.1, with 10.5 component-units and 28.5 term-units, for a total of 39 units. The general component efficiency is 10.5/10, or 105 percent; and the general term efficiency is 28.5/23, or 124 percent. General total efficiency is 39/33, or 118 percent. Consider fundamental cladogram 4.29.3, with 1.5 component-units and 0.5 term-units. Its general component efficiency is 1.5/10, or 15 percent. Its general term efficiency is 0.5/23, or 2 percent. And its general total efficiency is 2/33, or 6 percent. In cladograms with ambiguity, general efficiency may exceed 100 percent—indicating that the cladograms are, in some sense, too efficient. In cladograms with false information, both general information and general efficiency may achieve negative values (table 4.3).

OPTIMUM CLASSIFICATION

What is the optimum, or best, kind of classification? A variety of answers is possible, in accordance with the variety of possible taxonomic philosophies. According to one such philosophy, there is no optimum at all. The assumption that there is an optimum defines a philosophy that we call cladistics—a modern version of an old tradition in systematics,

Table 4.3. Analysis of General Information and General Efficiency of Cladograms

	Component Information					Term Information						Efficiency (%)		
	Total Units	Units That Are:			Total True	Total Units	Units That Are:			Total True	Total C + T True	Component	Term	Total
Cladograms		True	False	Ambiguous			True	False	Ambiguous					
4.29.1	11	10	0	1	10.5	34	23	0	11	28.5	39	105	124	118
4.29.2	7	7	0	0	7	12	12	0	0	12	19	70	52	58
4.29.3	4	2	1	1	1.5	19	3	7	9	0.5	2	15	2	6
4.29.4	5	1	4	0	-3	26	3	23	0	-20	-23	-30	-87	-70
4.29.5	3	2	1	0	1	9	5	4	0	1	2	10	4	6
4.30.1	2	0	2	0	-2	3	0	3	0	-3	-5	-20	-13	-15
4.30.2	4	4	0	0	4	6	6	0	0	6	10	40	26	30
4.30.3	2	0	2	0	-2	6	0	6	0	-6	-8	-20	-26	-24
4.30.4	4	4	0	0	4	9	9	0	0	9	13	40	39	39
General	10	10	0	0	10	23	23	0	0	23	33	100	100	100

namely the theory and practice of methods to resolve *the* natural system, or classification.

Discussion of optimality has generated a dispute between adherents of phyletic, phenetic, and gradistic philosophies, all of which are cladistic—under the assumption that any optimal classification has a cladistic aspect that may be represented by a cladogram, which is true for all attempts at hierarchical classification. Thus, there is a generally optimal classification—of which the cladistic aspect is represented in a general cladogram.

It is entirely possible, of course, that some methods are more efficient than others in estimating a general cladogram. To judge from Michener's bee study, his phyletic methods are more efficient than his phenetic methods. But his phenetic methods are not without value. They contribute replications and apparently generate random variation (i.e., noncombinable components) that may be discriminated as such and set aside. Without methods that generate random variation, there would be no need for cladistic analysis and synthesis, for their results would be intuitively self-evident in advance (which is near the truth anyway, for much of the general cladogram can be immediately apprehended by anyone inspecting the fundamental cladograms with an eye for the replications that they contain; in short, cladistic analysis and synthesis can operate at an intuitive level).

Michener (1977) provided three classifications of bees, one based on all variables, one based on larvae, and one based on external features of adults. He did not say so explicitly, but we assume that the classification based on all variables was offered by him as a general classification. The three classifications, plus an alternative one (figure 4.30.4) are analyzed in table 4.3. They differ widely in their information and efficiency. What are the conclusions to be drawn?

The basic question is about optimal classification: is classification to reflect a parameter? If yes, then one may judge one classification better than another in accordance with its agreement with some standard estimate of that parameter. If no, then one may not. To date, various parameters have been suggested: phyletic, phenetic, gradistic, etc., all of which are cladistic in the sense that each may be represented by a cladogram. We suggest the parameter estimated by the general cladogram—the cladistic parameter.

There is no necessary correlation between the cladistic parameter and any one discipline, such as phyletics, phenetics, or gradistics. In the case of Michener's bees, the general cladogram agrees better with Michener's phylogram than with three of his four phenograms, but it agrees better with one phenogram than with the other three. The agreement between estimates of the cladistic parameter and phyletics may prove general, at least in theory, but who knows? In another case, phyletic methods, applied ineptly but with the best of intentions, might generate pure noise; and phenetic methods, with a careful analysis of characters, pure signal.

GENERAL PATTERN

Traditionally conceived, systematics is the search for general pattern in the living world. During the history of systematics, pattern has been resolved piecemeal, and its resolution has never been claimed to be complete. Discovery of pattern has depended upon sampling, and confidence in the results has been a function of prediction and replication. Traditionally, sample size has been small.

With the advent of phenetic methods, stress was placed on large samples, in the hope that their results would be imperturbable with further sampling, and in the hope that the results would estimate a parametric "overall similarity." The recommendation was made that further samples be pooled with the original data, with the consequence that prediction dissolves into nonfalsifiability, which may be the Achilles' heel of purely phenetic philosophy. Hence, the time is ripe for consideration of small samples, the extent to which elements of cladistic pattern (components) are replicated therein, and the probability of their occurrence due to chance alone. These are the traditional virtues of systematics.

The obvious weakness of phenetics is its assumption of parametric "overall similarity"—the reason for the recommendation for pooling further samples with the original data (presumably resulting in a more reliable estimate of "overall similarity"). The usefulness of the idea of "overall similarity" would seem to consist only in its implication that, given estimates of the "overall similarities" for the species of a certain group (a phenogram), the estimates specify a cladogram, i.e., an estimate of the cladistic parameter. But there is no reason to believe that

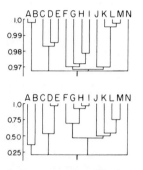

Figure 4.33. Two hypothetical phenograms that differ in their statements about levels of "overall similarity" but specify the same components.

either purported estimate depends for its existence upon the other. Even if there is no parametric overall similarity, there may yet be a cladistic parameter. Consider, for example, the hypothetical phenograms of figure 4.33. Each specifies estimates of the overall similarities for 14 taxa of a certain group. The estimates of the two phenograms differ in all particulars, but the cladistic aspect is the same for both.

Of interest is a comparison between the efficiency of the fundamental cladograms and the number of characters upon which each cladogram is based:

Cladogram	Characters	Efficiency
4.29.1	about 30	118 percent
4.29.2	54	58 percent
4.29.3	46	6 percent
4.29.4	144	−70 percent
4.29.5	25	6 percent

There seems to be no positive correlation, but there might be a negative one, with the implication that, with large numbers of characters processed by phenetic techniques, virtually all resolved components might be due to chance alone—a possibility enhanced by the findings of Mickevich (1978).

Suppose that a general cladogram is derived (figure 4.32). What does it mean? It specifies a nonrandom pattern of interrelationships. One may ask, what is the cause of the nonrandom pattern? Two possibilities come to mind: (1) the pattern is a methodological artifact; (2) the pattern reflects (and, therefore, estimates) an aspect (parameter) of the real world that is method-independent. Given these two possibilities, how might the truth be ascertained? Consider possibility (1): if the pattern is artifact, different methods, to the extent that they are truly independent, should produce different (randomly noncombinable) patterns. Consider possibility (2): if the pattern is method-independent, different methods, even if they are truly independent, should produce the same nonrandom pattern, or combinable parts of the same pattern, in addition to some random variation.

There is already available enough information to assert that different clustering methods produce fundamental cladograms that are to some extent different and to some extent the same. If the differences are artifact, and the samenesses method-independent, then the problem is only distinguishing signal from noise. Deriving a general cladogram merely specifies what is signal, given a number of fundamental cladograms that are to some extent the same, and to some extent different. As here conceived, its derivation amounts to asking, and answering in the affirmative and in detail, this question: given some fundamental cladograms, do they exhibit a nonrandom pattern of cladistic components?

One may ask, in addition, how this question might best be explored and answered. Suppose that a certain clustering method is applied to different data sets in order to generate various fundamental cladograms for the species of a certain group, one cladogram for each data set. A general cladogram might be derived. It could be an artifact of the clustering method. To explore the possible bias of a single clustering method, different methods might be applied, each method to a different data set, or all methods to the same data set. Another general cladogram might be derived. What might be expected in the way of agreement between the two general cladograms? The present study suggests that both general cladograms would estimate the same cladistic parameter.

Suppose that a general cladogram is derived (figure 4.32). What then? Is it artifact or knowledge of the real world? The bolder, and nontrivial, hypothesis is the latter. How might this hypothesis be falsified in a given case? The answer seems clear: by deriving a second, and independent, general cladogram from a different, and independent, set of fundamental cladograms. The prediction is that the two general cladograms, even though independently derived, will estimate the same cladistic parameter. The prediction is a null hypothesis: the two general cladograms will not significantly differ. The null hypothesis may be tested by sampling.

Suppose that the sampling is done, and that, in one's judgment, the two general cladograms differ significantly (we pass over the problem of what might constitute a significant difference). There seem to be four possibilities for an alternative: (1) there are two different cladistic parameters, each estimated by one general cladogram; (2) there is one cladistic parameter, estimated by one, or the other, or neither, of the general cladograms; (3) there is one cladistic parameter, estimated by the sameness (replication and combinability), if any, exhibited jointly by the two general cladograms; (4) there is no cladistic parameter, or parameters; and all pattern is artifact.

Possibility (1) seems purely ad hoc: "I still believe that there is one cladistic parameter (estimated by my first general cladogram), but now I believe that there is another." Possibility (2) also seems purely ad hoc: "I still believe that there is one cladistic parameter, but I have different estimates of it; I know not which to choose, but my first estimate might still be correct." Possibility (3) seems partly ad hoc, but is, nevertheless, interesting: "I still believe that there is one cladistic parameter, but I have revised my estimate of it; the revised estimate might prove more reliable." Possibility (4) seems trivial: "I no longer believe that there is a cladistic parameter, because my prediction was unfulfilled; and my hypothesis, falsified."

Possibility (3) seems unrejectable as an alternative, if there is some replication of components in the two general cladograms. The replicates, and components combinable with them, may be specified in a second-order general cladogram, i.e., a unique estimate of the cladistic parameter. Therefore, some replication between different general cladograms implies that possibility (3) is unrejectable, and that possibility (4) is merely false. We conclude that, even if the null hypothesis is falsified, the only alternatives are either partly ad hoc or trivial. Of the ad hoc possibilities, only (3) seems interesting.

We have asserted that possibility (4) seems trivial. It is a statement that replication of results (if any) is artifact, and noncombinability is an estimate of the real world. As such, it contradicts one of the premises of the falsification argument: namely, that the two general cladograms are, or should be, independently derived. If independence is assumed, and significant replication is the result, the result is *not* artifact. Significant artifact can mean *only* nonindependence in the derivation of the two general cladograms. Hence, possibility (4) is a misstatement. It purports to say something about the world, but in reality contradicts the premise of independence.

What then of the initial hypothesis: that a general cladogram is knowledge of the real world? Is it falsifiable? We conclude that it is not in a general sense, but that it is in a particular sense. In the general sense, all that a person can do is to search, but if one fails to find, the failure is no falsification. In the particular sense, a person may find one pattern rather than another.

What are the implications? The initial hypothesis can only be presumed. Once presumed, it may lead to particular null hypotheses that may be tested and, perhaps, falsified, and an alternative may be entertained. If this is the only implication, what is there in the way of a general theory of systematics? The answer seems clear: the theory of general cladograms. As here conceived, the theory includes phenetics, phyletics, gradistics, and any other theory that might be proposed, if its applications lead to results that might be rendered as hierarchically branching diagrams.

At present it is difficult to evaluate the completeness of the theory of general cladograms. Two areas of theoretical problems are evident: (1) levels of significance of difference between two or more general cladograms; (2) criteria of independence in derivation of general cladograms. We do not consider either area particularly important in a practical sense; such a judgment would presuppose that systematic data are necessarily so noisy in themselves that only mathematical treatment can isolate a nonrandom pattern. The history of systematics belies that presupposition. Systematic data are simply not in themselves necessarily noisy. So more practical significance may arise from a theoretical consideration of systematic data (characters). If some data sets are noisy, why so? Could it be that they include "characters" that are not general characters? Are there remedies? Could an explicit search for general characters suffice? Only time will tell.

EPILOGUE

A NOTE ON TERMINOLOGY

Persons may think it strange to use words beginning with "clado-" in a sense divorced from evolution and phylogeny. Their attention may be called to the definition of the term: "A combining form from Gr. *klados*, a shoot, branch, signifying branched or having branches." Persons may also doubt that cladistic analysis can be divorced from evolution and phylogeny. Their attention may be called to the numerous successful homology statements made by pre-evolutionary systematists, and to the fact that every homology statement implies one or more synapomorphy statements.

To state that a cladogram is a synapomorphy scheme invites the rejoinder that a cladogram must, therefore, be a phyletic concept. Not so, for by "synapomorphy" we mean "defining character" of an inclusive taxon. True, all defining characters, in the phyletic context, may be assumed to be evolutionary novelties. But making that assumption does not render it automatically true; nor does it change the characters, the observations on which the characters are based, or the structure of the branching diagram that expresses the general sense of the characters: i.e., that there exist certain inclusive taxa (components 1, 2, . . .) that have defining characters. That the taxa may be assumed to be "monophyletic" (derived from a single ancestral species) does not change the structure of the branching diagram, but merely adds the phyletic context of interpretation; hence the diagram becomes a phylogram. It is still atemporal in the sense that it is necessarily nonreticulate and that its branch points are mistakenly construed as speciation events. To equate "synapomorphy" and "defining character" invites the rejoinder that a branching diagram specifying only "poly-thetic" inclusive taxa is no cladogram, because no defining characters are present. Not so. Even if the inclusive taxa are assumed to lack defining characters (because none is yet known), that assumption does not change the fact that the branching diagram expresses the general sense of defining characters even in their absence: i.e., "if there were some defining characters, these are the inclusive taxa (components 1, 2, . . .) that the characters would define." In short, finding defining characters for each of the "polythetic" taxa would not change the structure of the branching diagram, but merely add replications of components.

A NOTE ON HISTORY

Rosa (1918) and Hennig (1950) may be considered among the many independent inventors of cladograms, at least in the phyletic context. Rosa's work is not widely known, and an excerpt of it is translated here, to illustrate the trend of his thought (even though Rosa's figure [Fig. 4.34] in this case is properly a tree, not a cladogram):

> . . . to the following scheme [Fig. 4.34], which represents the connections of affinity between the species of a group, such as they would be if the species were the result of dichotomous speciation. . . . Having before us 32 terminal species, represented by the black dots above, we would be able to make four groups (such as genera): A, B, C, D.
>
> It is clear that, even without paleontological knowledge of the connections, an adequate knowledge of the morphology of these species would suffice to indicate that genus B is more closely related to genus A than to genus C; and that, before grouping the 32 species into four genera, it would be necessary to group them

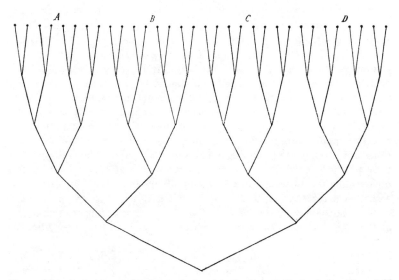

Figure 4.34. A cladogram (phylogram) showing the hypothetical relationships of 32 species of four groups (A, B, C, D). After Rosa (1918), figure on p. 138; also Rosa (1931), figure on p. 174.

into two "supergenera" or subfamilies: AB and CD. And within each genus it would be possible also to recognize subgenera and even smaller groups of more closely related species.

If this scheme corresponds to reality, one may conclude that the distinction between groups of equal taxonomic rank cannot be arbitrary; and, also, that the distinctions are not caused by gaps in the system, gaps produced by extinctions. Even in the absence of extinction, the distinctions would be quite clear. And while it might be arbitrary to consider group A a genus, rather than to consider group AB a genus, no good systematist would ever combine some species of group C with AB, and the other species of group C with D. (Rosa 1918:137–38, 1931:174–75, translated)

Mitchell (1901) and Hennig (1950) may be considered among the many independent inventors of the concepts of synapomorphy and of classification as a nested set of synapomorphies, at least in the phyletic context. Mitchell's work is not widely known, and an excerpt of it is presented here, to illustrate the trend of his thought:

In the description of the structure of an organ or anatomical part as it occurs in a large series of different forms, it becomes necessary from simple convenience to attempt some kind of valuation according to which the series of facts shall fall into definitely-named groups. When there is attempted the difficult passage from descriptive anatomy to morphology, it is necessary that the valuation and nomenclature should be in relation to the theory of descent with modification. I assume that birds were monophyletic in origin, and that the existing forms have branched out in diverging directions from the ancestral group. The members of this ancestral group, at the stage when they first might have been called birds, possessed an heritage of characters and tendencies, and these characters and tendencies have undergone modifications different in amount and nature in the different groups. The first business is to come to a decision as precise as possible as to the ground-plan, or archetype, the most ancestral condition of the structures under consideration. . . . I propose to call such a condition "arche-centric," implying that it represents a primitive, ancestral, or central condition, from which the conditions to be found in the other cases have diverged. It is obvious that the possession by two or more groups of birds of a character in its archecentric form cannot be an indication in itself that these groups are more closely related to one another than they are to groups possessing the character in another form; for if the diagnosis of archecentricity be correct, the condition has been present in all birds, and may be retained by any. . . .

When the ancestral condition is modified, it may be regarded as having moved outwards along some radius from the archecentric position. Such modified conditions I propose to call "apocentric." Again, it must be obvious that the mere apocentricity of a character can be no guide to the affinities of its possessor. . . . Before deciding as to the value of eutaxy [a putatively apocentric character of bird wings] in a natural classification, it would be necessary to decide whether the modification of the archecentric condition were a simple change that we might expect to occur in independent cases, or if it involved intricate and precisely combined anatomical changes that we could not expect to occur twice independently. In fact, having come to the conclusion that a character is apocentric, we must pass on to consideration of the problem whether or no the apocentricity be *uniradial* or *multiradial* [synapomorphic or convergent]. . . . Similar and common multiradial apocentricities, from which no direct argument of kinship is to be drawn, are to be found in conditions depending on the degeneration of a structure. . . . In the case of the alimentary canal, it is easy to set apart certain modifications as directly adaptive, and as therefore of no value when the character of an organ is being considered as an indication of the natural affinities of its possessor. For the apocentric modifications in question have been produced in different mammals as well as in different birds, and hence in birds these modifications must be multiradial and no indication of relationship. . . .

A complex apocentric modification of a kind that we cannot well expect to be repeated independently, and that may be designated as uniradial, must be the most certain guide to affinity. It happens frequently that such a modification forms a new centre around which new diverging modifications are produced. Such a centre I propose to call a "Metacentre," borrowing a convenient term from physics. It is obvious that the condition of a character, archecentric so far as the whole group of birds is concerned, is metacentric with regard to the common stock of birds and reptiles, and that the transformation of an apocentric character into a metacentre is simply an event in the general process of divergent evolution. I justify the nomenclature which I am proposing largely because it brings the valuation and classification of characters into line with our conception of the general processes of evolution. . . .

I trust that the ideas underlying this attempt at the valuation and nomenclature of characters, so far from being novel, are merely a codification of criteria in common employment among naturalists. I find, however, that such a codification was necessary when I tried to arrange systematically the modifications of the characters with which this memoir deals. So far as I have used them in my own work, I have found them illuminating, and I offer them in the conviction that the rigorous discipline which their use entails would prove of general utility. (Mitchell 1901: 178–82)

That anyone in particular could be recognized as the inventor of cladistic analysis is moot, for all of its formal properties are not yet clearly established. It is doubtful that either Mitchell, Rosa, or Hennig is the inventor of "cladistics" as a philosophy of classification, for all, or almost all, systematists seem to be, and seem ever to have been, "cladists." At least cladistic elements are discernible in systematic work of whatever age. However, Hennig's diagrams, as portrayed in his books and in the scientific papers of various European systematists (Brundin,

Schlee et al.), did provide a focus of general interest and an impetus to apply Hennig's type of diagram to the systematic study of various groups of organisms, both fossil and recent. And there is little doubt that Hennig's diagrams in particular, because of the widespread attention paid to them, have been a progressive factor beneficial to systematics generally. But the argument over "cladistics" has not centered on the diagram as much as on the "problem" of classification. In reality, no real "problem" ever existed. The argument developed, so it now seems, because of misapprehension on the part of the critics, due to their inability to understand the statements of the exponents, which were not, it is true, framed in their widest generality. The exponents can, perhaps, be forgiven for not realizing at once the full generality of cladistics and for not stating it clearly. It may prove more difficult to forgive certain of the critics—those who repeatedly failed to apprehend any generality at all.

A NOTE ON PHILOSOPHY

Since the advent of the so-called New Systematics, it has become popular to deprecate as "essentialistic" or "typological" the notions that species (and hence groups of them) have defining characters, and that it is the business of systematics to find them (see, for example, Mayr 1957; Hull 1965; Ghiselin 1969). The rationale for this deprecation seems to be that if evolution occurs, the characters of species (and hence groups) may change in the future; therefore, species and groups cannot be permanently characterized by means of a single character or set of characters such that the character or set is necessary and sufficient for membership in the species or group. The argument seems to rest on the misleading use of character states: it assumes that when a species is modified, and acquires a new apomorphic character (state), it no longer has, or is no longer recognizable as having, the original plesiomorphic character (state). In other words, according to this argument, we cannot use characters (such as fins) to define groups (such as Vertebrata), because some members of those groups (such as tetrapods) may acquire apomorphies (such as limbs). If one accepts the validity of ontogeny or outgroup comparison (i.e., parsimony) or any other possible test of hypotheses about character transformation, the argument is obviated. In this sense, systematists always have been, are, will be, and should be, typologists.

TIME

5

ONTOGENY, PHYLOGENY, PALEONTOLOGY, AND THE BIOGENETIC LAW

THE THREEFOLD PARALLELISM

It was argued in chapter 1 that the use of ontogenetic information is a direct, rather than indirect, technique of classification (because it does not depend on a prior, higher-level hypothesis of relationships), and can be justified by reference to parsimony rather than recapitulation (the theory that the sequence of developmental stages in ontogeny represents a vastly accelerated version of the sequence of adult forms in phylogeny). What, then, is the significance of Haeckel's so-called biogenetic law, that ontogeny recapitulates phylogeny? Perhaps Haeckel's principle might benefit from restructuring, as illustrated by the following example involving eye migration during the ontogeny of flatfishes.

Suppose there are two species of fishes, A and B. Species A has an eye on each side of the head (character x); species B has both eyes on the same side (character y). Which character is the more primitive? Without further information, no rational answer would be possible. But suppose we study the ontogeny of both species, and find that embryos of both species have character x, and that during the subsequent development of species B character x is transformed into character y. Informed by this ontogenetic character transformation, we may answer the original question: character x is more primitive than y.

In an evolutionary context, "primitive" means "ancestral." But we would not have observed character x to be ancestral in an evolutionary sense. We would have observed only that one character (x) is more general than another (y), in the sense that (x) occurs in two species and (y) in only one of them. Hence our answer, "character x is more primitive than y," is a hypothesis about the course of evolution—a hypothesis based on the observations that one character (x) is more general than another (y), and that during ontogeny the more general character (x) transforms into the other (y).

The biogenetic law may, therefore, be restated as follows: given an ontogenetic character transformation, from a character observed to be more general to a character observed to be less general, the more general character is primitive and the less general advanced.* Whether the biogenetic law has ever been stated in this way or not, the restated law seems open to falsification in at least two ways:

(I) If we assume to know the truth of evolution, and more particularly that character y really is more primitive than x, that knowledge would directly falsify the law.

(II) Suppose that we lack the truth of evolution, but we investigate two other species of fishes, C and D: species C with eyes on each side (character x′) and species D with eyes on the same side (character y′). Suppose, further, that we study the ontogeny of both species and find that species C undergoes a character transformation from character y′ to x′, and that on this basis we hypothesize that character y′ is primitive relative to x′.

Comparing the two examples, we note that according to one (species A and B), character x is primitive relative to y; according to the other (species C and D), y′ is primitive relative to x′. There seem to be only two implications:

(1) More than two characters are involved, such that
$$x \neq x' \text{ or } y \neq y'.$$
(2) Or the biogenetic law is falsified.

Although there seems to be no logical reason why the law could not be falsified, a preliminary search for a falsification has revealed none. For

---

*For ontogeny, the concept of relative generality is strictly an observational matter, as discussed below. The concept is exemplified in another statement of the biogenetic law (Crowson 1970:187): "when the immature stages of one kind of animal show resemblances to adults of some other kind, this should be taken as prima facie evidence that the first kind of animal has descended from ancestors resembling the second. The reverse relationship, implying neotenic evolution, may also be possible but is in general much less probable." Neoteny is also discussed below.

example, there seems to be no organism that, during ontogeny, exhibits the reverse eye migration, from an asymmetrical to a symmetrical arrangement. Thus, it is possible that the law is generally valid, at least as regards falsifications of type (II).

Falsifications of type (I) are another matter. The only folk who claim to know the truth of any evolution relevant in the present context are paleontologists. Not all paleontologists do so, of course, but there are enough of them so that discussion of the problem tends to become heated. In its most general form, the paleontological argument holds that the fossil record shows the course of evolution because it shows actual ancestor-descendant sequences (actual phylogenies). This general form of the argument is simply fallacious; stratigraphic sequence alone cannot indicate that two fossils belong to the same lineage (if it could, we might have to conclude that a fossil mammoth found only in one stratum is ancestral to a fossil cockroach found only in the next). Fossils must be ordered on the basis of systematic hypotheses, and since those hypotheses may always be incorrect, fossils so ordered cannot be said to show the truth, or the true history, of evolution. The notion that phylogeny can be read directly from the rocks is superstition and nothing more.

In its more specific form, the paleontological argument holds that the fossil record shows the course of evolution because it shows actual character transformations (actual character phylogenies); it postulates that, given two different but homologous characters, that character known to be geologically older is the more primitive. There is some apparent justification for the belief that this form of the argument has widespread support, for there are many statements that, read in an uncritical way, may be cited in support of such a belief. For example:

(1) "Its [paleontology's] methodology is the same as that of comparative morphology in general, and the added factor of time, although an invaluable aid in the determination of the direction of a morphological specialization, is more than offset by the disadvantages of having but one structural system, the skeleton, available for study" (Zangerl 1948:358–359).

(2) "Since we do not know from direct observation the direction in which a transformation . . . took place, we are dependent on accessory criteria here too. These are: (a) Criterion of geological character precedence. If in a monophyletic group a particular

character condition occurs only in older fossils, and another only in younger fossils, then obviously the former is the plesiomorphous and the latter the apomorphous condition of a morphocline . . ." (Hennig 1966:95).

It is clear, however, that Zangerl views the paleontological argument as subservient to his morphological concepts of "structural plan" and "morphotype," and Hennig, as subservient to his systematic concept of "monophyletic group." As viewed by these writers, the argument becomes an indirect technique of character phylogeny—one that presupposes a higher level phylogeny (or its morphological equivalent), and is, therefore, a technique at the same level as outgroup comparison.

There is no doubt that specific hypotheses about character transformation arise from many sources. There has been much argument about the relative value of different hypotheses, with particular reference to their source, as if their relative values could be ascertained by evaluation of their sources. It is an implication of the paleontological argument, for example, that a hypothesis based on the fossil record has greater validity than any other.

There is an alternative approach, namely a consideration of falsification. According to this approach, the source of a hypothesis is irrelevant. What is relevant is its potential falsifiability, and the source of its potential falsifiers. From this point of view, the paleontological argument means that the fossil record contains the only, or at least the best, potential falsifiers of a given hypothesis of character phylogeny. The argument is as follows: given a hypothesis that character x is primitive relative to y, the hypothesis is falsified if character y is known in fossils older than any fossils known to possess character x.

Of concern here is not only the potential falsification of some particular hypothesis about character phylogeny, but the potential falsification of the paleontological argument itself. Consider an example: the hypothesis that the asymmetrical condition of the eyes (character y) is primitive relative to the symmetrical condition (character x). Ontogenetic character transformations might be examined in search of a falsifier, namely a transformation from x′ to y′. Numerous falsifiers might be found. Alternatively, the fossil record might be searched for the earliest fossil to exhibit eyes. Likewise, a falsifier, in the form of a fossil with symmetrical eyes, might be found. If so, the hypothesis (that y is primitive to x) would be falsified by both ontogeny and paleontology.

Consider a subsequent example: the alternate hypothesis that the symmetrical condition of the eyes (x) is primitive relative to the asymmetrical (y). Ontogenetic character transformations might again be searched for a falsifier, namely a transformation from y' to x'. Suppose one were found. Comparing the two examples, we note that, as before, there are two implications:

(1) $x \neq x'$ or $y \neq y'$.
(2) Or the biogenetic law is falsified.

Alternatively, the fossil record might be searched for an earlier fossil that exhibits eyes. Suppose one were found that has asymmetrical eyes (y'''). There are three implications:

(1) $y \neq y'''$.
(2) Or the fossil record was not as complete as previously believed.
(3) Or the paleontological argument is falsified.

Considering the biogenetic law (the ontogenetic argument) and the paleontological argument as direct techniques of character phylogeny exposes both to falsification. As stated above, we know of no falsifier of the biogenetic law, but a thorough search of the literature pertaining to ontogenetic character transformations has not been made. Within the history of paleontology, there have been numerous apparent falsifiers of the paleontological argument; these have usually been rendered impotent as falsifiers by the ad hoc alternative that the fossil record was not as complete as previously believed. For these reasons, the biogenetic law may be a valid direct technique of character phylogeny, and the paleontological argument may be fallacious as a direct technique of character phylogeny. The paleontological argument seems fallacious because it is accepted in principle as nonfalsifiable (it is always protected from falsification by an ad hoc alternative that is always, and obviously, true). For the same reasons, the paleontological argument may be fallacious even as an indirect technique, that is, when considered subservient to morphological concepts such as structural plan or morphotype, or to systematic concepts such as monophyly.

Much paleontological effort is directed toward discovery of older and older fossils; as a result, they are found with some regularity. Their discovery is often announced with fanfare, and sometimes with claims that the newly discovered fossils subvert a previous evolutionary interpretation. Amidst the fanfare associated with discovery of such fossils, it is seldom realized that what is subverted is not only a previous evolutionary interpretation, but a previous application of the paleontological argument (the application that led to the previous evolutionary interpretation). The realization is seldom attained because of the ad hoc alternative, that the fossil record was not as complete as previously believed. The ad hoc alternative admittedly has appeal: how could its truth be doubted with the newly discovered fossils in hand?

Both the biogenetic law (the ontogenetic argument) and the paleontological argument reflect an expectation that nature is orderly. And, according to modern belief, the order has been imposed by a historical process rather than the will of a creator. To the extent that falsifiers might be found for either or both of these arguments, nature would appear unordered. That there are no, or few, apparent falsifiers of the biogenetic law implies that development, so far as known, is very orderly. This implication is no surprise; no biologist has recently claimed otherwise. That there are numerous apparent falsifiers of the paleontological argument implies that the fossil record, so far as known, is not very orderly. This implication also comes as no surprise: paleontologists have spoken clearly, with one voice as it were, about the "imperfections of the fossil record." Yet the fossil record has been invoked also as if it were revealed truth—not only about order in nature, but also about the truth of evolution, phylogeny, and character phylogeny.

It would be less than fair not to point out that paleontologists have almost always recognized limitations to the paleontological argument, but the hope has been expressed that the limitations might be transcended by adequate sampling. Consider, for example, an early statement of the argument by Bernard:

If we have sufficient material at our disposal, and if, on the other side, the chronologic order of the deposits is well established, we ought to be able to follow the transformations of all the types through the epochs, to determine whether any one form is derived by progression or regression from a more ancient one, to discover at what epoch and by what process the distinct groups, between which passage forms are wanting in living nature, came into existence. It will be understood that when the question concerns the establishing of relationships between genera and families, and especially between species of the same group, such a study can be entered on with some chance of success, when the materials for comparison are very abundant, when they come from numerous localities and are in a good state of preservation. (1895:183–84)

Bernard refers to "types," suggesting that he, like Zangerl, viewed the paleontological argument (Bernard's "principle of continuity") as subservient to comparative anatomy, and there are other reasons to suppose that he did:

The difficulty is sometimes even still greater, and the chronologic order of appearance is in complete discordance with what we might be permitted to expect from the method of Comparative Anatomy applied to Paleontology. If the ontogenic evolution is unknown, and if we can not bring in evidence any fact of regression, we are obliged to reserve our conclusion for the epoch when new evidence shall permit us to elucidate the question; until then we should formulate hypotheses with great reserve. (1895:185)

Bernard seems also to have been well aware of falsifiers of the paleontological argument:

It can be seen how dangerous it is to attach an exclusive importance to the order of appearance; every theory which in any way rests on these data, when the question concerns the primordial forms of large groups, incurs the risk of being ere long contradicted by some unexpected discovery. (*Ibid.*)

Bernard's statement of principles seems fairly to represent the historical mainstream of paleontological thought, namely that paleontology is basically comparative anatomy applied to fossils, but also imbued, perhaps, with the hope that truly adequate sampling might transcend certain limitations imposed by recent material alone.

What are the limitations that adequate sampling might transcend? There are at least two possibilities:

(1) Particular limitations imposed by the structural plans or monophyletic groups exemplified by recent organisms;
(2) General limitations imposed by the concepts of structural plan and monophyletic group, or their equivalents.

The structural plans and monophyletic groups exemplified by recent organisms have not prevented paleontologists from describing additional plans and groups based only on fossils. Examples among fishes include Heterostraci, Osteostraci, Thelodonti, Antiarchi, Arthrodira, Acanthodii, Osteolepiformes, and so on. The discovery of these plans and groups seems to have multiplied the problems of systematics, rather

than reduced them. There are more plans and groups of obscure
relationships, rather than fewer. And what of the plans and groups
exemplified by recent organisms? Have any of them disappeared as a
result of paleontological discoveries? We doubt that any significant
number of them has disappeared, even with the discovery of truly
intermediate fossils. From time to time, the plans and groups have been
redefined, often more narrowly and more carefully, and they have been
subdivided in a more complicated manner to allow for the inclusion of
fossils with more recent organisms in a single classification. Hence,
fossils generally complicate, rather than simplify, the systematic
landscape. If so, limitations of type (1) seem not in general to be
transcended by more complete sampling of the fossil record.

Limitations of type (2) are another matter. Transcending them
implies paleontology without morphological or systematic principles.
Such paleontology is implied, to one degree or another, by some persons who claim that theirs are truly adequate samples under rigid
stratigraphic control; recent examples have been furnished by Gould
(1969) and Gingerich (1976). That there is in some sense a paleontological argument or method independent of, or nonsubservient to,
comparative anatomy is implied, for example, by Gould (1973):

> Haeckel states . . . that he derived his evidence for phylogeny from three
> sources that Agassiz, in a pre-evolutionary context, has dubbed the "threefold
> parallelism"—ontogeny, comparative anatomy (sequences of living adults) and
> paleontology (the geological record of fossils). . . . I support . . . the pluralism
> of Haeckel. (pp. 322–23)

Are there, as Gould implies, three direct techniques: ontogeny (the
biogenetic law), comparative anatomy (outgroup comparison), and
paleontology (the paleontological argument)? Consider an example: the
hypothesis (based on an observed ontogenetic transformation from x to
y, several outgroups displaying x'', and some early fossils displaying x''';
in short, a threefold parallelism) that symmetrical eyes (x) are primitive
relative to asymmetrical eyes (y). An ontogenetic falsifier would be a
transformation from y' to x', with three implications:

   (I) $x \neq x'$ or $y \neq y'$.
  (II) Or the biogenetic law, the principle of outgroup comparison, and
      the paleontological argument are all falsified.
(III) Or the biogenetic law is falsified, but the principle of outgroup

comparison or the paleontological argument, or both, may be protected from falsification by the adoption of one or more of the ad hoc items (II) below.

An anatomical falsifier would be the discovery of one or more outgroups that display y'', with four implications:

(I) $y \neq y''$.

(II) Or the new outgroups are not outgroups, but ingroups.

(III) Or the biogenetic law, the principle of outgroup comparison, and the paleontological argument are all falsified.

(IV) Or the biogenetic law and the principle of outgroup comparison are falsified, but the paleontological argument may be protected from falsification by the adoption of the ad hoc item (II) below.

A paleontological falsifier would be a newly discovered earliest fossil with y''', with four implications:

(I) $y \neq y'''$.

(II) Or the fossil record was not as complete as previously believed.

(III) Or the biogenetic law, the principle of outgroup comparison, and the paleontological argument are all falsified.

(IV) Or the biogenetic law and the paleontological argument are falsified, but the principle of outgroup comparison may be protected from falsification by the adoption of the ad hoc item (II) above.

There are ad hoc possibilities for each type of falsification. If the ad hoc possibilities are disregarded, any falsifier suffices for all three principles together (ontogeny, anatomy, paleontology), if they are united in the concept of threefold parallelism. The adoption of Agassiz's concept, therefore, results in a predictable situation: falsification of any one of the principles falsifies the concept of threefold parallelism. The adoption of the concept of threefold parallelism can have only one of three outcomes:

(1) A refusal to accept falsification when it arrives.

(2) Acceptance of falsification of the concept of threefold parallelism, with no alternative principle in sight.

(3) Acceptance of falsification of the concept of threefold parallelism, but adoption of one or another ad hoc item in order to evade falsification of ontogeny, anatomy, or paleontology.

Has the concept of threefold parallelism no redeeming qualities? The concept seems to involve a strategy of confirmation rather than falsification. Its sense, after all, is that when ontogeny, anatomy, and paleontology all agree, the result is probably true. But agreement is not the same as truth; they could all agree, and their result could still be false. Suppose it were; how might the truth be learned? Unless some other principle could be invoked, a principle other than ontogeny, anatomy, and paleontology, falsification seems impossible. Given agreement among the three parallelisms, their result, which could be either true or false, seems unfalsifiable.

We do not wish to belabor the inadequacies of the concept of threefold parallelism; it has not had much popularity in recent years. No doubt, it has few, if any, adherents. But perhaps the concept should be revived. At present, there doesn't seem to be much hope for it in the context of falsification: it would be immediately falsified without ad hoc protection.

The most elementary ad hoc items are (I), rejections, or qualifications, of homology statements. Rejections and qualifications are possible for all three types of falsification. Adoption of item (I) in a particular case may evade falsification by invoking the concept of convergence. This concept is invoked with some frequency in the anatomical and paleontological types of falsification, but rarely, if ever at all, in the ontogenetic. The reason for the difference in frequency is, perhaps, a reflection of the frequency of appearance of apparent ontogenetic, anatomical, and paleontological falsifiers. The cause of the difference in frequency seems obscure, but perhaps is related to, or even caused by, the fact that mode of development is itself one of the most important criteria of homology. If real, the difference might be significant for an evaluation of the concept of "threefold parallelism," for the difference would indicate that two of the three parallelisms (anatomy and paleontology) are subject to ad hoc protection from falsification at a fundamental level. If true, the difference might be considered a falsification of the concept of threefold parallelism, under the assumption that all three parallelisms are, or should be, equally exposed to, rather than differently protected from, potential falsification.

The role of ad hoc items (I) is complicated by differences in existing concepts of homology. For example, parallel characters are sometimes considered homologous. Hence, for the anatomical type of falsification, there is possible an accessory ad hoc item, namely that the outgroup (the falsifier) really is an outgroup and that y really is homologous with y'' (y = y''), but that the homology is a parallel homology. Such parallels are sometimes referred to as "homoplasy" (Lankester 1870:39).

For the paleontological type of falsification, too, there is possible an accessory ad hoc item, namely that character y''' of the earliest fossil (the falsifier) really is homologous with character y, but that the homology is a parallel homology—an instance of homoplasy. The implication is that the "earliest fossil" is not really a relevant "earliest fossil."

It would seem that paleontology by itself can produce no criterion of relevance, except that of order of appearance in the fossil record. In order to discriminate relevant and irrelevant "earliest fossils," some accessory criterion is required. The accessory criterion usually adopted comes either from comparative anatomy or from systematics. The irrelevant "earliest fossil" is that fossil that belongs to some other structural plan or monophyletic group. How essential is the accessory criterion for the practice of paleontology? According to the traditional body of doctrine, as exemplified by Bernard, the accessory criterion is absolutely essential, for paleontology, by definition as it were, is comparative anatomy applied to fossils.

With respect to accessory ad hoc items, the three types of falsification may be compared. It is difficult to imagine any accessory ad hoc item for ontogeny. Consider, for example, the statement: x = x' and y = y', x really transforms to y, and in the falsifier y' really transforms to x', but either x and x', or y and y', or both, are parallel characters—an instance of homoplasy. This ad hoc alternative is possible conceptually, but it comes close to a nonsense statement: if x transforms to y, and y' transforms to x', there would be reason, according to the ontogenetic criterion of homology, to conclude either that $x \neq x'$, or that $y \neq y'$. In any case, we have never encountered such a statement, and doubt that it has ever been used to evade falsification of the biogenetic law.

That there are accessory ad hoc items for the anatomical and paleontological types suggests that two of the "threefold parallelisms" are protected from falsification at an additional fundamental level: once by convergence and again by homoplasy. If so, the concept of threefold parallelism might be considered falsified on two counts.

It seems that convergence and homoplasy must be accepted as falsifiers if the concept of threefold parallelism, or indeed any of the individual parallelisms (ontogeny, anatomy, paleontology), is to be exposed to falsification at all. The reason is simple: all falsifiers can be dismissed by invoking either convergence or homoplasy. Invocations of these concepts can be understood, therefore, in a general way as evasions of falsification. Because they may be invoked in any of the three types of falsification (ontogeny, anatomy, paleontology), their relative frequency becomes important, such that alternative arguments may be judged on the degree to which they minimize convergence and homoplasy, that is, on the degree to which they do not incorporate protection from falsification. It would appear, therefore, that the principle of parsimony can be understood in the context of falsification: an argument is parsimonious to the extent that it does not incorporate ad hoc items as protection against falsification. From this standpoint, the ontogenetic argument begins to emerge as the most parsimonious in the sense that it is least protected from falsification.

If items (I) and their accessory possibilities are disregarded, the three types may be compared with reference to their remaining ad hoc possibilities. For the ontogenetic type, there is no ad hoc item that seems relevant. For the anatomical and paleontological types, there is one ad hoc item for each, item (II).

For the anatomical type, the ad hoc item is a revised judgment of the systematic relationships of the outgroup (as an ingroup). For the paleontological type, the ad hoc item is a revised judgment of the previous adequacy of the fossil record, with respect to the earliest representative of some structural plan or monophyletic group. Such ad hoc judgments are commonly made, perhaps for one simple reason: to evade falsification of the principle of outgroup comparison or the paleontological argument. The former evasion presumably results in an advance in systematics—a more orderly arrangement. The latter evasion seems to have no inherent value, beyond the idea that, although the fossil record was not as complete as previously believed, now it is complete, or is at least more so.

The ontogenetic type seems peculiar, and different from the other two, in lacking a corresponding ad hoc item. If so, the difference might be considered a third falsifier of the concept of threefold parallelism.

These three general ad hoc possibilities (convergence, homoplasy, and revised judgments of systematic relationships or completeness of the fossil record) seem to be real differences between direct and indirect techniques of character phylogeny. That these differences emerge upon analysis of an example, based on the assumption of threefold parallelism, is surprising. Jointly, they seem to falsify the concept of threefold parallelism.

But the analysis also leads to something positive: a restatement of the biogenetic law in falsifiable form, with certain implications. One implication is that systematics and comparative anatomy (applied to fossils, too) are possible only to the extent that ontogeny is orderly. Another implication is that the concept of evolution is an extrapolation, or interpretation, of the orderliness of ontogeny.

If these conclusions are granted, what are their implications for the current debate in comparative biology? In a general sense, the nature of systematics is revealed by the interrelations of ontogeny, anatomy, and paleontology. If the decisive factor is the orderliness of ontogeny, and if comparative anatomy and paleontology are secondary disciplines, and if the theory of evolution itself is an extrapolation of ontogeny, well, then . . . what is left? Only one factor comes to mind, and that is space—biogeography.

This state of affairs is not necessarily the happiest possible. For ontogeny turns out to be too important, in the sense that evolutionary interpretations derived from the biogenetic law are nonfalsifiable, except with reference to ontogeny. If anatomy and paleontology merely order their data to conform with the framework provided by ontogeny and its orderliness, they do not have the possibility to falsify that framework. The framework gives direction, and anatomy and paleontology blindly follow. It would be better for science if the concept of threefold parallelism were valid. To some degree, perhaps, it can be rendered valid through critical reappraisal, so as to allow for potential falsification. But there will be no critical reappraisal if the concept is believed valid when in fact it is not—despite the testimony of those who might speak on its behalf.

What are the possibilities for reappraisal? Again, thoughts turn to space—to biogeography. It is space, after all, in which the factor form appears, and through which the factor time extends, so to speak. Hence, an alternative threefold parallelism can be imagined: space + time + form (Croizat 1964).

NEOTENY

During a period of history, the biogenetic law came under attack from various persons, among them Garstang (1922) and De Beer (1951), who argued, in effect, that neoteny is an apparent falsifier of the biogenetic law as conceived by Haeckel and others. With respect to the nature of evolutionary processes, there may be justification for their argument. But with respect to character phylogeny, is neoteny a falsifier? And if so, what does it falsify? The biogenetic law? Or, perhaps, anatomy and paleontology, too? Consider an example: species A with character x and species B with character y. Is it possible to imagine some hypothetical examples wherein neoteny would act as a falsifier? Here are three:

(1) Ontogeny: species B displays an ontogenetic character transformation from x' to y, implying that x is primitive relative to y. If it is assumed that $x = x'$, and that character x is a neotenic character of species A, then the biogenetic law is falsified.

(2) Anatomy: there is an outgroup with character x'', implying that x is primitive relative to y. If it is assumed that $x = x''$, and that character x is a neotenic character of species A, then character x'' is a neotenic character of the outgroup, and the principle of outgroup comparison is falsified.

(3) Paleontology: there is an earliest fossil with character x''', implying that character x is primitive relative to y. If it is assumed that $x = x'''$, and that character x is a neotenic character of species A, then x''' is a neotenic character of the earliest fossil, and the paleontological argument is falsified.

There are, of course, ad hoc alternatives to falsification for anatomy and paleontology. For anatomy, the ad hoc alternative is that the outgroup is an ingroup. For paleontology, the ad hoc alternative is that the fossil record is not as complete as previously believed. Acceptance of these ad hoc alternatives protects anatomy and paleontology from falsification in a case of neoteny. For ontogeny, no ad hoc protection seems available. It is therefore possible to claim that neoteny selectively falsifies the biogenetic law, but only by invoking ad hoc alternatives to protect anatomy and paleontology.

Disregarding ad hoc alternatives, and assuming we know the truth (that character x is a neotenic character of species A), we perceive that neoteny falsifies the biogenetic law, the principle of outgroup comparison, and the paleontological argument. Neoteny, therefore, would seem to falsify a more general principle: that of character phylogeny itself. This circumstance is paradoxical: if neoteny falsifies character phylogeny, how is knowledge of neoteny achieved? We propose no answer to this question, except to note that, in the well-known case of the axolotl, the knowledge was achieved by observing an ontogenetic transformation: metamorphosis of the axolotl into the adult salamander. In this case, the missing adult characters were finally discovered. Discovering them established that they previously had been missing. Could their absence have otherwise been detected?

A more profitable inquiry might be focused on the principle of character phylogeny—that general principle falsified by neoteny. In principle, what is character phylogeny? Basic to it, seemingly, is the idea that, given two homologous characters (x in one species and y in another), one character is more advanced than the other. Is there some other idea implied by character phylogeny? Presumably there is, namely a principle that relates observation and inference. Our observations indicate that different characters sometimes have different generality. Given characters of different generality, the inference is that the more general character is primitive in an evolutionary sense, and the less general character advanced. What is meant, then, by different generality?

(1) Different ontogenetic generality: the occurrence of character x (more general) in species A and B, in contrast to the occurrence of character y (less general) only in species B.

(2) Different anatomical generality: the occurrence of character x (more general) in species A and B and in an outgroup (in two structural plans or monophyletic groups), in contrast to the occurrence of character y (less general) only in species B (in only one of the two structural plans or monophyletic groups).

(3) Different paleontological generality: the occurrence of character x (more general) in species A and B and in the earliest fossil, in contrast to the occurrence of character y (less general) only in

species B. In this case, generality means no more than relative age: x appears to be older than y. Paleontology seems to be an exception to the concept that relative generality per se is an idea essential to character phylogeny. Perhaps the reason that paleontology is exceptional is that the earliest fossil is assumed to be ancestral, or at least the best available estimate of an ancestor (its characters are primitive by definition).*

If allowance is made for the exceptional nature of paleontology, what is the relation between characters of different generality, on the one hand, and, on the other, ontogeny, anatomy, paleontology, and neoteny? Consider an example: species A with character x and species B with character y, an ontogenetic character transformation from x' to y in species B, an outgroup with x'', and an earliest fossil with x'''. What are the differences in generality?

	Greater generality	Lesser generality
Ontogeny	x (if x = x')	y
Anatomy	x (if x = x'')	y
Paleontology	x (if x = x''')	y
Neoteny	x (if x = x', etc.)	y

There is one and the same difference in generality of the characters as considered in the contexts of ontogeny, anatomy, and paleontology. Even if character x were assumed to be a neotenic character of species A, the same difference in generality would be implied. There is another difference, however, that appears when the inferences of character phylogeny are compared with the assumption of neoteny:

---

*Paleontology is differently construed in the traditional body of doctrine—as comparative anatomy applied to fossils. If so, there are relevant and irrelevant earliest fossils, as determined with reference to a concept of structural plan or monophyletic group. If the relevant earliest fossil is considered as that earliest fossil that belongs to some other plan or group, then the principle of generality may be applied in paleontology as in anatomy. Yet, considered in another context (convergence and homoplasy, see above), the relevant earliest fossil is that earliest fossil that belongs to the same, not another, structural plan (or group): "organisms can be profitably compared morphologically only within the scope of their common structural plan" (Zangerl 1948:355). This seeming contradiction disappears if one conceives of a hierarchical arrangement of more or less inclusive structural plans, such that comparisons may be made within an inclusive plan (or group), but also from one subplan (or subgroup) to another.

	Primitive	Advanced
Ontogeny	x	y
Anatomy	x	y
Paleontology	x	y
Neoteny	x ⟶ y	no y

If neoteny is assumed, the transformation from x to y is primitive, not character x alone; the loss of character y is the advanced character. This situation is problematical, since the absence of a character is not a character.

If neoteny is to serve as a falsifier, how is knowledge of neoteny achieved? If, as in the case of the axolotl, it is achieved by observation in species A of an ontogenetic character transformation from x to y, the new observation changes the assessment of generality:

	Generality		
	Greater	Lesser	Same
Ontogeny	–	–	x and y
Anatomy	x	y	
Paleontology	x	y	

The new observation also changes the inferences of character phylogeny:

	Primitive	Advanced
Ontogeny	–	–
Anatomy	x	y
Paleontology	x	y
Neoteny	x ⟶ y	no y

With the new observation, in species A, of an ontogenetic character transformation from x to y, neoteny no longer falsifies ontogeny, at least with respect to species A and B. There remain, of course, the outgroup and the earliest fossil (with x″ and x‴). They, too, might be assumed to be neotenic; under this assumption neoteny would be a falsifier. But if the transformation from x to y were discovered in them, too, so as to establish neoteny for both, neoteny would no longer falsify anything:

	Generality		
	Greater	Lesser	Same
Ontogeny	–	–	x and y
Anatomy	–	–	x and y
Paleontology	–	–	x and y

Thus neoteny is a falsifier only when neoteny is assumed to be true in the absence of an ontogenetic transformation—in the absence of the best evidence that neotenic evolution has occurred.

This conclusion will be explored in a final series of considerations about the possible conflicts that might arise in character phylogeny when ontogeny, anatomy, and paleontology give different results. A common example will be used for most comparisons: species A with character x and species B with character y, an ontogenetic character transformation in species B from character x' to y, an outgroup (with either x'' or y''), and an earliest fossil (with either x''' or y'''). The comparisons are summarized in table 5.1, which shows the results of seven comparisons, of which the first six are variants of the common example.

In comparison 1, the ontogenetic inference is that character x is primitive relative to y; the anatomical inference (if the outgroup has character y'') is that character y is primitive relative to x. In comparison 2, the ontogenetic inference is that x is primitive relative to y; the paleontological (if the earliest fossil has character y'''), that y is primitive relative to x. In comparison 3, the anatomical inference (if the outgroup has character x'') is, or would be, the same as the ontogenetic, that x is primitive relative to y; the paleontological (if the earliest fossil has character y'''), that y is primitive relative to x. In comparison 4, the paleontological inference (if the earliest fossil has character x''') is, or would be, the same as the ontogenetic, that x is primitive relative to y; the anatomical (if the outgroup has character y''), that y is primitive relative to x. In comparison 5, the ontogenetic inference is that x is primitive relative to y; the anatomical (if the outgroup has character y''), that y is primitive relative to x; the paleontological (if the earliest fossil has y'''), that y is primitive relative to x. In comparison 6, the ontogenetic inference is that x is primitive relative to y; the anatomical (if the outgroup has character x''), that x is primitive relative to y; the paleontological (if the earliest fossil has character x'''), that x is primitive relative to y.

Of interest are the techniques falsified and the apparent falsifiers. Identification of them is possible only with reference to ontogeny. The term "apparent falsifiers" is used because the real falsifiers are the assumptions about neoteny. Under the assumption of no neoteny,

Table 5.1. Seven Comparisons Between Different Techniques of Character Phylogeny

Comparisons	Generality		Element Falsified		Apparent Falsifier	
	Greater	Lesser	No Neoteny	Neoteny	No Neoteny	Neoteny
1. Ontogeny	x	y	Anatomy	Ontogeny	Ontogeny	Anatomy
1. Anatomy	y	x				
2. Ontogeny	x	y	Paleontology	Ontogeny	Ontogeny	Paleontology
2. Paleontology	y	x				
3. Anatomy (+ Ontogeny)	x	y	Paleontology	Anatomy (+ Ontogeny)	Anatomy (+ Ontogeny)	Paleontology
3. Paleontology	y	x				
4. Paleontology (+ Ontogeny)	x	y	Anatomy	Paleontology (+ Ontogeny)	Paleontology (+ Ontogeny)	Anatomy
4. Anatomy	y	y				
5. Ontogeny	y	x	Anatomy + Paleontology	Ontogeny	Ontogeny	Anatomy + Paleontology
5. Anatomy	x	y				
5. Paleontology	y	x				
6. Ontogeny	x	y	None	Ontogeny + Anatomy + Paleontology	None	None
6. Anatomy	x	y				
6. Paleontology	x	y				
7. Anatomy	m	n	None	None	None	None
7. Paleontology	n	m				

ontogeny is nonfalsifiable, and is a general apparent falsifier; any technique that agrees with ontogeny (such as anatomy in comparison 3 or paleontology in comparison 4) is also a falsifier. The techniques falsified are those that conflict with ontogeny, either directly (comparisons 1, 2, 5) or indirectly (comparisons 3 and 4). Under the assumption of neoteny, ontogeny is automatically falsified, and any technique that conflicts with ontogeny is an apparent falsifier, as is anatomy (comparisons 1, 4, 5) or paleontology (comparisons 2, 3, 5); in addition to ontogeny, the techniques falsified are those that agree with ontogeny (comparisons 3 and 4). If there is no conflict (comparison 6), there is neither falsification nor apparent falsifiers under the assumption of no neoteny; there is falsification but no apparent falsifiers under the assumption of neoteny. If there is no ontogenetic information (comparison 7), there is neither falsification nor apparent falsifiers under either assumption.

If the real falsifiers are the assumptions about neoteny, then anatomy and paleontology in themselves cannot be recognized as apparent falsifiers, unless they can be related, through ontogeny, to the assumptions. In comparison 7, the anatomical inference is that character m is primitive relative to n; the paleontological inference is that n is primitive relative to m. Which technique is falsified and which technique is the falsifier? With no ontogenetic information, no answers are possible.

If the real falsifiers are the assumptions about neoteny, on what basis may such assumptions be made? For one or the other of these assumptions to be a falsifier, it must be made in the absence of an ontogenetic transformation (see above). If neoteny were assumed to be common, and no neoteny rare, or vice versa, one might argue that the more common (neoteny), or rare (no neoteny), should be assumed in a given case. Perhaps it is unimportant which is assumed in advance, if both assumptions are equally exposed to potential falsification. But such seems not to be the case.

If no neoteny is assumed and the assumption is false, how might the truth be discovered? The first hint would be a conflict between ontogeny and either anatomy or paleontology, or both. But even in a case of conflict, neoteny might not have occurred. If neoteny is assumed, and the assumption is false, how might the truth be discovered? The first hint would be the absence of conflict between ontogeny and either anatomy or paleontology, or both. But even in a case of no conflict (comparison 6), neoteny could have occurred.

Clearly, the mere presence or absence of conflict is not an infallible indication. Yet the difference between these indications can be viewed as decisive, on the grounds that no conflict is negative evidence, and that negative evidence is unsatisfactory as an indication. The assumption of no neoteny is therefore preferable on a priori grounds to the assumption of neoteny, in any given case; the only way to establish the existence of neoteny is to assume no neoteny and find a falsifier for this assumption. If the first hint of a falsifier is a conflict between ontogeny and either anatomy or paleontology, or both, there are various possibilities:

(1) Invoke convergence, so as to evade falsification of any of the three techniques (ontogeny, anatomy, paleontology).

(2) Or invoke homoplasy, so as to evade falsification of anatomy or paleontology, or both.

(3) Or consider the outgroup an ingroup, so as to evade falsification of anatomy.

(4) Or consider the fossil record incomplete, so as to evade falsification of paleontology.

(5) Or consider neoteny a possibility, so as to evade falsification of anatomy or paleontology, or both.

Items (3) and (5) seem to be the most interesting, inasmuch as item (3) presumably leads to an advance in classification, and item (5) opens up the possibility of further investigation. The other items seem to have little intrinsic interest.

How commonly do conflicts occur between any of the techniques of character phylogeny? Probably all systematists regularly confront all five possibilities, either in reading the literature or in the course of thought processes. Neoteny is perhaps the most interesting of the five. If it occurs, it falsifies ontogeny; and either anatomy or paleontology, or both, serves as an avenue for falsification.

As mentioned above, the best evidence for neoteny, namely an observed ontogenetic transformation, renders neoteny impotent (and with it, anatomy and paleontology) as a falsifier. When neoteny otherwise emerges as a possibility, it may do so under the guise of an ad hoc item protecting anatomy or paleontology, or both, from falsification. Crucial to the role of neoteny as a real falsifier, therefore, is the manner in which it is suspected and discovered in the absence of an ontogenetic transformation. The above discussion suggests that the first hint of neoteny is a conflict between ontogeny and either anatomy or paleontology.

Neoteny has been invoked for fishes, perhaps most commonly in species with an endoskeleton that fails to ossify (character x). During the ontogeny of the endoskeleton of vertebrates generally, the individual elements normally appear as cartilages that subsequently ossify (transformation from x' to y). Most vertebrates (potential outgroups) have a bony endoskeleton (character y''), with the exceptions of cyclostomes and chondrichthyans (character x''). The earliest relevant fossil, depending upon which one was chosen, might be cartilaginous (character x''') or bony (character y'''). Whether cartilage is primitive relative to bone, or vice versa, is generally considered a moot question.

If fishes such as *Salanx* are assumed to be teleosts, as they always have been, outgroup comparison with other teleosts having a bony endoskeleton suggests that retention of cartilage in *Salanx* is a neotenic character. But suppose that *Salanx* were assumed to be a cyclostome or shark (or that a cyclostome or shark were selected as an outgroup), and that the similarities of *Salanx* and teleosts were dismissed as convergence or homoplasy? Neoteny would emerge under the former assumption (that *Salanx* is a teleost) but not the latter (that *Salanx* is a cyclostome or shark). Why is the former assumption preferable? Presumably because neoteny is a more parsimonious interpretation than convergence and homoplasy. This judgment of parsimony, however, is dependent upon other characters. What are these other characters? And how have they been resolved? Presumably, they are the results of character phylogeny. In other words, in order to be assumed to be a teleost and not a shark, *Salanx* must possess some characters that are not themselves neotenic, and that are defining characters (synapomorphies) of teleosts. How have such characters been resolved? And what technique has served as a falsifier of alternative evolutionary interpretations of the non-neotenic defining characters of teleosts? The only general falsifier for non-neotenic characters is ontogeny—either by itself or in combination with anatomy and paleontology.

This is a curious situation. Ontogeny is falsifiable in a case of neoteny, even when no ontogenetic transformation is observed. But falsification depends upon a prior assumption that itself depends upon ontogeny in its role as falsifier. The argument seems always to come full circle, back to ontogeny and its orderliness. If so, systematics might be best construed, as Danser (1950) did, as classification of life cycles, rather than individual specimens or species, with ontogenetic transformations as the relevant characters. This is also in accord with the suggestion made above that all characters can be viewed as modifications of other characters (i.e., as parts of life cycles).

Two general falsifiers of the biogenetic law have been discussed here: contradictory ontogenetic character transformations and neoteny. The former is a falsifier in the broadest sense, but we are unaware of any example of it. Hence the biogenetic law may be valid in a broad sense. Neoteny is an apparent falsifier of a more general principle, that of character phylogeny; hence it is an apparent falsifier of the biogenetic law in its role as a direct technique, and also an apparent falsifier of anatomy and paleontology in their roles as indirect techniques of character phylogeny. But neoteny is an apparent falsifier in a narrow sense. Perhaps it is not a falsifier at all, but only a reflection of lack of information. One may doubt, for example, that any characters are truly lost, rather than transformed. Apparent loss may be an indication that the characters and transformations are merely poorly understood and, consequently, wrongly defined. The problem may thus be amenable to investigation through analysis of the nature of particular characters and their comparison.

SPACE

6

BIOGEOGRAPHIC HISTORY:
KINDS OF QUESTIONS

HISTORY

Historically, concern with systematics is older than concern with biogeography. Humans of every age could immediately observe a diversity of organisms everywhere that they lived. For their survival, humans faced the very practical necessity of identifying and classifying the organisms around them. Not until the eighteenth century was it learned that the diversity of organisms of one place might be, and indeed is, fundamentally different from that of another.

Carl Linnaeus (1707–1778) is generally credited with inaugurating the modern age of systematics—as distinct from the Medieval—with the publication of his various compendia listing all of the plants and animals known at the time. He was also one of the first naturalists to pay close attention to the native localities of the specimens that he collected and described. To some degree, he was aware that different species live in different places, but he was more interested in the systematic aspect, rather than the ecological or geographical aspects, which were less well known at that time. He nevertheless tried to form some interpretation of the geographical and ecological distributions of organisms. He supposed that all species of organisms were created, in pairs or singly, and lived with Adam and Eve in the Garden of Eden. Where the garden was located he was not sure:

To enter into the remainder of this subject with as much brevity as possible, I think myself not greatly in danger of error in laying down the following proposition: *That the Continent in the first ages of the world lay immersed under the sea, except a single island in the midst of this immense ocean; where all animals lived commodiously, and all vegetables were produced in the greatest luxuriance.* (1781:77)

Knowing that some organisms require a warm climate, and other organisms a cold climate, Linnaeus considered how both kinds of organisms could have survived on a single island:

First let us conceive Paradise situated under the Equator; and nothing further is requisite to demonstrate the possibility of these two indispensible conditions, than supposing a very lofty mountain to have adorned its beautiful plains. (1781:90)

By his reasoning, the organisms requiring a cold climate would have lived atop the mountain; and the ones requiring a warm climate, on the plains. Those requiring intermediate conditions would have lived at intermediate altitudes. Such a concept was modern by Linnaeus' standards, for it followed closely a recent discovery of Joseph Pitton de Tournefort (1656–1708):

"I found," says he, "at the foot of Mount Ararat those plants which were common in Armenia,—a little further those which I had before seen in Italy; when I had ascended somewhat higher such vegetables as were common about Paris; the plants of Sweden possessed a more elevated region; but the highest tracts of the mountain, next the very summit, was occupied by the natives of the Swiss and Lapland Alps." (1781:91)

Certain observations made by Linnaeus convinced him that the continents were growing in area as the sea receded from the shore, and he reasoned that such had occurred since the earliest times. With the growth of "Paradise"—the primordial island—he reasoned that the plants and animals migrated from there onto the newly emerged areas. And gradually, as the extent of land increased and became occupied by plants and animals, the world changed from its primordial state to the modern one.

Knowing that animals move about more freely than plants rooted firmly in one place throughout their life, he allowed that there was a problem:

Here we foresee that a difficulty will be started by some, how such a number of vegetables, such immense woods and thickets, those multitudes of flowers that cover every field and meadow, can have been disseminated over the world by single plants. (1781:94)

Observing that plants are capable of rapid multiplication through seeds, Linnaeus considered how

from a single central spot, a plant of a given species may be so disseminated as to be found in all parts of the world. (1781:95)

He considered various possibilities for seed transport, and he concluded:

We have seen the Winds, the Rains, the Rivers, the Sea, Heat, Animals, Birds, the structure of Seeds, and Seed Vessels, the peculiar Natures of Plants, and even Ourselves, contribute to this great work.—I have shewn, that any one single plant alone would have been able to have covered the face of the globe: I have demonstrated that the dry land has always been increasing, and dilating itself; and therefore once was infinitely less than it is at this present:—I have traced back the orders of animals and vegetables, and found them to terminate in individuals created by the hand of God. (1781:113-14)

The exposition of Linnaeus boils down to two fundamental ideas about the nature of the world: (1) one small area, or center, wherein species appear (through creation) and whence the species (2) move, or disperse, to other areas far and near as the areas become available for colonization by the species that can survive and propagate in them.

The exposition of Linnaeus is an early example of a type of explanation that has been very commonly advanced during the subsequent 200 years. This type has been termed center-of-origin/dispersal explanation. Because this type has been, and still is, so common in the history of biogeography, the example furnished by Linnaeus will be considered at greater length.

In science, it is always interesting to begin by doubting the truth of whatever proposition one wishes critically to examine. Consider the possibility that Linnaeus' center-of-origin/dispersal explanation might be wrong in some fundamental respect. If it were wrong, could its error be discovered through empirical observation?

In order to keep matters simple, consider first only those facts that might be learned about (1) what species exist, (2) where the species naturally occur, and (3) how the species might be interrelated, as expressed in some classification wherein related species are grouped together in supraspecific taxa such as genera, families, and orders. Does Linnaeus' explanation allow the possibility that there might be any number of species that live anywhere and that are interrelated in any way whatsoever? If so, his explanation, at least with regard to these three types of information, is unfalsifiable. His explanation could be wrong, but if it were, its error might never be discovered because it is not discoverable.

Linnaeus' explanation includes two auxiliary concepts about the nature of the world: (1) that species all were created at the same time and (2) all of the continental areas, save for one small island, were initially submerged beneath the sea. Suppose that one or both of these concepts were wrong, and that the errors could be discovered by investigating the ages of different species and the history of the continents. Even if the two auxiliary concepts proved erroneous, his center-of-origin/dispersal explanation could still be maintained, with only slight modifications of the two fundamental concepts: (1) a small area, or center, wherein species appear (through whatever causal process) and (2) whence the species move, or disperse, to other areas far and near (through whatever causal process).

This generalized center-of-origin/dispersal explanation might seem without any possibility of falsification, either with reference to empirical observations of species, of distributions of species, and of classifications of species, or with reference to causal processes of origin of species and of dispersal of species.

The generalized explanation is very close, but not as precisely stated, as Linnaeus' original, which was advanced by him to explain everything—all species, all of their distributions, and all of their classification. But, one may ask: is his explanation satisfactory? Does it really explain everything? Indeed, does it explain anything?

Answers to these questions, indeed all questions in science, are matters of judgment. One person might answer yes, and believe the answer to be correct. Another person might answer no, and believe with equal conviction that that answer is correct. Such differences in judgment always surround any live scientific problem. They pinpoint it as an area that might profitably be investigated in search of a solution: which judgment, if either, is correct?

To permit investigation of a problem, such as whether or not Linnaeus' explanation is true, his concept *must* suggest some additional observations that could be made and that would tend either to confirm or refute his explanation. Whatever Linnaeus might have intended as implications of his theory about species and their origins, his contemporaries and followers deduced that, if he were correct, different geographical areas of the earth should be populated by the *same* species to the extent that the different areas have the same physical conditions. Different mountaintops, if they were at the same height and latitude, and had the same kind of soil and climate, should have the same species of plants growing upon them. Different deserts, again at the same latitude, and with the same kind of soil and climate, should be populated by the same species of plants and animals.

Of course, not much was known in Linnaeus' time about the various plants and animals of the world. In the first edition of his *Systema Naturae*, for example, he listed only a few hundred kinds of plants and animals (modern estimates run into the millions). And most of what was known pertained to Europe and its immediate vicinity. Nevertheless, there was a clear expectation that distributions of organisms would be found to depend, elsewhere as in Europe, upon the physical factors of the environment.

This expectation was early contradicted by a contemporary of Linnaeus—George Leclerc, Comte de Buffon (1707–88). Having studied the mammals then known from the tropical parts of the Old World (Africa) and the New World (tropical America), he found not a single species in common between them. He wrote of his discovery as a "general fact, which first appeared very strange, and which no one before us even suspected, that no species of the torrid zone of one continent is found in the other" (1761:96, translated). Due credit must be given Buffon for calling attention to what seemed an anomaly within the context of his time: where, according to theory, there should be the same species, there is, in fact, not a single species in common. He referred to this anomaly—different species of mammals in the tropics of different continents—as a "general law." His doing so caused other investigators to focus on the problem. Within 50 years, Alexander von Humboldt (1769–1859) had compared the plants of Africa and South America; Pierre Latreille (1762–1833) had compared the insects; Georges Cuvier (1769–1832), the reptiles; and other investigators, other groups of organisms. All groups told the same story.

Curiously, Buffon did not question the center-of-origin/dispersal

explanation as a result of his discovery, for he believed that the Old
World was the original source area for the mammalian species now
found only in the New. What he questioned is the idea that all species
were independently created and immutable. In short, he adopted an
evolutionist point of view. His argument is complex, but seems
summarized in the following statement:

And when, because of revolutions on the globe or by the force of humans,
animals are forced to abandon their native country—when they have been
hunted or relegated to distant climes—their nature has undergone alterations so
large and so profound, that it is not recognizable at first glance; to recognize it
requires the most attentive inspection, experiment, and analogy. (1766:316,
translated)

It would go too far here to consider the factors that Buffon reasoned
were responsible for the "alterations so large and so profound" that
certain species had undergone. Suffice it to say that he believed that
factors external to the organisms—such as climate, food, and domesti-
cation—acted upon the organisms and caused their natures to change.
In this sense, physical factors of the environment were again brought
into play in order to explain the different distributions of organisms.
The external factors of South America are different from those of
Africa. When animals migrated from Africa to South America, they
encountered different conditions, and they changed accordingly.

Buffon's considerations of these matters were not the most consistent.
Whereas the species of the New World must have changed during or
after the emigration from the Old, the species of the Old World seem to
have been created, much in the fashion of Linnaeus, if not exactly all of
them together on a mountainous tropical island.

Buffon's reasoning on these matters is interesting. He starts with the
idea that the physical conditions of Africa are similar to those of South
America (both areas are located in the tropics, and extensive areas of
each are covered with rain forest). Accordingly, he might expect the
same species of organisms to occur in both places. He studies the facts of
the case, and finds that there is not one single species of mammal
common to both areas. He therefore concludes that *some* conditions
must be different after all. If the physical conditions are not so different,
then, well, . . . maybe different food supplies are responsible. Buffon
discovered what he might have construed as a falsifier of Linnaeus'

explanation. Yet, instead of rejecting the explanation, he modified it to include the idea that organisms might change, or evolve into new and different species, when moving, or dispersing, from their home, or center of origin, into other areas.

Buffon's modification is an example of adopting an ad hoc item that protects a theory from falsification—in this case the center-of-origin/ dispersal explanation of Linnaeus. To anyone familiar with the subsequent history of biogeography, Buffon's modification—the idea that organisms might change when dispersing into areas with conditions different from those of their homeland—is very familiar. The idea permeated biogeography for a period of about 100 years, from the publication of Darwin's *Origin of Species* in 1859 through the 1960s. Writers of this period frequently attributed the idea to Darwin's originality. Mayr, for example wrote that

when the young Darwin boarded the *Beagle* in 1831, he took for granted that the floras and faunas of all regions were the "products" of these regions and that faunas owed their characteristics to the local physical environment. But what he found, and he never tired of emphasizing it, was totally at variance with his preconceptions. For instance: "In the southern hemisphere, if we compare large tracts of land in Australia, South Africa, and western South America, between latitudes 25° and 35°, we shall find [these] parts extremely similar in all their conditions, yet it would not be possible to point out three faunas and floras more utterly dissimilar." (1965a:473; Darwin 1859:347)

And Mayr concluded:

These observations, and others to be discussed presently, destroyed for Darwin the idea, widespread since Buffon and earlier, that flora and fauna are "the product of a country" or, as we might say today, that factors of the physical environment determine the composition of faunas and floras. (*Ibid.*)

There is a difficulty here with Mayr's rendition of the history of these ideas. The comparison between Africa and South America, for example, was already an old one in the 1830s. James Prichard stated the following in 1836:

It would be easy to discover districts, situated respectively in North America and in Europe, or in equinoctial America, Africa, and Asia, in which all the same physical conditions exist, namely, a parallel temperature and elevation, a similar soil, and the same degrees of humidity in the atmosphere; yet the species of plants in these several districts will be far from being identical. The vegetable tribes will present, in each respectively, analogies of form and general character;

but few, if any, of the same species will be found in localities thus separated. (1836:50)

Prichard cited an essay of A.-P. de Candolle, published in 1820. In a book that appeared a few years before Prichard's, and from which Prichard learned of Candolle's essay, Charles Lyell stated:

> The luminous essay of Decandolle on "Botanical Geography" presents us with the fruits of his own researches and those of Humboldt, Brown, and other eminent botanists, so arranged, that the principal phenomena of the distribution of plants are exhibited in connexion with the causes to which they are chiefly referrible. "It might not, perhaps, be difficult," observes this writer, "to find two points, in the United States and in Europe, or in equinoctial America and Africa, which present all the same circumstances: as, for example, the same temperature, the same height above the sea, a similar soil, an equal dose of humidity, yet nearly all, *perhaps all*, the plants in these two similar localities shall be distinct. A certain degree of analogy, indeed, of aspect, and even of structure, might very possibly be discoverable between the plants of the two localities in question, but the *species* would in general be different. Circumstances, therefore, different from those which now determine the *stations*, have had an influence on the *habitations* of plants." (1832:68)

These quotations from Prichard and Lyell—ultimately from Candolle—show that Darwin was repeating a theme that had been enunciated and repeated during the forty years before 1859. The history of this period is, therefore, of some interest to anyone hoping to understand the subsequent history of biogeography or, indeed, the ideas of that discipline, even as they persisted until recent times.

The key to the period between Buffon and Darwin—indeed, the key to the subsequent history of biogeography—is Candolle's essay of 1820. It summarized, in very lucid terms, what was then known of plant geography, and it posed a problem that remained submerged and unsolved throughout the subsequent 150 years. Candolle's essay is divided into three parts: (1) the influence of external elements, (2) the relation between external elements and the study of stations and (3) the study of habitations. Candolle considered five external elements and their effects on plants: temperature, light, water, soil, and atmosphere. With respect to stations and habitations, Candolle stated:

> By the term *station* I mean the special nature of the locality in which each species customarily grows; and by the term *habitation*, a general indication of the country wherein the plant is native. The term station relates essentially to climate, to the terrain of a given place; the term habitation relates to

geographical, and even geological, circumstances. The station of *Salicornia* is in salt marshes; that of the aquatic *Ranunculus*, in stagnant freshwater. The habitations of both of these plants is in Europe; that of the tulip tree, in North America. The study of stations is, so to speak, botanical topography; the study of habitations, botanical geography. (1820:383, translated)

The terms "stations" and "habitations" are not familiar in modern usage. In this combination Candolle used them for the first time. His usage was followed by Lyell, but not by Darwin. Alfred Wallace (1823–1913) used the terms, but he abbreviated "habitation" to "habitat," which today is understood in a different sense—that of Candolle's term "station." Even so, the terms, as used by Candolle, have modern counterparts: ecological and historical biogeography. Ecological biogeography is the study of stations (Candolle's "botanical topography"); historical biogeography, the study of habitations (Candolle's "botanical geography"). Candolle considered that "the confusion of these two classes of ideas is one of the causes that have most retarded the science and that have prevented it from acquiring exactitude" (*Ibid.*). To anyone familiar with the twentieth-century literature of biogeography, Candolle's words are provocative. If ecological and historical biogeography were clearly distinguished so long ago, how is it that they have subsequently and so often been confounded? Turning to a book entitled "Biogeography," even of a very recent date, one is not sure quite what to expect, but the extremes are easy enough to specify: pure ecology, pure (historical) biogeography, or a mixture of the two in any proportion. And whatever the nature of the book, its author is apt to believe that its subject matter explains whatever it is that is worth explaining about the distribution of organisms.

The distinction between stations and habitations was important to Candolle, because he believed that these two classes of facts (ecological-topographical and historical-geographical) relate to two different classes of causes: " Stations are determined uniquely by physical causes actually in operation, and . . . habitations are probably determined in part by geological causes that no longer exist today" (1820:413, translated).

In order to appreciate Candolle's concern with "geological causes," one must first understand what Candolle believed had been caused. What was his view of the world? Interestingly, Candolle never travelled far beyond France, although he explored that country's plants, growing in their native localities, in great detail; hence his views of the world depended partly upon his own studies of specimens brought to Europe from elsewhere, and upon the reports of explorers and collectors—many of which had been inspired by Buffon's comparison of tropical mammals. Alexander Humboldt (1816) reported:

These studies of the law of the distribution of forms lead naturally to the question whether there exist plants common to both continents—a question that inspires all of the more interest, for it relates directly to one of the most important problems of zoology. It has long been known, and it is one of the most elegant results of the *geography of animals*, that no mammal, no terrestrial bird, and as it appears from the studies of M. Latreille, almost no insect is common to the equatorial regions of the two worlds. M. Cuvier is convinced after careful study, that this rule applies even to reptiles.

As for phanerogamous plants (with the exception of species of *Rhizophora* and *Avicennia*, and some other littoral plants), Buffon's law seems exact for dicotyledonous species.

Generally in tropical America, only monocotyledonous plants, and among them only Cyperaceae and Graminae, are common to the two worlds. These two families are exceptions to the general law that we examine here, the law which is so important in the history of the catastrophes of our planet, and according to which, organisms of the equatorial regions differ essentially in the two continents. (1816:234, translated)

Pierre Latreille (1817) reported:

The totality, or a very large number of arachnids and insects, of which the native countries have the same temperature and soil, but which are separated by large distances, is composed, in general, of different species. The majority of animals [spiders and insects] also differ specifically, when their native countries, with identical soil and temperature, are separated from each other, no matter what the difference in latitude, by natural barriers, interrupting the communications of these animals, or rendering them very difficult. . . . Consequently spiders and insects, and even reptiles of America, of Australia, cannot be confused with animals of the same classes that inhabit the Old World. (1817:43, translated)

These reports indicate that investigators such as Humboldt and Latreille were inspired by, and were essentially repeating, Buffon's early "experiment" with tropical mammals, and obtaining the same result. Reading their reports, even at this late date, suggests that they felt some principle of great generality was slowly being discovered. Humboldt referred to "the law of the distribution of forms" and, again, to "Buffon's law," by which he meant merely that tropical organisms were different in the two continents. Could this "law" have been construed as a falsifier of

Linnaeus' explanation? And if so, what particular part of the explanation would it have falsified? No more than a tentative answer can be suggested here, based on the assumption that, inherent in Linnaeus' explanation is the idea that the entire world biota is a single system, the history of which dates from one center of origin for all species. The early "experiment" of Buffon, if its result had been viewed as a falsifier, would have meant that there were, not one, but *two* apparently independent biotic systems, one in the Old World and the other in the New, each with its own history. Although Buffon, by invoking the idea that species change while they disperse, was able to retain Linnaeus' explanation in modified form, the later investigators, each of them, had an additional opportunity to recognize the same falsifier—as represented by the distributions of the groups that they studied.

It was Candolle who recognized most clearly that a radically new view of the world was forming, and his statement of it far surpasses in its generality the repeated comparisons of Africa and South America. He wrote:

From all of these facts, one may deduce that there are *botanical regions;* and by this term I denote whatever areas that, with the exception of introduced species, have a certain number of plants that to them are peculiar, and that can be called truly *aboriginal.* (1820:410, translated)

By "botanical regions" Candolle meant not only Africa and South America, but many more besides. Each of the regions he recognized carried with it the implication of a long and independent history. There was, in short, not a single system for the world biota, nor two of them, but many more—what would be called today "areas of endemism," wherein one would expect to find species of plants and animals that occur natively nowhere else on earth. After stating that his determinations of "botanical regions" were not yet complete, Candolle listed twenty such regions, and added that any island isolated enough from a continent to have its own flora would be, in effect, another botanical region. His list reads:

(1) northern Asia, Europe, and America; (2) Europe south of region (1) and north of the Mediterranean; (3) Siberia; (4) the Mediterranean area; (5) eastern Europe to the Black and Caspian Seas; (6) India; (7) China, Indochina, and Japan; (8) Australia; (9) south Africa; (10) east Africa; (11) tropical west Africa; (12) Canary Islands; (13) northern United States; (14) northwest coast of North

America; (15) the Antilles; (16) Mexico; (17) tropical America; (18) Chile; (19) southern Brazil and Argentina; (20) Tierra del Fuego. (1820:411, translated)

Candolle's list of botanical regions is remarkable, coming as it did at a time when other investigators were focused on Africa and South America. The list is remarkable also because its regions are immediately recognizable, even today, as real areas of endemism for terrestrial organisms, plant and animal alike, and even for aquatic and marine organisms as well. The significance of Candolle's list was, perhaps, most clearly perceived by Charles Lyell (1797–1875), who had the opportunity to visit Candolle in Switzerland during the late 1820s. In 1832 Lyell wrote:

Decandolle has enumerated twenty great botanical provinces inhabited by indigenous or aboriginal plants; and although many of these contain a variety of species which are common to several others, and sometimes to places very remote, yet the lines of demarcation are, upon the whole, astonishingly well defined. Nor is it likely that the bearing of the evidence on which these general views are founded will ever be materially affected, since they are already confirmed by the examination of seventy or eighty thousand species of plants. (1832:71)

These botanical regions, of course, are the "habitations" of Candolle. As a statement about the nature of the world, the regions lead directly to an implication, or expectation, very different from that based on Linnaeus' explanation (different geographical areas of the earth should be populated by the *same* species to the extent that the different areas have the same physical conditions). Instead of expecting the same species in areas of the same physical conditions, one would expect different species in different botanical regions—even in areas having the same physical conditions. Mountaintops in Siberia would be expected to have plant species different from those of mountaintops of any other region—even if the physical conditions were the same in the two regions.

Because Candolle's list of regions implied a different, indeed contradictory, expectation, it expressed a view of the world radically different from that of Linnaeus. But Candolle's view grew out from that of his predecessor. This growth of knowledge and transformation of ideas about the nature of the world are well expressed by Candolle:

The first explorers always thought that they found plants of their home in faraway countries, and they delighted in giving the plants the same names. But as soon as specimens were brought back to Europe, the illusion dissipated for the vast majority of the plant species. Even when examination of dry specimens left some doubt, horticultural investigation generally removed the doubts, leaving only a very small number of species common to different regions. (1820:403, translated)

So clear was Candolle's view of these matters that he boldly declared:

Botanists know that in general the plants from these twenty regions are different, the ones from the others, such that when one finds in the writing of an explorer that the plants of one of these regions have been found in another, one must, before supposing such to be true, study the specimens from the two areas with particular care. (1820:412, translated)

Another difference between Candolle and his predecessors concerns cosmopolitan species—that "very small number of species common to different regions." Such cosmopolitan species had been accepted, indeed expected, under Linnaeus' explanation. They were looked for but not found by Buffon, Humboldt, Latreille, and others. Under Candolle's view they became exceptions to the rule—to what Humboldt had termed "Buffon's law"—as extended by Candolle to apply not merely to two regions (Africa and South America), but to twenty or more of them, including all parts of the world. Candolle's discussion of cosmopolitan species began: "Before some degree of importance is attached to this small number of species common to widely separated regions, it is appropriate to examine the various means by which seeds are able to be transported from one country to another" (1820:403, translated).

His discussion concluded: "If one now reflects on the continuous action of the four causes of seed transport that I have listed—water, wind, animals, and humans—one will find, I believe, that they are sufficient to explain the small number of plant species that are the same in various continents" (1820:410, translated). Candolle's discussion of the means by which seeds might have been dispersed is remarkable. First, its substance is much the same as Linnaeus' discussion of the same subject. But whereas Linnaeus believed that dispersal from "Paradise" explains *all* present distributions of organisms, Candolle believed that dispersal from one region to another explains only *a small fraction*—the small, but anomalous, number of plant species that are the same in different regions. Second, Candolle's discussion contrasts with Buffon's. For Buffon, dispersal of organisms allows them to colonize new areas with different conditions, and, therefore, creates the possibility of their changing, or evolving, into different species. For Buffon, dispersal explains why there are *different* species in different areas; for Linnaeus and Candolle, dispersal explains why there are the *same* species in different areas.

What can be said by way of an evaluation of these different views regarding dispersal of organisms and what it means? The answer to this question is not of interest merely because of what it might say about a remote period in the history of biogeography—the period from Buffon to Darwin. The answer will be of interest also for what it says about the period from Darwin to the present. Why this is so will be considered below. For the present it is enough to focus on the views of these three early naturalists, and attempt to come to a judgment about the meaning and significance of their views in their own times.

The sense of the whole discussion seems best analyzed in relation to two alternative views of the world: one, shared by Linnaeus and Buffon, that all species have a history traceable to one center of origin (Linnaeus' "Paradise" or Buffon's "Old World"), from which species dispersed to colonize other areas; the other, held by Candolle, that each species has a history traceable to one of many regions, wherein species were confined by barriers that were generally effective in preventing dispersal.

Candolle's view was based on knowledge of the distributions of many plant species, and, of course, on what he knew about animal distribution, too. This knowledge must have seemed to him sufficient to falsify the older view of a single center of origin. Once he had accepted the alternative, of many different regions, each with its independent history, Candolle had no need to invoke dispersal to explain the many different regions, but to explain only the small number of species common to different regions.

There is one similarity in all three views: that dispersal explains whatever seems to contradict the basic idea held by each author. For Linnaeus, the basic idea was the small island; for him the seeming contradiction was supplied by the wide distribution of many species. What explained this anomaly? *Dispersal* from "Paradise." For Buffon, the basic idea was an implication of the above: areas with similar physical conditions should have the same species; for him the seeming contradiction was supplied by the tropical mammals—different species in similar environments. What explained this seeming anomaly?

Evolution arising as a by-product of *dispersal* from the Old World to the New. For Candolle, the basic idea was the existence of numerous regions, each with its own native flora; for him, the seeming contradiction was supplied by species common to different regions. What explained this seeming anomaly? *Dispersal* of species from one region to another.

How might these different views be evaluated? There seem to be only two possibilities: either by evaluating the truth of the basic idea of each view, or by evaluating their implications. Among their implications are the different roles of dispersal, as conceived by the three authors.

One possible question crucial to an evaluation is: does Candolle's list of regions refute the notion of a single center of origin for all species? Of interest is that in 1838 Candolle added another twenty regions to his list, making a total of 40. A few of his older regions he subdivided. Region 17 (tropical America), for example, he replaced with four new regions: Colombia, Peru, Guiana, and Central America. Most of his new regions, however, were islands such as the Falklands, Aleutians, Juan Fernandez, Madeira, the Azores, Madagascar, Zanzibar, Mauritius, New Zealand, New Caledonia, the Societies, Tristan de Cunha, Saint Helena, and Hawaii. The items on his revised list, like the first, are immediately recognizable as *bona fide* areas of endemism by any modern standard. And this revised list might be considered Candolle's definitive statement on the subject. Does the revised list refute the notion of a single center of origin?

Discussed above were the possibilities for falsification of the single center-of-origin explanation of Linnaeus. There seemed to be very few possibilities—beyond the implication that different geographical areas of the earth should be populated by the *same* species to the extent that the different areas have the same physical conditions. This expectation, surely, was refuted by Candolle, if not previously by Buffon and those investigators who followed his lead in comparing Africa and South America.

Charles Lyell, who commented on this situation, was ready to credit Buffon with accomplishing the entire feat:

Although in speculating on "philosophical possibilities," said Buffon, the same temperature might have been expected, all other circumstances being equal, to produce the same beings in different parts of the globe, both in the animal and vegetable kingdoms, yet it is an undoubted fact, that when America was discovered, its indigenous quadrupeds were all dissimilar from those previously known in the old world. . . .

These phenomena, although few in number relatively to the whole animate creation, were so striking and so positive in their nature, that the French naturalist caught sight at once of a general law in the geographical distribution of organic beings, namely, the limitation of groups of distinct species to regions separated from the rest of the globe by certain natural barriers. (1832:87)

Here Lyell gives too much credit to Buffon. His summary of what he calls Buffon's "general law in the geographical distribution of organic beings" is, nevertheless, succinct and clear, even if, as a generalization, it is best credited to Candolle and Lyell himself.

Reading this early literature of biogeography, one is tempted to see a progression, and real progress in knowledge, from Linnaeus' rather simple and pious view, through Buffon's early investigations, to the comprehensive summaries of Candolle and Lyell. If the sense of this early history is the gradual attainment of a new view of the world, the view can be only that the world is subdivided into some number of areas of endemism—Candolle's regions.

However the world was viewed by these early authors, all of them agreed that the present state of the world arose through a process of historical development, which for Linnaeus and Buffon involved wholesale geographical change through emergence of the continents from the sea. Coming later than the other early authors, Candolle was in the best position to assess the nature of the processes of historical development. Yet Candolle was unable to decide whether species did change under the influence of the environment, as Buffon suggested; whether new species might arise through hybridization; or whether they were more or less permanent. He did believe that regions—his "habitations"—could not be explained by external factors alone. He wrote:

So far I have tried to prove that habitations, considered in their totality, appear to be determined by temperature. Without doubt, it is necessary to combine, with temperature, the considerations deduced from a study of stations; it is clear, for example, that the more a certain country is sandy, the more one will find plants of the sand growing there, etc. But even when one gives to these causes all of the latitude possible, does one do justice to the best-known facts? It is this that I doubt; it is this that requires a new discussion. (1820:402, translated)

Unfortunately, his "new discussion" did not lead very far beyond the idea that unknown geological causes were somehow, at least in part, responsible, and that species seemed to have something of a permanent nature. His discussion concluded:

I see that there exists in living things some permanent differences that cannot be accounted for by any of the known causes of variation; these are the differences that constitute the species. These species are distributed over the globe partly after laws that can be deduced from the combination of the principles of physiology and physics, and partly from the principles that pertain to the origin of things—principles about which we know nothing. Such is, in summary, the point where geographical botany is forced to stop its enquiries. (*Ibid.*)

The period from Buffon to Candolle—roughly 1750–1850—was the golden age of French achievement in systematics and biogeography, which benefited from the colonialistic activities of that nation during that time. New areas of the earth were explored for treasure and resources. To further these endeavors, scientists were encouraged to collect and study specimens of plants, animals, and minerals, in the hope that knowledge of new worlds would aid in the future exploitation of their resources. Hence specimens from remote areas tended to accumulate in Paris, and other European capitals, where they formed the beginnings of the large museums, and in particular the present Muséum National d'Histoire Naturelle in Paris. Much new knowledge, indeed, was acquired during this period, and the natural world began to emerge in the rough outlines of its full complexity in human understanding, which in this way benefited from the otherwise dubious ambitions behind colonialistic endeavor. French science, however, was not destined to continue in its position of dominance, at least not in the area of natural history.

The reasons behind the rise and decline of the scientific activities of nations are manifold and subject to diverse interpretation—a subject much too complex to be considered here. It is enough to note that early developments in biogeography were a virtual monopoly of French culture and enterprise. As a consequence, many of the publications of this period were in French, one of the first modern languages to replace Latin as the medium of scientific and scholarly discourse—one prime example, perhaps, being Buffon's multivolume treatise, *Histoire Naturelle*, published during the period 1749–1804. Several of the early French publications, as well as timely reports of the activities of their scientific academies, were translated into other European languages, primarily German and English, and must have served in these countries as an impetus for scientific advance on their own nationalistic level— conducted and published in the living language of the country wherein the investigators lived and worked.

Whatever the reasons, activity in the areas of systematics and biogeography waned in France but intensified in Britain during the last half of the nineteenth century. To anyone native to the English-speaking world of even recent times, the scientific figures of that period assumed awesome proportions, and the string of their names seemed never ending. Lyell, Murchison, Owen, Darwin, Hooker, Wallace, Bates, and Huxley are names that long continued to capture interest, imagination, and emulation, and to influence lives, thoughts, and activities.

Most attention, of course, has been given to Darwin as a result of his book, *On the Origin of Species,* which no doubt did much to focus the attention of at least the English-speaking world on matters pertaining to systematics and biogeography—matters that seemed to be enlightened by Darwin's writings on evolution by means of natural selection. Yet it is often the case that the flurry of excitement that attends and follows the exploits of the heroes of a nation or culture also dims those of others. Within the English-speaking world there has been a tendency, all too palpable, for commentators, educators, historians, and scientists alike, in the area of biology, to date the beginnings of their science with the year 1859, in the belief that, somehow, all that was worthy in the eras before Darwin was either recast or subsumed in his writings and those of his successors. Too many are there who have argued all too persuasively that Darwin and the discoveries that followed him solved whatever problems that might have lingered as a result of the earlier activities of Linnaeus, Buffon, and Candolle.

Darwinism and neo-Darwinism, as historical forces, are much too complex for treatment here, beyond noting that there are persons who proclaim that, even 100 years after the fact, there are still benefits to be gained from continuing the "Darwinian revolution" in biology (Mayr 1971). But also relevant here, perhaps, is the attitude of the one person who, more than any other, was responsible for igniting and continuing whatever Darwinian "revolution" there was—if indeed there ever was one. In 1858, the year preceding the publication of the *Origin,* Thomas Huxley revealed his expectations of the events to come:

Wallace's impetus seems to have set Darwin going in earnest, and I am rejoiced to hear we shall learn his views in full, at last. I look forward to a great revolution being effected. Depend upon it, in natural history, as in everything else, when the English mind fully determines to work a thing out, it will do it better than any other. I firmly believe in the advent of an English epoch in science and art, which will lick the Augustan (which, by the bye, had neither science nor art in our sense, but you know what I mean) into fits. (L. Huxley 1900:171)

Huxley's is a revealing statement, and his insight into the movement of history was generally correct. The spotlight was then about to focus on Britain.

DISTRIBUTION

Biogeography, as a topic for discourse or discussion, is in some ways like religion: both topics lend themselves to ever more complicated treatment in the abstract, which is apt to border even on the miraculous, but which is apt to crumble in confrontation with concrete facts of life. It is well here to pause briefly to consider some facts characteristic of the distribution of organisms, if not of religions. Indeed, how are organisms distributed?

Consider first the distribution of a single species, which might have been observed in various localities, such as the sea star *Solaster endeca,* known from scattered collections in the North Atlantic, North Pacific, and Arctic Oceans (figure 6.1). In this case, the species might be expected to occur in some at least of the intervening areas of these oceans, and, in fact, its distribution might be fairly continuous throughout the oceans of the far northern hemisphere.

Consider next the distribution of the Eagle Owl, *Bubo bubo* (figure 6.2). Exact localities are not represented by dots (as in figure 6.1), but the entire range of the species is represented by darkened areas on the map. A few small areas, in Sweden, the Arabian peninsula, and the Japanese archipelago, indicate populations that might be isolated from the main one by intervening areas of unfavorable habitat—areas wherein eagle owls might simply not survive even if they were transported there.

Consider finally the distribution of the sea cucumber, *Holothuria atra,* known from many places throughout the warmer oceans of the world (figure 6.3). These three examples show that the distributions of single species vary considerably, one from another, in their extent and in

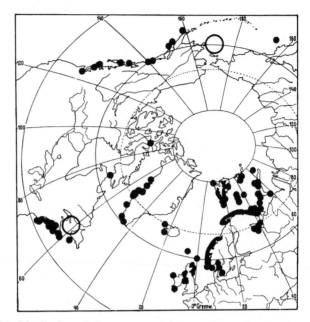

Figure 6.1. Distribution of the sea star, *Solaster endeca.* After S. Ekman (1935), Tiergeographie des Meeres (Leipzig: Akademische Verlagsgesellschaft), figure 145, p. 257.

the areas of the world that they occupy. Not much more could be said for any one or all of them except the obvious: that the organisms occur where they can survive and reproduce in accordance with their particular environmental requirements. One does not find, nor would one expect to find, the sea cucumber living with the Eagle Owl, nor either of them living with the sea star.

What is true for single organisms is true also for groups of species. Consider the genus of prawns, *Spirontocaris,* with about 40 species, occupying about the same area as that of the sea star, *Solaster endeca* (figure 6.4); the family of loaches, Cobitidae, with about 150 species, occupying somewhat, but not much, more than the area of the Eagle Owl, *Bubo* (figure 6.5); and, finally, the genus of fiddler crabs, *Uca,* with about 60 species, occupying about the same area as that of the sea cucumber, *Holothuria atra* (figure 6.6). Again, not much beyond the obvious can be said about these additional distributions. It would, of course, come as no surprise to find specimens of *Spirontocaris* and

Figure 6.2. Distribution of the Eagle Owl, *Bubo bubo*. After M. Fogden (1973), Fishing owls, eagle owls and the Snowy Owl *Ketupa, Scotopelia, Bubo, Nyctea,* figure on p. 73. Map by Geographical Projects. From J. A. Burton, ed., Owls of the World: Their Evolution, Structure and Ecology, pp. 61–93 (New York: A and W Visual Library). Copyright © 1973, Eurobook Ltd. By permission of E. P. Dutton.

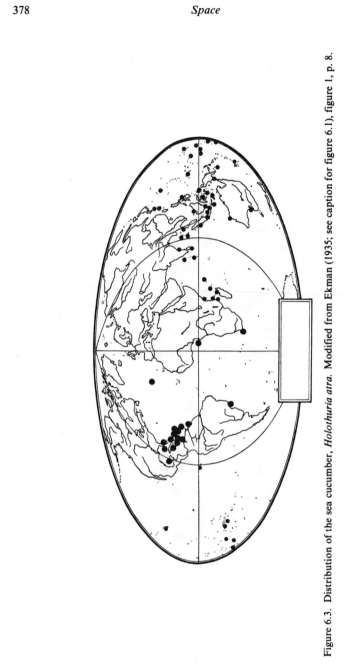

Figure 6.3. Distribution of the sea cucumber, *Holothuria atra*. Modified from Ekman (1935; see caption for figure 6.1), figure 1, p. 8.

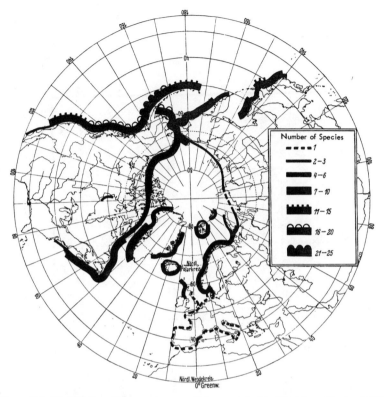

Figure 6.4. Distribution of the prawn genus *Spirontocaris*. Modified from Ekman (1935; see caption for figure 6.1), figure 124, p. 224.

Solaster in the same localities; or specimens of *Cobitis* and *Bubo*; or specimens of *Holothuria* and *Uca*.

Distributional summaries of species and groups are acquired slowly, as observations, records, and specimens accumulate in collections, sometimes over a period of a hundred years or more. Accurate distributions for very few species, or groups, were available in the early days of biogeography. Today, of course, there are more, but still not very many in absolute numbers. Even when all of the available information is gathered together, and plotted on a map, for a single species, or group, the result is apt to be a large area of geography with a few dots sprinkled over the surface. Impressionistic as it might be, such a summary is very often the best that can be had, unless, of course,

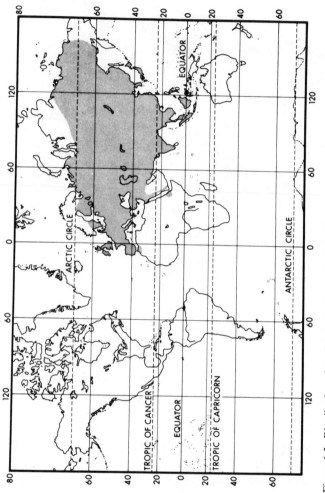

Figure 6.5. Distribution of the fish family Cobitidae. Modified from J. S. Nelson (1976), Fishes of the World (New York: Wiley), map 21, p. 334.

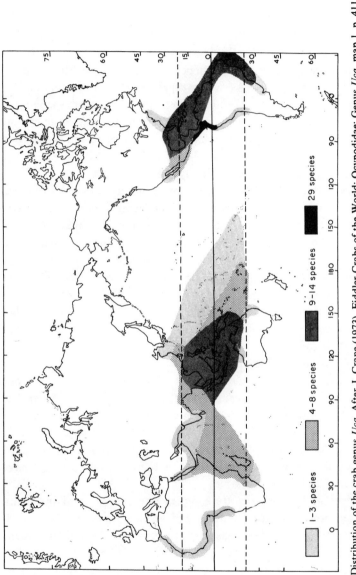

Figure 6.6. Distribution of the crab genus *Uca*. After J. Crane (1973), Fiddler Crabs of the World: Ocypodidae: Genus *Uca*, map 1, p. 411. Copyright © 1975, Princeton University Press. Reprinted by permission of Princeton University Press.

additional data are collected. Additional data, however, can never produce perfect accuracy, but only a summary that is relatively more complete.

What is important is not the degree of completeness per se, but what the data, however scanty they might be, suggest about the distribution of a certain species or group. A few dots representing localities for *Solaster endeca, Bubo bubo,* or *Holothuria atra* might be enough to suggest that the first is a cold-water marine organism of the northern hemisphere; the second, a terrestrial organism widespread in Eurasia; and the third, a warm-water marine cosmopolite. These concepts would be essentially correct, even if they had been initially suggested by very few data.

As portrayed by diagrams in the scientific literature, distributions are, more often than not, depicted as *concepts* rather than as detailed and factual summaries of localities exactly rendered. Figure 6.7, for example, shows a concept of the distribution of the trout-perch, *Percopsis omiscomaycus,* in North America (inset), as well as a fairly detailed summary of the localities known for this species in the State of Ohio. But even the many localities in Ohio are only more or less exact, within the resolution of the map, and only more or less complete. Other localities doubtless could have been, or could be, found if someone searched for them.

One factor underlying all summaries of distributions, and rendering them all inaccurate, is a dynamic aspect. Distributions are apt to change with time—even the time during which the data have been collected. The Ohio localities for the trout-perch, for example, were accumulated over a period of 100 years. The present distribution of the species in Ohio might not be very accurately reflected by the numerous recorded localities, simply because the details of the distribution might have changed.

Changes in distributions are complex phenomena involving many causal factors, not the least important of which is human activity, which in the last few hundred years has dramatically altered the details of plant and animal distribution, as well as some of its more general aspects. The most dramatic influences, perhaps, stem from human introductions of species into areas wherein they are not native. One example is the muskrat, *Ondatra,* native to North America, but introduced into Europe earlier in this century. Its spread in Czechoslovakia was

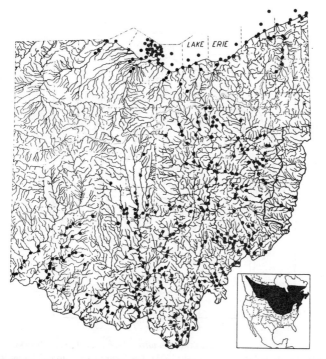

Figure 6.7. Distribution of the fish, *Percopsis omiscomaycus*, in the State of Ohio and (inset) its distribution within North America. From M. B. Trautman (1957), The Fishes of Ohio, map 21, p. 464. Copyright © 1957, Ohio State University Press (Columbus). All rights reserved. Used by permission of the author and the publisher.

recorded, from an initial population of five individuals transported in 1905 from North America to the village of Dobris near Prague (figure 6.8). Within 50 years, subsequent introductions and expansion of range caused a very widespread distribution (figure 6.9).

An interesting, if not the most interesting, aspect of distribution is that which is denoted by the term vicariance, and displayed by the members of a group of related species or other taxa. Bumblebees (*Bombus*) of the subgenus *Cullumanobombus*, for example, include four species, ranging through the northern hemisphere. One species, *B. silantjewi*, occurs in isolated, or disjunct, populations (figure 6.10). The species of this group are said to be vicariously, or allopatrically, distributed, meaning that the species occupy different areas of the globe. Where species tend to have a

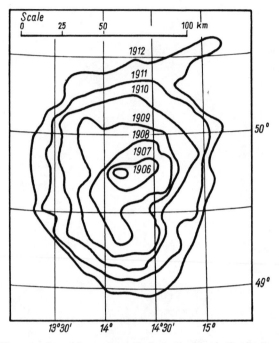

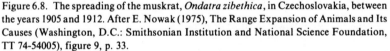

Figure 6.8. The spreading of the muskrat, *Ondatra zibethica*, in Czechoslovakia, between the years 1905 and 1912. After E. Nowak (1975), The Range Expansion of Animals and Its Causes (Washington, D.C.: Smithsonian Institution and National Science Foundation, TT 74-54005), figure 9, p. 33.

restricted, as opposed to a widespread, distribution, they are often said to be "endemic" to the areas they occupy.

Vicarious, or allopatric, distributions are common phenomena among closely related taxa, such as species within a genus, or subspecies within a species. Vicarious distributions are less common among distantly related taxa, such as genera within a family, families within an order, or orders within a class. Distantly related taxa tend to be sympatric, meaning that they occupy the same area. The catfish family, Siluridae, for example, is broadly sympatric with the loach family Cobitidae (figure 6.11). Both groups are members of the higher taxon Ostariophysi. Another group, more closely related to the Cobitidae than to the Siluridae, is the sucker family, Catostomidae (figure 6.12), which,

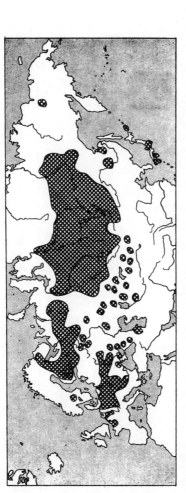

Figure 6.9. Eurasian distribution of the natively North American muskrat, *Ondatra zibethica*, as a result of human introduction beginning in 1905. After G. de Lattin (1967), Grundriss der Zoogeographie (Stuttgart: Fischer), figure 18, p. 50.

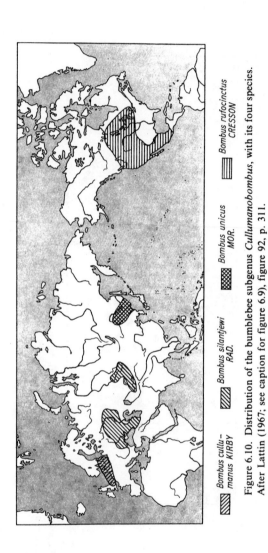

Figure 6.10. Distribution of the bumblebee subgenus *Cullumanobombus*, with its four species. After Lattin (1967; see caption for figure 6.9), figure 92, p. 311.

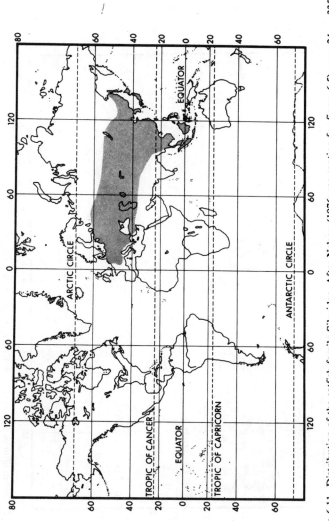

Figure 6.11. Distribution of the catfish family Siluridae. After Nelson (1976; see caption for figure 6.5), map 24, p. 335.

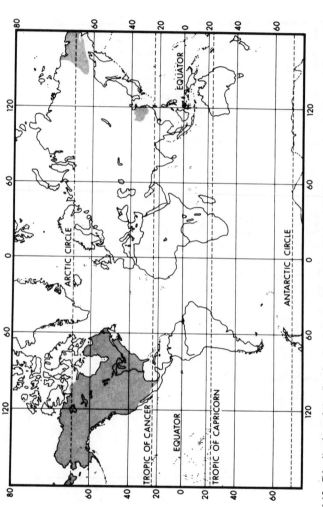

Figure 6.12. Distribution of the fish family Catostomidae. After Nelson (1976; see caption for figure 6.5), map 20, p. 333.

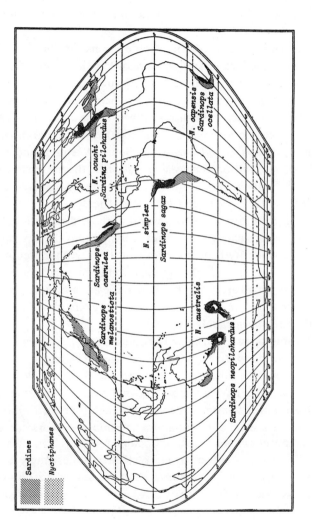

Figure 6.13. Distribution of species of sardines (*Sardina* and *Sardinops*) and species of the euphausiid crustaceans of the genus *Nyctiphanes*. Modified from J. W. Hedgpeth (1957), Marine biogeography. In J. W. Hedgpeth, ed., Treatise on Marine Ecology and Paleoecology, 1:359–82 (New York: Geological Society of America, Memoir 67), figure 12, p. 372.

although present both in North American and Eurasia, overlaps the distribution of the Cobitidae only in one small area of China. The Cobitidae and Catostomidae are members of one subdivision (Cypriniformes) of the Ostariophysi. Even as families, they are vicariously distributed.

Patterns of vicarious distributions of different groups are, like distributions of single taxa generally, more or less similar or different. Consider a case of similarity (figure 6.13). Shown are the distributions of the four species of the euphausiid genus *Nyctiphanes*, and the six species of pilchards (fishes, sometimes called sardines, of the closely related, if not synonymous, genera *Sardina* and *Sardinops*). *Nyctiphanes australis* occurs with *Sardinops neopilchardus* in South Australia and New Zealand; *N. simplex*, with *S. caerulea* and *S. sagax* in the eastern Pacific; *N. couchi*, with *S. pilchardus* in the eastern Atlantic-Mediterranean; and *N. capensis*, with *S. ocellata* in southern Africa.

Relative to marine distributions generally, these distributions are fairly restricted, approaching in size those distributions that are called "endemic." The similarity between them is of the sort that eventually leads, if it is observed in still other groups, to the concept of areas of endemism—fairly small areas that have a significant number of species that occur nowhere else. In this particular case, the pilchards indicate six areas; and the euphausiids, only four.

If these distributions are considered as indicators of possible areas of endemism, which group is the more informative? The answer to this question is obvious: the pilchards. The northwest Pacific could be an area of endemism, even if *Nyctiphanes* does not occur there. And there could be two areas of endemism in the eastern Pacific (one north and one south), even if one species of *Nyctiphanes* (*N. simplex*) occurs in both.

It is not the purpose here to list, or document a list, of the areas of endemism for the world—either marine or terrestrial—but simply to call attention to the basic *ideas* pertaining to plant and animal distribution, as they were developed in the pre-Darwinian period. The maps of figures 6.1–6.13 are, of course, from the recent scientific literature. The *facts* of distribution that they contain are more abundant than they were in earlier times. But greater abundance of facts is not the same as progressive change in basic ideas.

REGIONS

With these factual matters as a background, the accomplishments of the early period of biogeography may be better, more easily, and more clearly understood. One of the last publications of Candolle is a case in point. In a detailed consideration of the distributions of the plants of the family Compositae (1838), in which he listed 40 areas of endemism for land plants of the world, he wrote:

These regions were not established *a priori*, for I have recognized as regions only those areas that can be defined naturally, and in which I have noted many endemic species. With reference only to the family that I have chosen as an example, the results of my tabulation are that, of 8523 known species, no more than 562 have been found in more than one region. This number, however, is an overestimation of the actual number of species: firstly, because the same species, for example *Gnaphalium luteoalbum*, is sometimes found in 3, 4, or 5 regions [and would have been counted in my table as 2, 3, or 4 of the 562 occurrences]; and secondly, I have been obliged to count as common to several regions those species cited by diverse authors and, consequently, of dubious identity. Without risk of error, therefore, the number can be reduced to less than 500, which would mean that 6% of the species of Compositae are found in more than one region; or, in other words, that about 94% of the species cited for each region are, on the average, endemic there and occur nowhere else. (1838:9, translated)

These 500 widespread, or nonendemic species, Candolle considered exceptions to his generalization about the global distribution of land plants: that there are 40 or more real botanical regions exemplified by botanical distribution generally. He considered these exceptions in greater detail:

Among continental areas, plants are much more easily able to pass from one area to another, such that the distribution of a species is often very large, extending over two or more regions. Also, one is sometimes obliged to recognize regions that are too large, because the available facts are not precise enough to warrant exact subdivision. Thus, I have considered all of tropical Africa as one "region"—this one with an area of several million square kilometers. This "region" surely will be subdivided into several when it is better known. With respect to Compositae, however, the error that might result from recognizing it is small, for only 62 species of this family are known there. These considerations apply, although to a lesser degree, to Brazil, to China and surrounding areas, and also to Central Asia, which I have only mentioned in the table [as region no. 40], because I know of no species of Compositae that lives there. If one examines the 500 Compositae that occur in two or more regions, one quickly perceives that almost all of them occur in regions that: 1. are contiguous (Europe and the Orient, the Orient and Siberia); 2. are separated by discontinuous and irregular arms of the sea (Siberia and northern America), by oceans that are, perhaps, younger than the vegetation (Central Europe and North Africa); 3. appear to have been influenced by human activity, voluntary or not, resulting in the introduction of plants from one region into another—which is known fairly certainly for *Erigeron canadense, Xanthium macrocarpum*, and *Bidens leucantha* (which are not cited in the ancient floras of those parts of Europe wherein the plants grow today in great abundance), and which is probably true for many others, such as *Cnicus benedictus* (introduced into austral America), *Guizotia oleifera* (cultivated in India and Abyssinia), etc.

If we disregard these causes of error, and a few cases of doubtful identification, we find that the number of species of Compositae that actually seem to be native to widely separated regions is singularly small and doubtful. (1838:10, translated)

Candolle listed eight such species, noting their widely scattered distributions. Then he continued:

We see, then, that in one of the families that is the best known, eight of 8500 species seem to be exceptions to the general laws of the botanical regions of the world. We note also that these eight species have seeds that are so small and so numerous, and mainly grow so near the coasts, that it seems difficult not to believe that they were transported by humans or by physical causes such as water and wind.

This result is all the more remarkable, because the family Compositae is one of those for which dispersal of species seems the most easy: they are robust, able to accommodate themselves to different climates; each year they produce a large number of seeds that germinate with ease; the seeds are small and almost always provided with hooks that might aid in long-distance transport. It is such a family for which we find so few species distributed beyond their regions or regions contiguous among themselves! (1838:11, translated)

It is not possible here to do justice to the subsequent history of biogeography, if only because the publications on this subject are too numerous to be read, digested, and integrated—a fact true even for the period, up to about 1900, during which British authors were very active. Two books by Scharff (1899, 1912), who made a real attempt to summarize the literature for two large areas of the world (Europe and America), give some indication of the complexity of the subject, which is, perhaps, the main reason why biogeography changed its character during this period. On the one side, biogeography fell more and more into the hands of successive generations of taxonomists, who were becoming more and more narrowly specialized; and, on the other, into the hands of theorists preoccupied, for a time at least, with the problem of causal explanation of the facts of distribution as revealed by the studies of taxonomists. Increasingly, however, the activities of these two groups diverged, and communication between them became difficult. The increasingly specialized taxonomists lost their ability meaningfully to generalize about life, its distribution, and its history; and the theorists consequently had fewer and fewer generalizations to theorize about. Perhaps the best that can be said for this period is that biogeography suffered progressive degeneration, until a large number of the scientists supposedly active in this field became fixed in their adoration of what must have seemed to them the ultimate pinnacle of achievement made manifest by Darwin.

This period of stasis is what is commonly called the Darwinian era, which lasted for about 100 years. Its spirit was, perhaps, best captured by a recent commentator on biogeography (Fleming 1958):

> My chief qualification to present a lecture on this topic is that I am a thorough-going and unashamed Darwinian in my beliefs and in my approach to Natural History. Having said this, I can retire from the scene, except as a mouthpiece for Darwin and his contemporaries, for my intention is to present for your consideration an address composed very largely of quotations from Darwin's letters and books, and New Zealand illustrations of the principles he did so much to establish, some of them from work with which I have been associated. (p. 65)

With respect to biogeography, the roots of the Darwinian era extend back into the early activities of Linnaeus, Buffon, and Candolle. In Britain, these early activities were summarized in seminal works by Lyell (1832), Swainson (1835), and Prichard (1836), but it was Lyell who brought them together most forcibly, and it was Lyell who, ironically, prepared the way for their eventual effacement in the events, and the effects of the events, of the Darwinian revolution.

The heritage of this early period might be summed up by the term "Buffon's law," which by the time of Candolle and Lyell had achieved a significant degree of generality. According to this law, life is distributed as, and in, a large number of regions, or areas of endemism. Further investigation was expected, both by Candolle and Lyell, to reveal more and more regions and more and more endemic species—a task that was eventually accomplished, not directly in response to Candolle's or Lyell's impetus, but accidentally, as it were, by generations of taxonomists who were working in an ostensibly Darwinian tradition. Also, further investigation was hoped, at least by Candolle if not by Lyell, to reveal *the* generally valid, causal explanation of the existence of regions—again a task that was eventually accomplished, not in response to Candolle's or Lyell's impetus, but accidentally, as it were, by geologists studying the nature of the ocean floor in the 1960s.

Lyell accepted the reality of Candolle's regions, but believed that, to some extent at least, they were capable of causal explanation with the facts then on hand. He theorized that the number of living species is in equilibrium. He believed that some species had suffered extinction in the past and that, therefore, there should be a creative principle responsible for the emergence of new species, such that the equilibrium could be maintained. New species, he imagined, were created one at a time, in one region or another.

Lyell's concept of "creation" is not miraculous creation, but creation according to natural law and process. He was not specific about the nature of creative laws and process, but he did argue against the idea, as expounded by Buffon and Lamarck in particular, that the process involved evolution—or transmutation, as it was then called—of species. Although he never said as much, Lyell's statements on these matters amount to the belief that a species—by some natural process to be sure—arises from thin air, albeit within the confines of an area of endemism, wherein the new species becomes native upon its creation.

Lyell's treatise on biogeography, published as volume 2 of his *Principles of Geology* (1832), was the definitive statement of the subject in England for the middle years of the nineteenth century. Because Lyell allowed, and even argued for, "creation" of species, he conformed to the outlook of the times. And naturalists who enthused over his work felt free to write not of areas of endemism arising through natural processes, but of "centers of creation" arising through the working of divine Providence.

One of the more remarkable items in the pre-Darwinian period is a paper by Alfred Wallace (1855), entitled *On the Law which has regulated the Introduction of New Species.* Wallace believed that his "law"—newly discovered by himself—

explains and illustrates all the facts connected with the following branches of the subject:—1st. The system of natural affinities. 2nd. The distribution of animals and plants in space. 3rd. The same in time, including all the phaenomena of representative groups, . . . 4th. The Phaenomena of rudimentary organs. (1855:186)

Wallace referred to Lyell's *Principles*, noting Lyell's arguments to the effect that species become extinct. Wallace then stated:

To discover how the extinct species have from time to time been replaced by new ones down to the very latest geological period, is the most difficult, and at the same time the most interesting problem in the natural history of the earth. The present inquiry, which seeks to eliminate from known facts a law which has determined, to a certain degree, what species could and did appear at a given epoch, may, it is hoped, be considered as one step in the right direction towards a complete solution of it. (1855:190)

Wallace's argument, in effect, was a commentary on Lyell's vague and unspecific remarks about "creation" of new species. Wallace also used the term "creation," but for him it was not "creation" from thin air, but "creation" from pre-existing species. He wrote:

Such phaenomena as are exhibited by the Galapagos Islands, which contain little groups of plants and animals peculiar to themselves, but most nearly allied to those of South America, have not hitherto received any, even a conjectural explanation. The Galapagos are a volcanic group of high antiquity, and have probably never been more closely connected with the continent than they are at present. They must have been first peopled, like other newly-formed islands, by the action of winds and currents, and at a period sufficiently remote to have had the original species die out, and the modified prototypes only remain. In the same way we can account for the separate islands having each their peculiar species, either on the supposition that the same original emigration peopled the whole of the islands with the same species from which differently modified prototypes were created, or that the islands were successively peopled from each other, but that new species have been created in each on the plan of the pre-existing ones. (1855:188)

Wallace's commentary on causal principles bears some similarity to Buffon's argument about dispersal of species into new areas. But Wallace did not consider *why* species might change when colonizing new areas until a few years later when he and Darwin together published identical views on the subject—views that, as far as biogeography is concerned, were a rebirth in Britain of Buffon's causal explanation under the guise of a supposedly newly discovered principle, natural selection.

Wallace's 1855 paper, however, is remarkable, being written, as it was, when Wallace was busy collecting natural-history specimens in Borneo. The background of the paper was reasonably presented by Wallace:

The great increase of our knowledge within the last twenty years, both of the present and past history of the organic world, has accumulated a body of facts which should afford a sufficient foundation for a comprehensive law embracing and explaining them all, and giving a direction to new researches. It is about ten years since the idea of such a law suggested itself to the writer of this paper, and he has since taken every opportunity of testing it by all the newly ascertained facts with which he has become acquainted, or has been able to observe himself. These have all served to convince him of the correctness of his hypothesis. Fully to enter into such a subject would occupy much space, and it is only in consequence of some views having been lately promulgated, he believes in a wrong direction, that he now ventures to present his ideas to the public, with only such obvious illustrations of the arguments and results as occur to him in a place far removed from all means of reference and exact information. (1855:185)

Wallace's 1855 paper makes interesting reading even today. The entire text, short as it is, is too long to reproduce here, but Wallace summarized his argument thus:

The following propositions in Organic Geography and Geology give the main facts on which the hypothesis is founded.

Geography
1. Large groups, such as classes and orders, are generally spread over the whole earth, while smaller ones, such as families and genera, are frequently confined to one portion, often to a very limited district.
2. In widely distributed families the genera are often limited in range; in widely distributed genera, well-marked groups of species are peculiar to each geographical district.
3. When a group is confined to one district, and is rich in species, it is almost invariably the case that the most closely allied species are found in the same locality or in closely adjoining localities, and that therefore the natural sequence of the species by affinity is also geographical.
4. In countries of a similar climate, but separated by a wide sea or lofty mountains, the families, genera and species of the one are often represented by closely allied families, genera and species peculiar to the other.

Geology
5. The distribution of the organic world in time is very similar to its present distribution in space.
6. Most of the larger and some small groups extend through several geological periods.

7. In each period, however, there are peculiar groups, found nowhere else, and extending through one or several formations.

8. Species of one genus, or genera of one family occurring in the same geological time are more closely allied than those separated in time.

9. As generally in geography no species or genus occurs in two very distant localities without being also found in intermediate places, so in geology the life of a species or genus has not been interrupted. In other words, no group or species has come into existence twice.

10. The following law may be deduced from these facts: —*Every species has come into existence ₚoincident both in space and time with a pre-existing closely allied species.* (1855:185–86)

One element that recurs in many of the items of Wallace's list is the notion of "affinity" (or relationship) among taxa. Wallace considered that patterns of affinities among taxa, viewed either geographically or geologically (stratigraphically), are best explained by "creation" of new species from old ones. Geographically, this means that new species arise within the spatial distributions of the old species; stratigraphically, the new arise within the temporal distributions of the old.

Wallace's view of the world—what he thought needed explanation— was not very different from that of Candolle and Lyell because of Wallace's emphasis on patterns of taxonomic affinity. Candolle, for example, was well aware of the items summarized under Wallace's heading of "Geography" (items 1–4). Candolle even introduced the term "endemic" to describe those genera and families for which, in his words, "all of the species are native to a single region (I call them, by analogy with the medical language, *endemic genera*)," as contrasted with other genera and families, of wider distribution, for which "the species are distributed over the whole world (I call them, by similar analogy, *sporadic genera*)" (1820:412, translated). The items summarized under Wallace's heading of "Geology," in contrast, were not known in Candolle's time.

Wallace conceived of his "law" as an explanation of certain facts about spatial and temporal distributions of organisms—both fossil and recent. In particular, his "law" purportedly explained in general terms *how* geographical and stratigraphic patterns came into existence. What is important is to realize that these facts—the patterns of affinities— were known prior to Wallace's attempt to explain them with reference to a theoretical principle about the creation of new species.

Consider again the bumblebees of the subgenus *Cullumanobombus* (figure 6.10). By Wallace's reasoning, the distributions of the four species would have originated by derivation, ultimately, from the distribution of a single "antitype" (or ancestral species), which might have been either restricted in distribution or widespread throughout the northern hemisphere. In the former case, dispersal from the originally restricted area of distribution of the ancestral species would be necessary to account for the different distributions of the recent species. The bumblebee distributions would, viewed in this way, be exactly comparable with the Galapagos distributions as interpreted by Wallace in the passage quoted above.

RELATIONSHIPS OF REGIONS

Beneath the surface of Wallace's argument about the origin of new species is another idea bearing more directly on the empirical data of biogeography—an idea that derives from patterns of affinity. The idea is not simply affinity among taxa, but affinity among the areas occupied by the taxa. This idea was not developed much by Wallace, but it was considered a few years later by Philip Sclater (1858):

An important problem in Natural History, and one that has hitherto been too little agitated, is that of ascertaining the most natural primary divisions of the earth's surface, taking the amount of similarity or dissimilarity of organized life solely as our guide.

The world is mapped out into so many portions, according to latitude and longitude, and an attempt is made to give the principal distinguishing characteristics of the Fauna and Flora of each of these divisions; but little or no attention is given to the fact that two or more of these geographical divisions may have much closer relations to each other than to any third. (pp. 130–31)

There are some problems with Sclater's notions about areas (as delimited by latitude and longitude) and relationships of areas (as determined by similarity or dissimilarity), as will be discussed below. But his was one of the early statements about this problem (interrelationships of areas). The problem was soon explored in a practical way by Wallace, who in 1863 published an analysis of the interrelationships of four areas: Sumatra, Java, Borneo, and the Asian mainland. His analysis concludes:

Java possesses numerous birds which never pass over to Sumatra, though they are separated by a strait only 15 miles wide, and with islands in mid-channel. Java, in fact, possesses more birds and insects peculiar to itself than either Sumatra or Borneo, and this would indicate that it was earliest separated from the continent; next in organic individuality is Borneo, while Sumatra is so nearly identical with the peninsula of Malacca in all its animal forms, that we may safely conclude it to have been the most recently dismembered island. (1863:228)

Wallace's concept of the interrelationships of these four areas may be expressed in a branching diagram (figure 6.14), which is easily interpreted to suggest the idea of a historical development of a once-continuous land that was divided initially into two fragments: one, Java; and the other, a continuous Borneo + Sumatra + Asia. Next, the latter was divided into two fragments: one, Borneo; and the other, a continuous Sumatra + Asia. Finally, Sumatra and Asia were divided.

Viewed in the historical context, Wallace's concept (figure 6.14) has the characteristics of a tree, rather than a cladogram; and a tree wherein the "ancestors" are specified—not as single areas, but as areas in combination. Interestingly, the "ancestors" (SU + AS, etc.) specify the components of a cladogram of areas. It is of interest to view the concept as a cladogram—that is, as a general summary of knowledge. Wherein, one may ask, is the knowledge represented by component SU + AS? by component BO + SU + AS? by component JA + BO + SU + AS?

Wallace's summary of these matters, quoted above, gives no answers to these questions, not even to the first: even though most species found in Sumatra are found also on the Asian mainland, there might be more such species shared by Borneo and the mainland, or by Java and the mainland. Even if not, are shared species the knowledge represented by components? As will be argued below, the answer is no.

Wallace, in fact, based his conclusions not on shared species, but rather on the relative numbers of endemic species. There is an obvious

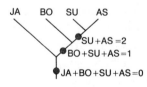

Figure 6.14. Alfred Wallace's concept of the historical interrelationships of four areas: Java (JA), Borneo (BO), Sumatra (SU), and the Asian mainland (AS). Based on Wallace (1863).

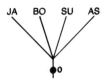

Figure 6.15. The correct cladogram for Wallace's four areas, based on the information supplied by him (relative numbers of endemic species, which merely suggest, or confirm, that there are four such areas of endemism).

problem with this argument. Firstly, one may ask, what is necessary in order to suggest the existence of an area of endemism? The answer is obvious: one endemic species. But is one endemic species enough to prove such to be true? The answer is a matter of judgment. If one is not enough, are two? Or ten? Or twenty? Clearly, the discovery of additional species endemic, say, to Java, can only confirm that Java is an area of endemism. Each newly discovered endemic species is confirmatory but redundant information.

If endemic species merely establish the identity of an area of endemism, what kind of information might serve to indicate the relationships between different areas of endemism? This question will be considered below and, in detail, in the next two chapters. It suffices for the moment to note that, on the information supplied by Wallace (relative numbers of endemic species), the correct cladogram for the four areas is merely the most general cladogram possible, i.e., the tertiary cladogram (figure 6.15). It should not be concluded, therefore, that Wallace's concept (figure 6.14) is wrong, for his concept is, after all, one of the possibilities subsumed in the tertiary cladogram.

It is pointless to argue that because Wallace was premature in his judgment of the history of these areas, he was fruitlessly blundering. He was, in fact, personally breaking new ground in his attempts, of which this was one, to use distributional information about organisms as a basis for inferring the history of landmasses. His was not the first such attempt, but his certainly was an early one, which came at an interesting moment in the history of biogeography. A considerable amount of knowledge had already been acquired about distributions of plants and animals, about areas of endemism, about earth history, about the possibilities of organic evolution, such that Wallace might have posed the questions of the preceding paragraphs. Of course, he did not, but, then, he did not have the advantage of hindsight.

Wallace, in fact, was not directly concerned with areas of endemism and their interrelationships. Nor was Sclater in his 1858 paper, in which he argued that there are, as evidenced by birds, six major zoological regions of the world. In retrospect, both Sclater and Wallace missed an opportunity to deal decisively with the matter of area interrelationships. That the opportunity was missed may be appreciated by analysis of Wallace's example. Wallace did not supply the exact numbers of the species endemic to Java, Borneo, Sumatra, and Asia. Consider, therefore, some hypothetical numbers of endemic species of birds and insects—numbers for which the relative magnitude follows the logic of Wallace's argument:

	Endemic Species	
	Birds	Insects
Java	4	4
Borneo	3	3
Sumatra	2	2
Asia	1	1

The number of different cladograms possible for 10 species is about 35 million. Consider two cladograms of the ten species that are perfectly

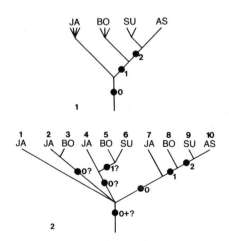

Figure 6.16. Two cladograms for 10 hypothetical species endemic to Java, Borneo, Sumatra, and Asia, with an analysis of their components (0, 1, 2).

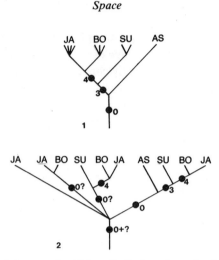

Figure 6.17. Two cladograms for 10 hypothetical species endemic to Java, Borneo, Sumatra, and Asia, with an analysis of their components (0, 3, 4).

consistent with Wallace's concept (figure 6.16); and two others (figure 6.17), consistent among themselves, but which would lead to a different concept of the interrelationships of the four areas (figure 6.18).

Given cladograms specifying the interrelationships of the endemic species, such as those of figure 6.16, the components of the cladograms may be examined and compared with the components of Wallace's concept (figure 6.14). For convenience, the components are numbered as follows:

Component 0:   JA + BO + SU + AS
Component 1:   BO + SU + AS
Component 2:   SU + AS

Cladogram 6.16.1 includes each of these three components. Cladogram 6.16.2 includes seven components as follows:

Components	Occurrence
0	4
1	2
2	1

Component 0 occurs four times; component 1, twice; component 2, once. Interestingly, component 0 is manifested in four different ways; and component 1, in two different ways:

Figure 6.18. An alternative concept of the interrelationships of four areas: Java, Borneo, Sumatra, and Asia (cf. figure 6.14).

Component 0		Component 1	
Manifestation	Occurrence	Manifestation	Occurrence
10 species	1	2 species	1
2 species	1	3 species	1
3 species	1		
4 species	1		

It is not readily apparent, and seems counterintuitive, that all of these manifestations are really of the same components. Hence they will be analyzed in further detail in the sequence listed above, beginning with component 0.

Ten species: all ten species are interrelated, indicating that the areas are interrelated in some way (figure 6.19.1).

Two species: two species are interrelated, indicating that Java and Borneo are related in some way (figure 6.19.2).

Three species: three species are interrelated, indicating that Java, Borneo, and Sumatra are interrelated in some way (figure 6.19.3).

Four species: four species are interrelated, indicating that Java, Borneo, Sumatra, and Asia are interrelated in some way (figure 6.19.4).

It is clear that the first and last items (ten and four species) are exactly the same. A moment's reflection shows that the other two items (two and three species) are subsumed in the first and fourth. If items 1 and 2 are added, for example, the result is merely item 1 (following the arrow between figure 6.19.1 and figure 6.19.2); this is because item 2 merely indicates that Java and Borneo are related in some way, not that they are more closely related to each other than to Sumatra or Asia (at least a 3-taxon cladogram would be required to indicate that). Similarly, if items 2 and 3 are added, the result is item 3; if items 3 and 4 are added, the result is item 4. Components of this sort (manifestations of component

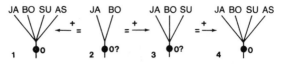

Figure 6.19. Four manifestations of the same component (0). If items 1 and 2 are added, the result (following the arrow between them) is item 1; 2 + 3 = 3; 3 + 4 = 4.

0) are uninformative with respect to the interrelationships of areas, as may be appreciated in comparison with the informative components.

Component 1: Borneo, Sumatra, and Asia are more closely related among themselves than any one of them is to Java (figure 6.20.1).

Component 2: Sumatra and Asia are more closely related among themselves than either of them is to Borneo (figure 6.20.2).

Components 1 and 2 may be added and, together with component 0, produce a fully resolved, and maximally informative, cladogram of the areas (figure 6.20.3).

The two manifestations of component 1 (figure 6.16) may be similarly analyzed (figure 6.21).

Two species: two species endemic to Borneo and Sumatra, respectively, are more closely related among themselves than either is to a species endemic to Java, indicating that Borneo and Sumatra are more closely interrelated among themselves than either is to Java (figure 6.21.1).

Three species: three species endemic to Borneo, Sumatra, and Asia, respectively, are more closely interrelated among themselves than any one of them is to a species endemic to Java, indicating that Borneo, Sumatra, and Asia are more closely interrelated among themselves than any one of them is to Java (figure 6.21.2).

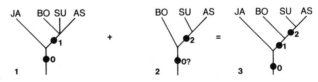

Figure 6.20. Addition of informative components (1, 2), which results in a maximally informative cladogram incorporating both components.

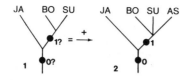

Figure 6.21. Two manifestations of the same component (1). If items 1 and 2 are added, the result (following the arrow between them) is item 2.

The complexity of this sort of analysis seems forbidding at first glance, but in reality the sense of it is easily grasped intuitively after a little practice. Consider once again the two cladograms of figure 6.16, and suppose that one of them (say, 6.16.1) fitted Wallace's birds; and the other (figure 6.16.2), his insects. With these cladograms Wallace would have had independent evidence (from birds *and* from insects) that his concept (figure 6.14) is correct, so far as could be determined from the facts concerning the distribution and relationships of these organisms. Consider, also once again, the two cladograms of figure 6.17, and suppose that they, not those of figure 6.16, fitted Wallace's birds and insects. With these cladograms he would have had independent evidence that his concept (figure 6.14) is wrong. One further supposition is, perhaps, worth consideration: that cladogram 6.16.1 fitted Wallace's birds; and cladogram 6.17.1, his insects. In that case, he would not have been able to conclude anything at all.

Wallace, of course, had no cladograms for the species he believed to be endemic to these areas. One might be tempted to conclude that Wallace had not yet penetrated the nature of the biogeographical problem that he posed about the interrelationships of these areas. This conclusion would be enhanced by the fact that Wallace later modified his views on these matters. He wrote (1876):

These various indications enable us to claim, as an admissible and even probable supposition, that at some epoch during the Pliocene period of geology, Borneo, as we know it, did not exist; but was represented by a mountainous island at its present northern extremity, with perhaps a few smaller islets to the south. We thus have a clear opening from Java to the Siamese Peninsula; and as the whole of that sea is less than 100 fathoms deep, there is no difficulty in supposing an elevation of land connecting the two together, quite independent of Borneo on the one hand and Sumatra on the other. This union did not probably last long; but it was sufficient to allow of the introduction into Java of the *Rhinoceros javanicus*, and that group of Indo-Chinese and Himalayan species of mammalia and birds which it alone possesses. When this ridge had disappeared by subsidence, the next elevation occurred a little more to the east, and produced the union of many islets which, aided by subaerial denudation, formed the present island of Borneo. It is probable that this elevation was sufficiently extensive to unite Borneo for a time with the Malay Peninsula and Sumatra, thus helping to produce that close resemblance of genera and even of species, which these countries exhibit, and obliterating much of their former speciality, of which, however, we have still some traces in the long-nosed monkey and *Ptilocerus* of Borneo, and the considerable number of genera both of mammalia and birds confined to two only out of the three divisions of typical Malaya. The subsidence which again divided these countries by arms of the sea rather wider than at present, might have left Banca isolated, as already referred to, with its proportion of the common fauna to be, in a few instances, subsequently modified.

Thus we are enabled to understand how the special relations of the *species* of these islands to each other may have been brought about. To account for their more deep-seated and general zoological features, we must go farther back. (I:358–59)

Wallace's modified view seems hopelessly complex and wildly speculative, but it can be rendered easily enough into a tree diagram defined, in part, by two components: 0 and 1 (figure 6.22). The cladogram corresponding to the tree diagram is that of figure 6.23, which differs from Wallace's original concept only in lacking component 2 ("SU + AS" of figure 6.14). Interestingly, Wallace abandoned component 2 because species he thought endemic to Borneo were later found in Sumatra and elsewhere:

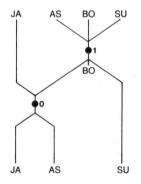

Figure 6.22. Alfred Wallace's revised concept of the historical interrelationships of four areas (Java, Borneo, Sumatra, Asia), with two components indicated (0, 1). Based on Wallace (1876).

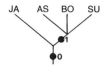

Figure 6.23. The cladogram corresponding to Wallace's revised concept of the inter-relationships of four areas (cf. figure 6.22).

In birds we hardly find anything to differentiate Borneo and Sumatra in any clear manner. *Pityriasis* and *Carpococcyx*, once thought peculiar to the former, are now found also in the latter. . . . The Malay Peninsula is perhaps the best known, but it is probable that both Sumatra and Borneo are quite as rich in species. With the exception of the genera noted above, and two or three others as yet found in two islands only, the three districts we are now considering may be said to have an almost identical bird-fauna, consisting largely of the same species and almost wholly of these together with closely allied species of the same genera. There are no well-marked groups which especially characterise one of these islands rather than the other, so that even the amount of speciality which Borneo undoubtedly exhibits as regards mammalia, is only faintly shown by its birds. (1876, I:355)

In this way Wallace was true to his original belief that *relative numbers* of endemic taxa were in themselves, or in part, sufficient to indicate *relative degrees* of interrelationships between areas. Consider again,

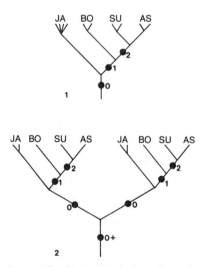

Figure 6.24. Two cladograms for 10 hypothetical species endemic to Java, Borneo, Sumatra, and Asia, with an analysis of their components (0, 1, 2).

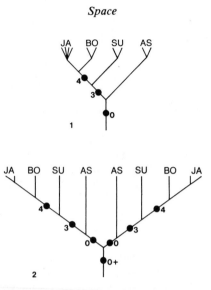

Figure 6.25. Two cladograms for 10 hypothetical species endemic to Java, Borneo, Sumatra, and Asia, with an analysis of their components (0, 3, 4).

therefore, certain hypothetical numbers of endemic species of birds and insects that are evenly divided between Borneo, Sumatra, and Asia (numbers simulating Wallace's revised data on this subject):

	Endemic Species	
	Birds	Insects
Java	4	4
Borneo	2	2
Sumatra	2	2
Asia	2	2

There are, as before, about 35 million possible cladograms for ten taxa. Consider two cladograms of the ten species that are perfectly consistent with Wallace's original concept (figure 6.24; cf. figures 6.14, 6.16); and two other cladograms, consistent among themselves, but which would lead to a different concept of the interrelationships of the four areas (figure 6.25; cf. figures 6.17, 6.18).

From the above considerations, one may conclude that relative numbers of endemic species allow for different cladograms of the interrelationships of the species. By themselves, relative numbers of endemic species are uninformative with respect to the nature of the interrelationships between species and, consequently, to the nature of the interrelationships between the areas of endemism. If these conclusions are accepted, then Wallace's views on these matters, in this particular case, may be judged accordingly. In effect, Wallace proposed a hypothesis about area-interrelationships—a hypothesis that leads to certain predictions about the interrelationships of endemic species. Specifically, his earlier view (figure 6.14) predicts two components (1, 2) in any cladogram of species-interrelationships; and his later view (figure 6.23) predicts only one component (1) in any such cladogram. Precisely, his earlier view (figure 6.14) predicts:

(1.) Given four species endemic to the four areas, the species endemic to Borneo, Sumatra, and Asia, respectively, are more closely related among themselves (and will be found to be so) than to a species endemic to Java.

(2.) Given three species endemic to Borneo, Sumatra, and Asia, respectively, the species endemic to Sumatra and Asia are more closely related among themselves (and will be found to be so) than to a species endemic to Borneo.

And his later view predicts the first, but not the second, of the above two items.

7

BIOGEOGRAPHIC PATTERNS: COMPONENT ANALYSIS

ERRORS IN PREDICTION

Prediction, as considered in the preceding chapter, is easily grasped in its generality, according to which there should be a correspondence between species-relationships and area-relationships. Whether or not this correspondence actually exists, or can be discovered, is a matter of interest to current investigation. Final answers have not yet been obtained on these points of information about the world of living things. Wallace never considered these questions, nor did most of the biogeographers since Wallace's time. The questions are implications of vicariance biogeography, which is a development primarily of the last twenty-five years.

Why these questions were so long unasked is itself a question that cannot be answered very satisfactorily, beyond the obvious fact that Wallace and other biogeographers after him were busy asking and exploring different questions. Wallace, for example, was interested in the relative numbers of endemic species. Also, careful analysis of branching diagrams (cladograms) is another development of the last twenty-five years. Finally, there is an additional reason (sampling error)—aside from the problem of correctly assessing species-relationships—that might have been important. This reason pertains directly to the testing of specific predictions. Consider once again cladogram 7.1.2, showing the interrelationships of 10 hypothetical species endemic to Java, Borneo, Sumatra, and Asia. Suppose that there are 10 such species, really interrelated as specified by the cladogram. Suppose also that an investigator wishing to test the predictions implied by Wallace's original concept (figure 7.1.1) samples these 10 species and chooses, by chance, species 6, 7, 8, and 10. Suppose finally that he arrives at the correct assessment of the species-relationships (figure 7.2). He would find two components (5!, 6!) that apparently contradict the predictions. In this case, the apparent contradiction would stem only from sampling error, not from incorrect assessment of species-relationships. Thus, the investigator would really have discovered components 0 and 1, but would have misidentified them, because of chance alone, as new components contradicting the predictions implied by the original concept. Sampling error of this sort might have been very common in the past, and may be very common even now. If so, its presence might have totally frustrated efforts to investigate the possible correspondence between species- and area-relationships.

Sampling error results from missing knowledge about species and their occurrences, either because certain species (for example, species 1–5 and 9 of figure 7.1.2) are omitted from consideration, or are unknown or uncollected. The species could even be extinct. At present, it is not possible to state the real difficulties caused by sampling error, for these types of problems have received little investigation. Nevertheless, the

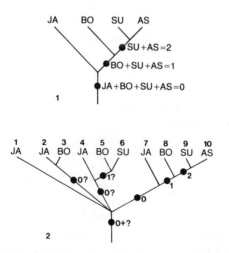

Figure 7.1. Cladograms of (1) Wallace's concept of the interrelationships of four areas (cf. figure 6.14), and (2) 10 hypothetical species endemic to those areas (cf. figure 6.16.2).

investigator is not totally at the mercy of sampling error. In this case, for example, the investigator might compare the original concept (figure 7.1.1) and the result (figure 7.2) and, reasoning that the result might be due to sampling error, infer that component 5! = component 0 and that component 6! = component 1. These inferences would lead to certain secondary predictions:

1. That species 6 is related to yet undiscovered species endemic to, or occurring in, Java, Borneo, or Asia (species 4 and 5 of figure 7.1.2).
2. That species 7, 8, and 10 are related to a yet unknown species endemic to, or occurring in, Sumatra (species 9 of figure 7.1.2).

Cases of sampling error, therefore, may lead directly to a secondary set of predictions that, in effect, overcome the sampling error—provided, of course, that the initial concept is correct, and that the interrelationships of the known species are correctly assessed. The secondary set of predictions is different from the preceding, or primary, predictions, for the secondary set is not falsifiable. Consider, for example, prediction 1: that species 6 is related to yet undiscovered species endemic to, or occurring in, Java, Borneo, or Asia. Given the prediction, one might search for the yet undiscovered species. Finding the species would, of course, fulfill the prediction. But failure to find the species would not refute the prediction, because the prediction asserts only that there are yet undiscovered species. The prediction asserts nothing about when such species might be found, by whom they might be found, or what kind of searching in exactly what place would be sufficient to find them. Hence, the secondary predictions have, at most, only heuristic value. If followed up, they might result in the discovery of previously unknown species, or species of previously unknown, or unsuspected, relationships.

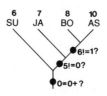

Figure 7.2. Cladogram of four hypothetical species endemic to Sumatra, Java, Borneo, and Asia, with an analysis of its components (5!, 6!). The components are apparent falsifiers of Wallace's concept, but in reality stem from sampling error (cf. figure 7.1). Component 5! = component 0; component 6! = component 1.

Because the secondary predictions have some potential value, their origin may be considered in greater detail.

	Original concept (figures 7.1.1, 7.1.2) components		Result found (figure 7.2) components
0:	JA+BO+SU+AS	0:	JA+BO+SU+AS
1:	BO+SU+AS	5!:	JA+BO+AS
2:	SU+AS	6!:	BO+AS

How might components 5! and 6! be inferred, because of sampling error, to represent components 0 and 1? If sampling error arises because of missing information, the missing information can be added to components 5! and 6!, in order to convert them into the components of the original concept.

Result found		Element added		Conversion	
5!: JA+BO+AS	+	SU	=	0:	JA+BO+SU+AS
6!: BO+AS	+	SU	=	1:	BO+SU+AS
6!: BO+AS	+	JA	=	0:	JA+BO+SU+AS

The only possibility for component 5! is adding to it element SU, thereby converting component 5! into component 0. There are two possibilities for component 6!: adding SU results in component 1; adding JA results in a manifestation of component 0. Interestingly, adding element SU to component 6! also adds SU to component 5! Hence both components (5!, 6!) that apparently contradict the original concept are converted into components of the original concept by the simple addition of SU to component 6!—in other words, by the discovery of a species endemic to, or occurring in, Sumatra, and the correct determination of its relationships.

The mere discovery of a species endemic to Sumatra does not by itself cause the conversion of the falsifying components, for there is another aspect of sampling error, exemplified by an incorrect or imprecise statement of the relationships of the species. This aspect of sampling error may also be considered to be due to lack of information—the information necessary for a correct and fully informative statement of the relationships of the species. Suppose, for example, that species 9 (with relationships as shown in figure 7.1.2) is discovered, and that it is to be added to cladogram 7.2. There are 10 different ways it might be added (figure 7.3), only one of which (7.3.10) is correct and fully informative in all of its details.

Four of the 10 cladograms (7.3.1–4) have the falsifying components unchanged (5!, 6!). In six cladograms (7.3.5–10), component 5! converts to component 0. In three of them (7.3.5–7), component 6! remains: in one of these (7.3.6), a new falsifying component appears (7!); in another (7.3.7), component 1 appears. In the remaining three cladograms (7.3.8–10), component 6! converts to component 1: in one of these (7.3.9), a new falsifying component appears (8!); in another (7.3.10), component 2 appears.

A newly discovered Sumatran endemic would have one of the ten relationships shown in figure 7.3. If it were species 9, with its closest

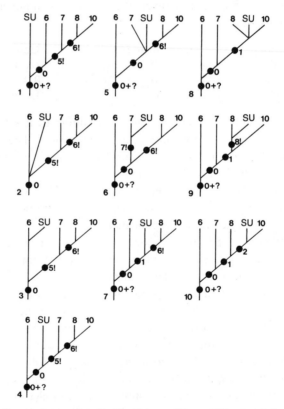

Figure 7.3. Ten cladograms, each of which specifies a different relationship for a hypothetical and newly discovered species endemic to Sumatra (cf. figure 7.2).

relationship with species 10 (figure 7.3.10), certain of the cladograms would be erroneous, not merely in lacking information (7.3.8), but in containing falsifying components when in reality there is none.

Number of cladograms	Falsifying components per cladogram
2	0
3	1
5	2

A summary of the ten cladograms shows that in five of them (7.3.5, 7.3.7-10) there is a reduction (to less than two) in the number of falsifying components. In two of them (7.3.8, 7.3.10) there is total reduction (to zero).

Within the limits of the problem under consideration, there is no way to increase the number of falsifying components, unless the relationships of the species are misconstrued. Thus, no other species (of species 1-5 of figure 7.1.2) can be added to the cladograms of figure 7.3, in order to increase the number of falsifying components, unless the species added is incorrectly placed on the cladogram (if, for example, the SU species of figure 7.4.2 is species 9). If, however, additional species are sampled, the number of falsifying components may increase, because of sampling error, even if the relationships are correctly construed. Consider for example, the possibility that cladogram 7.1.2 is the result of sampling from a larger and more complex cladogram (figure 7.5). In this case, the SU species of figure 7.4.2 would represent species 11 (figure 7.5) in its correct position on the cladogram, and the number of apparently falsifying components would be raised to three.

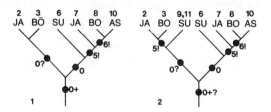

Figure 7.4. Cladograms with falsifying components (cf. figures 7.1, 7.5). In cladogram 1 the falsifying components result from missing species (cf. figure 7.1). In cladogram 2 there is an additional falsifying component that in one circumstance (9, cf. figure 7.1) results from an incorrect statement of relationships; in another circumstance (11, cf. figure 7.5), from missing species.

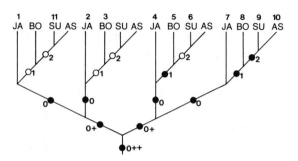

Figure 7.5. An interpretation that would allow the falsifying components of cladograms 7.4.1 and 7.4.2 to result from missing species rather than from incorrect statements of relationships.

The effect of incorrect statements of relationships cannot be assessed in the absence of some idea of what the correct relationships are. Suppose, for example, that a group of four taxa of birds is first studied with the results shown in figure 7.6.1, and that a group of four taxa of insects is then studied with the results shown in figure 7.6.2. The insects would supply two falsifying components (3!, 4!). If the same result is expected for both groups, but the results conflict, one might suspect that the conflict is due to sampling error. Alternatively, if different results are expected for different groups, and the results conflict, one would have no reason to doubt the truth of the results even if they conflict. What is an indication of error under the first expectation is no indication of error under the second expectation.

There is reason to believe that statements of relationships are sometimes erroneous, and that erroneous statements of relationships, when they are made, are indicated by conflicts like that between figures 7.6.1 and 7.6.2. Given such a case of conflict and possible error, what can

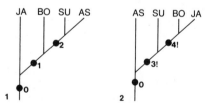

Figure 7.6. Two cladograms: one for 4 taxa of hypothetical birds (1); the other for 4 taxa of hypothetical insects (2).

be done in order to investigate the possibility? The results could be rechecked in the hope that any errors might be discovered. If no errors are discovered, what then? Some additional sample of information about these taxa might be taken, in order to allow for an independent set of statements of relationships to be made. If the new results duplicate the old, one would have greater confidence that the results, even if they conflict, are not due to sampling error. And one might infer that there are two kinds of area-relationships—one kind indicated by birds and another kind indicated by insects.

Alternatively, one may turn toward a third group of organisms, for example fishes. If there is a group of 10 taxa of fishes, distributed and interrelated as shown in figure 7.1.2, it would tend to agree with the results obtained from the birds; alternatively, if there is a group of 10 taxa of fishes, distributed and interrelated as shown in figure 6.17.2, it would tend to agree with the results obtained from insects. In the first case, there would be reason to conclude that if either the bird or the insect results were wrong, the insects were probably wrong; in the second case, the birds.

The logic of the above argument seems straightforward enough, but the problem may be analyzed under the assumption that all errors are due to missing species (the first aspect of sampling error), rather than from incorrect statements of relationships. Thus, if the fish results are those of figure 7.1.2, they may be interpreted as having resulted from missing species interrelated and distributed according to each of two possible patterns: one pattern specified by figure 7.6.1 and the other pattern specified by figure 7.6.2 (figure 7.7).

A comparison of the two interpretations shows the degree to which the hypothetical results obtained from fishes are economically explained by different concepts of area-relationships. In one case (7.7.1) the results require 12 items of error; in the other case (7.7.2), the results require 36 items of error (table 7.1). On the basis of this comparison, one may conclude that the hypothetical results are more economically explained by the former concept rather than the latter; and that, if there is a unique explanation for all organisms (birds, insects, and fishes), the explanation involves the former, not the latter, concept of area-relationships. One may also reach the same conclusion by merely counting the number of times a concept has to be replicated to account for the data. The results obtained from fishes that can be explained by

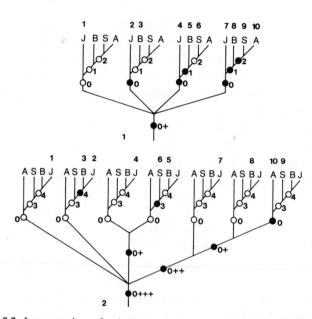

Figure 7.7. Interpretations of a cladogram for 10 taxa of hypothetical fishes (cf. figure 7.1.2): one (1) based on the concept of area-relationships displayed by the hypothetical birds of figure 7.6.1; the other (2), on the concept of area-relationships displayed by the hypothetical insects of figure 7.6.2.

Table 7.1. Items of Error

	Cladogram 7.7.1		Cladogram 7.7.2	
	Occurrences	*Totals*	*Occurrences*	*Totals*
Missing species:				
JA	0		3	
BO	1		4	
SU	2		5	
AS	3	6	6	18
Missing components:				
0+++	–		0	
0++	–		0	
0+	0		0	
0	1		6	
1	2		–	
2	3		–	
3	–		6	
4	–	6	6	18
Totals		12		36

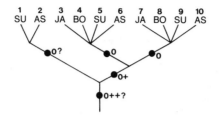

Figure 7.8. Cladogram of certain taxa of birds: 1–2, *Rhopodytes tristis elongatus* and *R. t. longicaudatus*; 3–6, *Otus spilocephalus angelinae*, *O. s. luciae*, *O. s. stresemani*, *O. s. vulpes*; 7–10, *Strix leptogrammica bartelsi*, *S. l. leptogrammica*, *S. l. myrtha*, *S. l. maingayi*. Data from J. L. Peters (1940), Check-list of Birds of the World, 4:52, 88–89, 157–58 (Cambridge, Mass.: Harvard University Press).

four replications of one concept (figure 7.7.1) require seven replications of the other concept (figure 7.7.2).

As has been noted above, information necessary for resolving area-relationships was generally not available to Wallace, and is scarcely available even today. Data on bird distributions, for example, are abundant, but, with respect to the interrelationships of the taxa, there is virtually no available information that would resolve area-relationships. Some examples, chosen at random from Peters' *Check-list of Birds of the World* are shown in figure 7.8. A comparison of the same two interpretations of the bird data is shown in figure 7.9. The data are explained with equal efficiency by both concepts. In other words, the bird data are uninformative with respect to area-relationships (table 7.2).

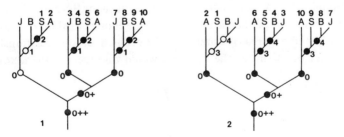

Figure 7.9. Interpretations of a cladogram for certain taxa of birds (cf. figure 7.8): one (1) based on the concept of area-relationships displayed by the hypothetical birds of figure 7.6.1; the other (2), on the concept of area-relationships displayed by the hypothetical insects of figure 7.6.2.

Table 7.2. Items of Error

	Cladogram 7.9.1		Cladogram 7.9.2	
	Occurrences	*Totals*	*Occurrences*	*Totals*
Missing species:				
JA	1		1	
BO	1		1	
SU	0		0	
AS	0	2	0	2
Missing components:				
0++	0		0	
0+	0		0	
0	1		0	
1	1		–	
2	0		–	
3	–		1	
4	–	2	1	2
Totals:		4		4

A third aspect of sampling error is exemplified by the possibility that a missing species might already have been found and identified not as a previously unknown species, but misidentified as a species previously known and thought to be endemic to one area or another. Consider again the hypothetical example of contradictory cladograms for birds and insects (figure 7.6). Suppose again that a group of fishes is studied, in which there are only three species, one of which occurs in two areas (figure 7.10).

Finding that, for example, a certain taxon thought to be endemic to Java (or Borneo) occurs also in Borneo (or Java) would reduce by one the number of taxa thought to be endemic to Java (or Borneo). Provided that the number of endemic taxa is still large enough to allow the area to be considered an area of endemism, the problem of area-relationships would not be changed (the problem would disappear if the taxon in

1   2 3
JA+BO SU AS

●?

Figure 7.10. A hypothetical group of three fish species, one of which occurs both in Java and in Borneo.

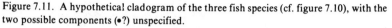

Figure 7.11. A hypothetical cladogram of the three fish species (cf. figure 7.10), with the two possible components (•?) unspecified.

question had been the only endemic, for finding it elsewhere would have reduced the list of endemic taxa to zero).

If the widespread taxon (species 1 in Java and Borneo) can be placed in an informative cladogram (e.g., figure 7.11), the cladogram may be analyzed under either of two assumptions about the nature of the widespread taxon.

ASSUMPTION 1

Under the assumption that the Javanese and Bornean occurrences of species 1 will never be distinguished as separate taxa, it might then be argued that whatever is true of the one occurrence is also true of the other occurrence. In concrete terms (figure 7.11), if species 2 in Sumatra and species 3 in Asia are more closely related among themselves than to species 1 in Java (or Borneo), then species 2 and 3 are more closely related among themselves than to species 1 in Borneo (or Java). Under this assumption, the species cladogram yields one component (2; figure 7.12.2), which allows for three dichotomous area-cladograms (figure 7.13). In this case, the hypothetical fishes would tend to confirm the birds (figure 7.6.1), not the insects (figure 7.6.2).

The species-cladogram (figure 7.11) may be interpreted as having resulted from missing species interrelated according to each of the 15 possible dichotomous patterns of area-relationships (figures 7.14 and 7.15). In figures 7.14.1–3, there is one species missing from each interpretation (one species missing jointly from Java and Borneo). And there is one component missing from each interpretation: component (1) from 7.14.1; component (9) from 7.14.2; and component (4) from 7.14.3. In figures 7.15.1–12, there are five species and five components missing from each interpretation (table 7.3). Some interpretations allow for alternative placement of either the Javanese (7.15.1–2) or Bornean (7.15.10–11) occurrences of species 1 (figure 7.16).

Figure 7.12. Analysis of the area-components (2) implied by the species-cladogram (1) for the three fish species (figure 7.11).

Under assumption 1, therefore, the species-cladogram is sufficient to divide the 15 possible dichotomous area-cladograms into two sets: one of 3, and the other of 12, area-cladograms. The former set includes those area-cladograms that are economical interpretations relative to those of the latter set. Component analysis of the species-cladogram yields only one informative area-component (2). The widespread taxon (species 1) by itself contributes no area-component.

A similar analysis may be done for a different species-cladogram having the widespread taxon (species 6) in a different position (figures 7.17 and 7.18). Again, the widespread taxon contributes no area-component; there is only one area-component resolved (1); and there are three possible economical dichotomous area-cladograms (figure 7.18).

Interestingly, combining the two species-cladograms (figures 7.19.1–2) is equivalent to combining their implied area-cladograms (figures 7.19.3–4). The combination yields two area-components (1 and 2), which specify as economical only one of the 15 possible dichotomous area-cladograms. Hence, under assumption 1, two 3-taxon 4-area species-cladograms are sufficient to specify a fully resolved (dichotomous) area-cladogram—if each species-cladogram yields a different area-component, and if the area-components are combinable in a single cladogram. If the area-components are noncombinable, no informative area-cladogram need result (e.g., figure 7.20).

There are 15 possible dichotomous cladograms for four areas (figure

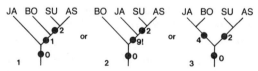

Figure 7.13. The three dichotomous area-cladograms allowed by the component analysis of figure 7.12.

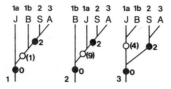

Figure 7.14. Three economical interpretations (cf. figure 7.15) of the species-cladogram for the three fish species (figure 7.11).

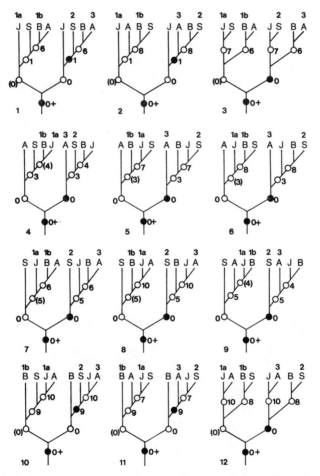

Figure 7.15. Twelve noneconomical interpretations (cf. figure 7.14) of the species-cladogram for the three fish species (figure 7.11).

Table 7.3. Items of Error

	Area Cladograms 7.14.1–3				Area Cladograms 7.15.1–12												
	Occurrences			Totals/ Cladogram	Occurrences												Totals/ Cladogram
	1	2	3		1	2	3	4	5	6	7	8	9	10	11	12	
Missing species																	
JA } BO	1	1	1		3	3	3	3	3	3	3	3	3	3	3	3	5
SU	0	0	0		1	1	1	1	1	1	1	1	1	1	1	1	
AS	0	0	0	1	1	1	1	1	1	1	1	1	1	1	1	1	
Missing components																	
0+	–	–	–		0	0	0	0	0	0	0	0	0	0	0	0	
0	0	0	–		2	2	1	1	1	1	1	1	1	2	2	1	
1	0	–	0		1	1	–	–	–	–	–	–	–	–	–	–	
2	0	0	–		–	–	–	2	2	2	–	–	–	–	–	–	
3	–	–	1		–	–	–	2	2	–	–	–	–	–	–	–	
4	–	–	–		–	–	–	–	–	–	2	2	2	–	–	–	
5	–	–	–		2	–	2	–	–	–	2	2	–	–	–	–	
6	–	–	–		–	–	2	–	–	–	–	–	–	–	2	–	
7	–	–	–		–	–	–	–	2	2	–	–	–	–	–	2	
8	–	–	–		–	2	–	–	–	–	–	–	–	1	1	–	
9	–	1	–		–	–	–	–	–	–	–	–	–	2	–	–	
10	–	–	–		–	–	–	–	–	–	–	2	–	2	2	2	
		Total:		1 2											Total:		5 10

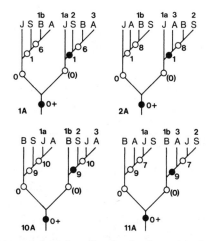

Figure 7.16. Four interpretations that allow for the alternative placement of Java (1A, 2A) and Borneo (10A, 11A; cf. figure 7.15).

7.21). There are 18 possible 3-taxon 4-area species-cladograms (figure 7.22). Each such species-cladogram allows for three economical area-cladograms (table 7.4). Any combination of two or more species-cladograms will specify some number of economical area-cladograms, which may be determined by comparing the lists of three area-cladograms specified by the species-cladograms. For example, the two species cladograms of figure 7.19 are numbers 1 (7.19.1) and 16 (7.19.2), under the assumption that A=Java, B=Borneo, C=Sumatra, and D=Asia; in combination they specify as economical only area-cladogram 1 (7.19.5). The two species-cladograms of figure 7.20 (cf. figures 7.12 and 7.13) are numbers 1 (7.20.1) and 5 (7.20.2), which in combination specify as equally economical six different area-cladograms: 1 (7.13.1), 11 (7.13.2), 13 (7.13.3), 2, 8, 14.

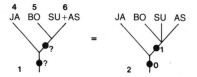

Figure 7.17. Analysis of the area-components (2) implied by the species-cladogram (1) for a different group of three species.

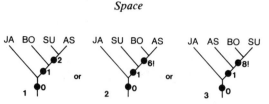

Figure 7.18. The three dichotomous area-cladograms allowed by the component analysis of figure 7.17.

Cases involving widespread taxa are to some extent ambiguous under assumption 1, for they permit some number of alternative dichotomous cladograms. For two of the four cases considered above (figures 7.12.1, 7.17.1), there is no problem in graphic representation, for each component analysis yields a secondary (partially resolved) area-cladogram (figures 7.12.2 and 7.17.2). In one case involving two widespread taxa (figure 7.19), there is no problem in representation, for the two component-analyses (figures 7.19.3–4) are combinable in one primary (dichotomous) area-cladogram (figure 7.19.5). In another case involving two widespread taxa (figure 7.20), there is a problem in representation, for the two different component-analyses yield a tertiary (noninformative) area-cladogram (figure 7.20.5). In this case, the tertiary cladogram means only that there is no informative area-cladogram that is an economical interpretation of the two species-cladograms. The tertiary cladogram is not, however, an exact summary, or representation, of the two conflicting component-analyses, inasmuch as the tertiary cladogram would allow for all possible area-cladograms, even though the component-analyses include only components 2 and 6 (figures 20.3–4). The component-analyses do not imply as economical

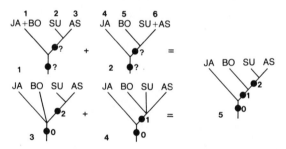

Figure 7.19. The combination of two different species-cladograms (1–2), or two different area-component-analyses (3–4), each of which combination gives the same fully resolved area-cladogram (5).

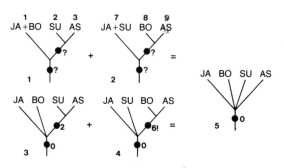

Figure 7.20. The noncombination of two different species-cladograms (1–2), or two different area-component-analyses (3–4), each of which noncombination gives an uninformative area-cladogram (5).

interpretations those based on area-cladograms containing components 8 (BO+SU) and 10 (JA+AS), and those area-cladograms containing components 4 (BO+JA) and 7 (JA+SU), except in combination with components 2 (SU+AS) and 6 (BO+AS), respectively. The component-analyses, in short, allow for 6 of 15 possible area-cladograms. Graphically, two secondary area-cladograms are required to specify these six possibilities.

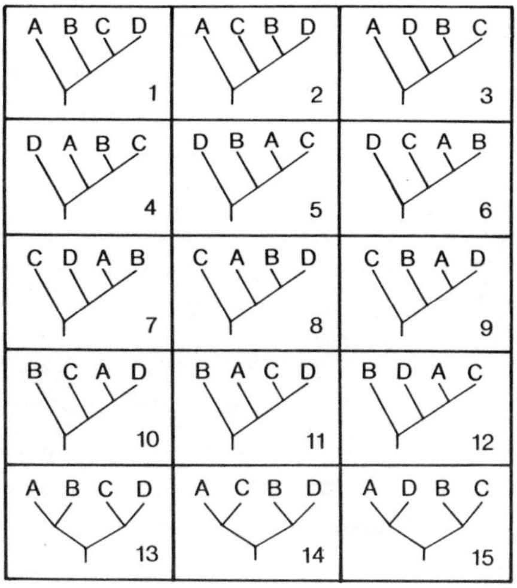

A B C D	A C B D	A D B C
1	2	3
D A B C	D B A C	D C A B
4	5	6
C D A B	C A B D	C B A D
7	8	9
B C A D	B A C D	B D A C
10	11	12
A B C D	A C B D	A D B C
13	14	15

Figure 7.21. The 15 possible area-cladograms (four areas) that are fully resolved (dichotomous).

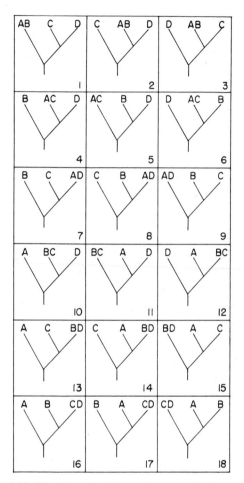

Figure 7.22. The 18 possible 3-taxon 4-area species-cladograms.

An alternative scheme of graphic representation is possible with reference to networks. The component-analysis of figure 7.12.2 is equivalent to a network with only three of its possible five roots (figure 7.23.2), each with its own pathway through the network, and each yielding one of the dichotomous area-cladograms of figure 7.13. The component-analysis of figure 7.17.2 is equivalent to three networks (figures 7.24.2–4), each with one root and pathway, and each yielding one of the dichotomous area-cladograms of figure 7.18. In combination

Table 7.4. Assumption 1

Area-cladogram[b]	\multicolumn{18}{c}{3-taxon 4-area species-cladograms[a]}

Area-cladogram[b]	1	2	3	4	5	6	7	8	9	10	11	12	13	14	15	16	17	18
1	–																	
2					–											–	–	–
3			–										–			–		
4			–	–								–						
5			–		–										–			–
6		–		–		–												
7		–		–				–										
8								–				–		–				
9		–						–	–									
10										–								
11		–									–							–
12				–														
13	–																	
14					–									–	–		–	
15									–		–							

[a]Numbered as in figure 7.22.
[b]Numbered as in figure 7.21.

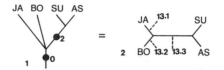

Figure 7.23. A component-analysis (1; cf. figure 7.12), and its equivalent network with three roots specified (2; cf. figure 7.13).

networks are themselves equivalent to a less informative network with only one of its possible seven roots (figure 7.24.5; see below for an explanation of the various possible roots), which nevertheless allows for three different pathways through the network.

Inasmuch as the component-analyses may be combined to yield a fully resolved area-cladogram (figures 7.19.3–5; 7.25.1–3), their equivalent networks may also be combined to give the same result (figures 7.25.4–6). The combination of the networks (figures 7.25.4–5) involves merely the preservation of roots and pathways, if any, that are common to the two networks. In the first network (7.25.4) there are three roots, each with one pathway; in the second network (7.25.5) there is one root with three pathways, one of which is the same as in the first network (JA; cf. figures 7.23.2 and 7.24.2). The combination of the networks thus reduces the possible roots and pathways to one (figure 7.25.6).

The conflicting component-analyses of figure 7.20 are equivalent to two networks (figures 7.26.3–4), without any common roots and pathways. They may nevertheless be combined in a modified network with five of its possible seven roots (figure 7.26.5), one of which (JA)

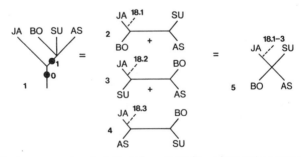

Figure 7.24. A component-analysis (1; cf. figure 7.17); its equivalent three networks (2–4), each with one root specified (cf. figure 7.18); and their equivalent modified network with one root specified (5).

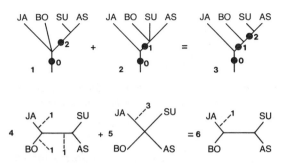

Figure 7.25. The combination of two different component-analyses (1–2) into a fully resolved area-cladogram (3); with the equivalent combination of the corresponding networks (4–5) into a network with a single root and pathway (6) specifying the same area-cladogram (3).

allows for two alternative pathways, making a total of six, through the network. Additional notation is necessary to eliminate certain combinations and redundancies. Thus, for each root the appropriate pathways through the network may be designated by small letters (a, b, c, d), with the result that the same root-location may yield different cladograms (e.g., 1 and 4) in accordance with different pathway-notations. That the two conflicting component-analyses (figures 7.26.1–2) might be represented by some such modified network (figure 7.26.5) is not particularly interesting or important in itself, because the modified network is not much of a simplification over the two networks equivalent to the component analyses (figures 7.26.3–4). What might be important is the possibility that conflicting component-analyses are informative in their combination, which cannot itself be accurately designated by a single

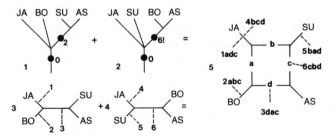

Figure 7.26. Two conflicting component-analyses (1–2; cf. figure 7.20); their equivalent networks (3–4); and their combination in a modified network with five roots and six network-pathways specified (5).

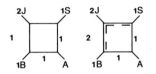

Figure 7.27. Modified networks (cf. figure 7.26), in which the details of pathways are omitted (1) or are rendered as short lines that specify the initial directions of the pathways through the network (2).

secondary or tertiary area-cladogram. The information can be understood, and rendered graphically in a general way, as a restricted set of the possibilities allowed by the modified network. By omitting the details about the pathways, such a network can be rendered in a simple way (figure 7.27.1). Or the pathways may be indicated in a rudimentary form by short lines that specify their initial directions through the network (figure 7.27.2).

ASSUMPTION 2

Under the assumption that the Javanese and Bornean occurrences of species 1 might in the future be distinguished as separate taxa, then whatever is true of the one occurrence might not be true of the other occurrence. In concrete terms (figure 7.28), if species 2 in Sumatra and species 3 in Asia are more closely related among themselves than to species 1, the relationship might be true only for the Javanese occurrence but not the Bornean (figure 7.28.2), or for the Bornean but not the Javanese (figure 7.28.3). Under this assumption, the species-cladogram (figure 7.28.1) yields two possibilities for component-analyses, but each of the possibilities includes only three of the four areas of concern (figures 7.28.2–3).

One possibility is analyzed in figure 7.29, which shows that there are five economical dichotomous area-cladograms for the four areas. The second possibility is analyzed in figure 7.30, which shows that there are also five economical dichotomous area-cladograms for the four areas. Comparison of the two possibilities shows that three area-cladograms are repeated (figures 7.29.2 and 7.30.4; 7.29.3 and 7.30.3; 7.29.4 and 7.30.2), giving a total of seven different area-cladograms, of which three

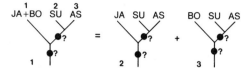

Figure 7.28. Preliminary analysis into two possibilities for area-components (2-3) implied by the species-cladogram (1) of the three fish species (cf. figure 7.12).

are repeated once each. The seven different area-cladograms reduce to three different component-analyses (figure 7.31) that jointly specify the seven cladograms. In this case, too, the hypothetical fishes would tend to confirm the birds (figure 7.6.1), not the insects (figure 7.6.2), for one or both components of the bird cladogram occurs in five of the seven area-cladograms, whereas the two components of the insect cladogram occur in none of the seven area-cladograms.

The species-cladogram (figures 7.11 and 7.28.1) may be interpreted as having resulted from missing species interrelated according to each of the 15 possible dichotomous patterns of area-relationships (figures 7.32 and 7.33). In figures 7.32.1-7, there is one species missing from each interpretation (one species missing jointly from Java and Borneo). And there is one component missing from each interpretation: component (9) from 7.32.1; (7) from 7.32.2; (10) from 7.32.3; (1) from 7.32.4; (8) from 7.32.5; (6) from 7.32.6; (4) from 7.32.7. In figures 7.33.1-8, there are five species and five components missing from each interpretation (table 7.5). Each interpretation of figure 7.33 allows for alternative placement of either the Javanese or Bornean occurrence of species 1; examples for two interpretations are given in figure 7.34.

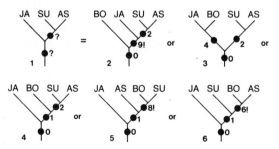

Figure 7.29. Analysis of one possibility (1) for the area-components, in the form of five alternative area-cladograms (2-6).

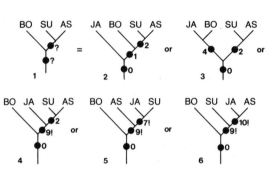

Figure 7.30. Analysis of the second possibility (1) for the area-components, in the form of five alternative area-cladograms (2–6).

Under assumption 2, therefore, the species-cladogram is sufficient to divide the 15 dichotomous area-cladograms into two sets: one of seven, and the other of eight. The former set includes those area-cladograms that are economical interpretations relative to those of the latter set. Component-analysis of the species-cladogram yields four components (figure 7.31), of which three are informative (1, 9, 2), but there is conflict among two of them (1 and 9). The widespread taxon (species 1) by itself contributes no area-component that is informative; hence component 4 is not really informative.

A similar analysis may be done for a different species-cladogram having the widespread taxon (species 6) in a different position (figure 7.35). Again the widespread taxon (species 6) contributes no area-component; there are nine area-components, all of which are informative, but there are several conflicts between them; and there are seven possible dichotomous area-cladograms (figures 7.36 through 7.38).

Combining the two species-cladograms (figures 7.39.1–2) is equivalent, as before, to combining their implied area-cladograms (figures 7.31 and 7.38). Under assumption 2, the combination yields only one area-

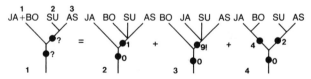

Figure 7.31. Summary of the component-analysis (both possibilities) of the species-cladogram (1) for the three fish species (cf. figure 7.12).

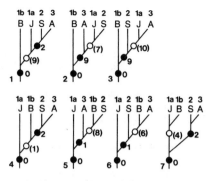

Figure 7.32. Seven economical interpretations (cf. figure 7.33) of the species-cladogram for the three fish species.

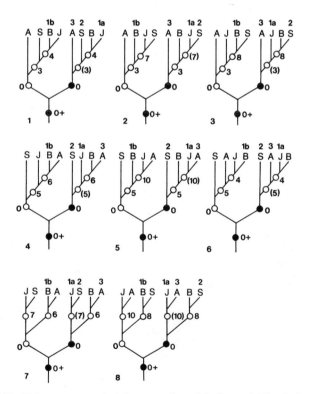

Figure 7.33. Eight noneconomical interpretations (cf. figure 7.32) of the species-cladogram for the three fish species.

Table 7.5. Items of Error

	Area Cladograms 7.32.1-7 Occurrences							Totals/ Cladogram	Area Cladograms 7.33.1-8 Occurrences								Totals/ Cladogram
	1	2	3	4	5	6	7		1	2	3	4	5	6	7	8	
Missing species																	
JA	1	1	1	1	1	1	1		3	3	3	3	3	3	3	3	
BO	0	0	0	0	0	0	0		1	1	1	1	1	1	1	1	
SU	0	0	0	0	0	0	0		1	1	1	1	1	1	1	1	
AS	0	0	0	0	0	0	0	1	1	1	1	1	1	1	1	1	5
Missing components																	
0+	0	0	–	0	0	–	–		0	0	0	0	0	0	0	0	
0	–	0	–	0	0	0	0		1	1	1	1	1	1	1	1	
1	0	–	–	–	–	0	–		–	–	–	–	–	–	–	–	
2	0	–	–	0	–	–	0		2	2	2	–	–	1	–	–	
3	–	–	–	–	–	–	1		2	–	–	–	–	–	–	–	
4	–	–	–	–	–	–	–		–	–	–	2	2	2	2	–	
5	–	–	–	–	–	–	–		–	–	–	2	2	2	–	–	
6	–	–	–	–	–	1	–		1	2	–	2	–	–	2	–	
7	–	1	–	–	–	–	–		1	–	2	–	–	–	2	2	
8	–	–	–	–	1	–	–		–	–	2	–	–	–	–	2	
9	1	0	–	1	1	–	–		–	–	–	–	–	–	–	–	
10	–	–	1	–	–	–	–	1	–	–	–	2	2	–	–	2	5
						Total:		2							Total:		10

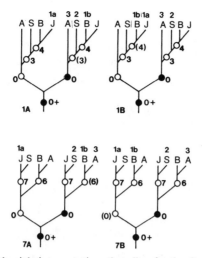

Figure 7.34. Two of the eight interpretations that allow for the alternative placement of Java and Borneo (cf. figure 7.33).

component (1), which allows for three of the 15 possible dichotomous area-cladograms (cf. figure 7.19). Interestingly, substituting a different species-cladogram (figure 7.39.5), which conflicts under assumption 1 (figure 7.20), yields the same area-component (1). Combining all three species-cladograms gives the same result (figure 7.40). Substituting a different third species-cladogram (figure 7.41.3), which also conflicts under assumption 1, yields an additional area-component (2) and a fully resolved (dichotomous) area-cladogram (figure 7.41.4). In some cases (e.g., figure 7.42), only two 3-taxon 4-area species-cladograms are necessary to yield a fully resolved area-cladogram (figure 7.42.3).

There are 15 possible dichotomous cladograms for four areas (figure 7.21). There are 18 possible 3-taxon 4-area species-cladograms (figure 7.22). Each such species-cladogram allows for seven economical area-

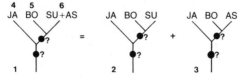

Figure 7.35. Preliminary analysis into two possibilities for area-components (2–3) implied by the species-cladogram (1) for a different group of three species (cf. figure 7.17).

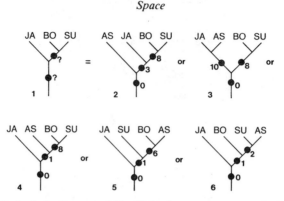

Figure 7.36. Analysis of one possibility (1) for the area-components, in the form of five alternative area-cladograms (2–6).

cladograms under assumption 2 (table 7.6). Any combination of two or more species-cladograms will specify some number of economical area-cladograms, which may be determined by comparing the lists of seven area-cladograms specified by the species-cladograms. For example, the three species-cladograms of figure 7.40 are numbers 1 (7.40.1), 16 (7.40.2), and 5 (7.40.3), under the assumption that A = Java, B = Borneo, C = Sumatra, and D = Asia; in combination they specify area-cladograms 1, 2, and 3 as economical interpretations. The three species-cladograms of figure 7.41 are numbers 1 (7.41.1), 16 (7.41.2), and 4 (7.41.3), which in combination specify area-cladogram 1 (7.41.4). The two species-cladograms of figure 7.42 are 1 (7.42.1) and 18 (7.42.2), which in combination specify area-cladogram 13 (figure 7.42.3).

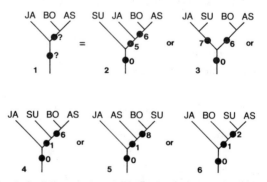

Figure 7.37. Analysis of the second possibility (1) for the area-components, in the form of five alternative area-cladograms (2–6).

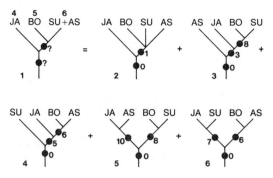

Figure 7.38. Summary of the component-analysis (both possibilities) of the species-cladogram (1) for the three fish species (cf. figure 7.17).

The entries in table 7.6 (the digits 1 and 2) reflect the redundancy in the different possibilities of component analysis. For example, in figures 7.29 and 7.30, certain area-cladograms are repeated in the analysis of species-cladogram 1 (figure 7.28.1): 1 (7.29.4, 7.30.2), 11 (7.29.2, 7.30.4), and 13 (7.29.3, 7.30.3). In figures 7.36 and 7.37, other area-cladograms are repeated in the analysis of species-cladogram 16 (figure 7.35): 1 (7.36.6, 7.37.6), 2 (7.36.5, 7.37.4), and 3 (7.36.4, 7.37.5).

Cases involving widespread taxa are to some extent ambiguous under assumption 2, for they permit some number of alternative dichotomous cladograms. For all of the cases considered above (e.g., figures 7.31 and 7.38), there is a problem in graphic representation of component-

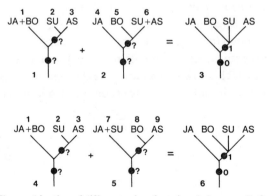

Figure 7.39. The combination of different pairs of species-cladograms (1–2; 4–5), each of which yields a secondary area-cladogram with one component (1).

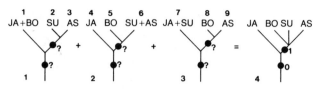

Figure 7.40. The combination of three different species-cladograms (1–3), which yields a secondary area-cladogram with one component (1).

analyses, for each analysis yields a suite of cladograms that conflict among themselves and that cannot be combined in a single cladistic structure that is fully informative.

An alternative scheme of graphic representation is possible with reference to a modified network. Such a network allowing for all 15 possible dichotomous cladograms is shown in figure 7.43.1. There are seven possible roots: four roots, each with three different pathways through the network (3J, 3S, 3B, 3A) and three roots with unique pathways (1, 1, 1). As noted above, the pathways may be specified by their initial directions, as indicated by the short lines within the square. The three roots with unique pathways (1, 1, 1) divide the four areas into two groups of two each. The root within the square yields JA and BS; the root to the side yields JS and BA; the root below yields JB and SA. Because three pathways per root are the maximum for four areas, and include all possibilities, indication of the pathways is unnecessary, and the modified network may be simplified to that in figure 7.43.2, which is fully informative of all roots and pathways.

The component-analysis of figure 7.31 is equivalent to a modified network with three of its possible seven roots, which allow seven pathways through the network (figure 7.44). The component-analysis of figure 7.38 is equivalent to a modified network with five of its possible seven roots, which allow seven pathways through the network (figure 7.45, wherein the initial direction for roots S and A is indicated by short

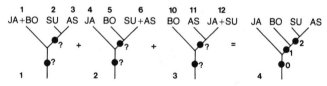

Figure 7.41. The combination of three different species-cladograms (1–3), which yields a fully resolved area-cladogram (4).

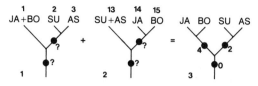

Figure 7.42. The combination of two different species-cladograms (1–2), which yields a fully resolved area-cladogram (3).

lines extending toward J). In combination, the two component-analyses (networks) are themselves equivalent to a network with only one of its possible seven roots (figure 7.46.3), which nevertheless allows for three different pathways through the network (cf. figures 7.39.1–3). The combination of the networks (figure 7.46) involves merely the preservation of roots and pathways, if any, that are common to the two networks.

Another combination involves the species-cladograms of figures 7.39.4 and 7.39.5 (figure 7.47), which yields the same result (cf. figure 7.39.6). Combining all three species-cladograms (figure 7.40) also gives the same result (figure 7.48). A different combination of three species-cladograms (figure 7.41) yields a network with a unique root and pathway, or in other words a single dichotomous cladogram (figure 7.49). Finally, the combination of two species-cladograms that under component-analysis yield a dichotomous area-cladogram (figure 7.42) gives the same result with networks (figure 7.50).

The relation between assumptions 1 and 2 may be demonstrated graphically with reference to networks. Consider a species-cladogram of 3 taxa in four areas (figure 7.51.1). Under assumption 1, the component-analysis may be represented as an area-cladogram (figure 7.51.2) or a modified network with three roots and three pathways (figure 7.51.3). Under assumption 2, the same species cladogram (figure 7.52.1) yields two possibilities for component-analysis (figures 7.52.2–3). If each possibility is represented as a modified network (figure 7.53), and if the two possibilities are compared for common elements (figures 7.54.1–2), the result (figure 7.54.3) is the same as the component analysis under assumption 1 (figure 7.51.3). The relation is the same for 3-taxon 4-area species-cladograms having the widespread species (6) in the other position in the cladogram (figures 7.55 through 7.58).

Table 7.6. Assumption 2

Area-cladogram[b]	3-taxon 4-area species-cladograms[a]																	
	1	2	3	4	5	6	7	8	9	10	11	12	13	14	15	16	17	18
1	2	—	—	1	1	—	—	1	1	2	—	—	2	—	—	2	—	—
2	1	1	—	—	1	1	—	—	—	2	—	—	2	—	—	2	—	—
3	—	—	1	—	2	—	—	—	2	2	—	2	2	—	—	2	—	—
4	1	—	—	—	—	2	1	—	2	1	—	2	1	—	1	1	1	1
5	—	—	2	—	—	2	—	2	1	—	—	2	—	1	2	—	—	1
6	—	2	2	—	2	2	—	2	—	—	—	1	—	2	1	—	—	2
7	—	2	2	1	—	—	2	2	—	—	2	—	—	2	—	—	—	2
8	—	2	—	2	—	—	2	—	—	—	2	—	—	2	—	—	2	1
9	—	1	—	2	—	—	2	—	—	—	—	—	—	1	—	—	2	—
10	2	—	—	1	—	—	1	—	—	—	—	—	—	—	—	—	2	1
11	1	—	—	—	—	—	—	1	—	—	—	—	—	—	2	—	1	—
12	1	—	—	1	—	1	1	1	—	—	—	1	—	—	1	—	—	1
13	2	2	1	—	1	—	—	—	—	1	1	—	1	1	—	—	1	2
14	1	1	1	1	2	1	—	—	—	—	—	—	—	—	2	1	1	—
15	—	1	1	1	—	1	—	—	2	—	2	—	1	—	—	1	1	—

[a]Numbered as in figure 7.22.
[b]Numbered as in figure 7.21.

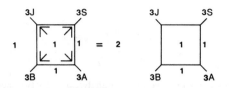

Figure 7.43. Modified networks allowing for all 15 possible dichotomous cladograms for four areas. In one (1) the three possible pathways for each of four roots (J, B, S, A) are specified by the short lines indicating the initial directions through the network. In the other (2) the pathways are not indicated but are understood to include all possibilities.

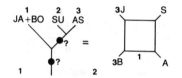

Figure 7.44. A species-cladogram (1) and its component-analysis in the form of a modified network with three roots and seven pathways (2).

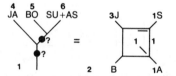

Figure 7.45. A species-cladogram (1) and its component-analysis in the form of a modified network with five roots and seven pathways (2). For two pathways (S and A), the initial direction (toward J) is indicated by the short lines within the square.

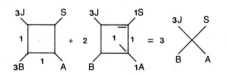

Figure 7.46. The combination of the component-analyses (1-2) of two species-cladograms (cf. figures 7.39. 1-2, 7.44, and 7.45), which yields a modified network (3) with one root and three pathways (cf. figure 7.39.3).

Space

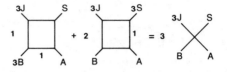

Figure 7.47. The combination of the component-analyses (1–2) of two species-cladograms (cf. figures 7.39.4–5), which yields a modified network (3) with one root and three pathways (cf. figure 7.39.6).

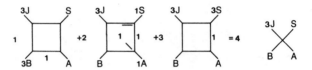

Figure 7.48. The combination of the component-analyses (1–3) of three species-cladograms (cf. figures 7.40.1–3), which yields a modified network (4) with one root and three pathways (cf. figure 7.40.4).

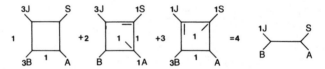

Figure 7.49. The combination of the component-analyses (1–3) of three species-cladograms (cf. figures 7.41.1–3), which yields a network (4) with one root and one pathway (cf. figure 7.41.4).

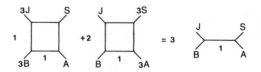

Figure 7.50. The combination of the component-analyses (1–2) of two species-cladograms (cf. figures 7.42.1–2), which yields a network (3) with one root and one pathway (cf. figure 7.42.3).

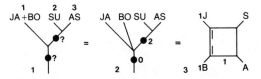

Figure 7.51. A species-cladogram of three taxa (1) and its component-analysis under assumption 1, in the form of an area-cladogram (2) and a modified network with three roots and three pathways (3).

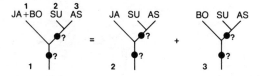

Figure 7.52. A species-cladogram of three taxa (1) and the two possibilities for its component-analysis under assumption 2, each in the form of an area-cladogram of three areas (2–3).

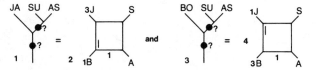

Figure 7.53. The two possibilities for component-analysis under assumption 2 (1, 3), each in the form of a modified network with three roots and five pathways (2, 4).

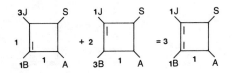

Figure 7.54. The combination of the two possibilities for component-analysis under assumption 2 (1–2), which yields a modified network with three roots and three pathways (3; cf. figure 7.51.3).

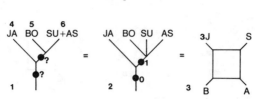

Figure 7.55. A species-cladogram of three taxa (1) and its component-analysis under assumption 1, in the form of an area-cladogram (2) and a modified network with one root and three pathways (3).

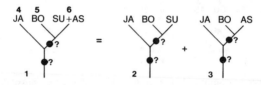

Figure 7.56. A species-cladogram of three taxa (1) and the two possibilities for its component-analysis under assumption 2, each in the form of an area-cladogram of three areas (2–3).

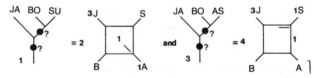

Figure 7.57. The two possibilities for component-analysis under assumption 2 (1, 3), each in the form of a modified network with three roots and five pathways (2, 4).

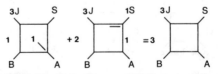

Figure 7.58. The combination of the two possibilities for component-analysis under assumption 2 (1–2), which yields a modified network with one root and three pathways (3; cf. figure 7.55.3).

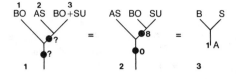

Figure 7.59. A species-cladogram of three taxa (1) and its component-analysis under assumption 1, in the form of an area-cladogram (2) and a network with one root and one pathway (3).

WIDESPREAD TAXA AND REDUNDANT DISTRIBUTIONS

Assumptions 1 and 2 emerge in new guises in the analysis of species-cladograms when all areas are represented and one or more taxon is widespread. Consider species-cladogram 7.59.1. Under assumption 1, its component-analysis yields component 8 (figure 7.59.2), which can also be represented as a network with one root and one pathway (figure 7.59.3). Consider species-cladogram 7.60.1. Under assumption 1, its component-analysis yields no informative component (figure 7.60.2); the analysis can also be represented as a network with three roots and three pathways (figure 7.60.3).

Under assumption 2, the results are different. Species-cladogram 7.59.1 (figure 7.61.1) yields two possibilities for component-analysis (figures 7.61.2–3). One possibility is uninformative (figure 7.61.2), leaving the other (figure 7.61.3), which includes all three areas, as the result (figure 7.61.4), which can be represented as a network with one root and one pathway (figure 7.61.5). Species-cladogram 7.60.1 (figure 7.62.1) also yields two possibilities for component-analysis (figures 7.62.2–3). Likewise, one possibility is uninformative (figure 7.62.3), leaving the other (figure 7.62.2), which includes all three areas, as the

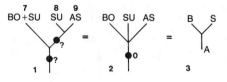

Figure 7.60. A species-cladogram of three taxa (1) and its component-analysis under assumption 1, in the form of an area-cladogram (2) and a network with three roots and three pathways (3).

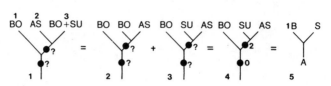

Figure 7.61. A species-cladogram of three taxa (1), the two possibilities for its component-analysis under assumption 2 (2–3), and the definitive component-analysis in the form of an area-cladogram (4) and a network with one root and one pathway (5).

result (figure 7.62.4), which also can be represented as a network with one root and one pathway (figure 7.62.5).

The results of such component-analyses under assumption 1 (figures 7.59.2–3 and 7.60.2–3) differ among themselves. The results under assumption 2 (figures 7.61.4–5 and 7.62.4–5) do not differ among themselves, but do differ from the results under assumption 1.

For 3-taxon 3-area species-cladograms having one widespread taxon in two areas, there are 18 possibilities (figure 7.63). Under assumption 1, six possibilities (1–2, 7–8, 13–14) yield informative component-analyses; and 12 possibilities (3–6, 9–12, 15–18) yield uninformative component-analyses (figure 7.64). Under assumption 2, all 18 possibilities yield informative component-analyses: possibilities 1–6 yield component 2; possibilities 7–12 yield component 6; and possibilities 13–18 yield component 8 (figure 7.64). In no case are the analyses the same under assumptions 1 and 2.

Each species-cladogram may be interpreted as having resulted from missing species according to each of the three possible dichotomous patterns of area-relationships (figures 7.65 and 7.66). For species-clado-grams yielding informative component-analyses under assumption 1 (figures 7.65.1 and 7.65.5), one interpretation (figures 7.65.2 and 7.65.8) is more economical than the other two interpretations (figures 7.65.3–4

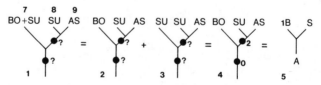

Figure 7.62. A species-cladogram of three taxa (1), the two possibilities for its component-analysis under assumption 2 (2–3), and the definitive component-analysis in the form of an area-cladogram (4) and a network with one root and one pathway (5).

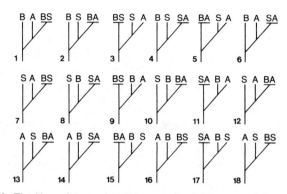

Figure 7.63. The 18 possible species-cladograms for three areas and three taxa, one of which occurs in two areas.

and 7.65.6–7), but the economical interpretation does not include the informative component (component 8 for species-cladogram 7.65.1, and component 6 for species-cladogram 7.65.5). For the economical interpretations, there are three missing species and three missing components (figures 7.65.2 and 7.65.8); for the noneconomical interpretations of the same species-cladograms, there are six missing species and six missing components (figures 7.65.3–4 and 7.65.6–7). For species-cladograms yielding no informative component-analysis under assumption 1 (figures 7.65.9–12), all three interpretations include three missing species and three missing components. One interpretation (7.65.11)

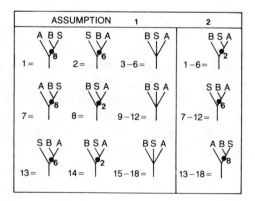

Figure 7.64. Summary of the component-analyses of the 18 species-cladograms of figure 7.63, under assumptions 1 and 2.

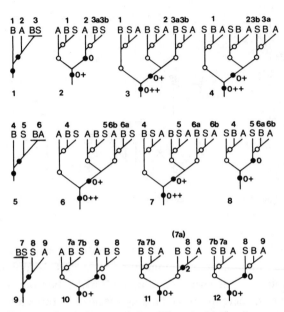

Figure 7.65. Three species-cladograms (1, 5, and 9), each with three interpretations under assumption 1 (2–4, 6–8, and 10–12). For species-cladograms 1 and 5, one interpretation is economical (2 and 8), and two interpretations are noneconomical (3–4; 6–7). For species-cladogram 9, all interpretations are equally economical (10–12).

allows for the alternative placement of the Bornean occurrence of one taxon (7a in parentheses).

All of the 3-taxon 3-area species-cladograms (figure 7.63) are informative under assumption 2 (figure 7.66). Of the three interpretations possible for each species-cladogram, one interpretation is economical and the other two interpretations are noneconomical. All economical interpretations include the informative component. For the economical interpretations (figures 7.66.3, 7.66.7, and 7.66.11) there are no missing species and no missing components; for the noneconomical interpretations (figures 7.66.2, 7.66.4, 7.66.6, 7.66.8, 7.66.10, and 7.66.12), there are three missing species and three missing components. Each noneconomical interpretation allows for alternative placement of one of the two (but not both) of the occurrences of the widespread taxon: 3a or 3b (figures 7.66.2 and 7.66.4); 6a or 6b (figures 7.66.6 and 7.66.8); 7a or 7b (figures 7.66.10 and 7.66.12).

For 3-taxon 3-area species-cladograms having two widespread taxa, each in two areas, there are 33 possibilities (figures 7.67–7.69). Of the first thirty, only six are informative under assumption 1, but under assumption 2 each of the thirty yields an informative component (figure 7.68). The three other possibilities are informative under both assumptions (figure 7.69): under assumption 1 they yield an informative component; but under assumption 2, they permit a component-analysis in the form of a network with two roots and two pathways—in effect one half-component.

For 3-taxon 3-area species-cladograms having three widespread taxa, each in two areas, there are 15 possibilities (figure 7.70). All are uninformative under assumption 1. Under assumption 2, nine yield a half-component (figures 7.70.1–9), and six yield an informative component (figures 7.70.10–15).

For 3-taxon 3-area species-cladograms having one widespread taxon in three areas there are nine possibilities (figure 7.71), all of which are

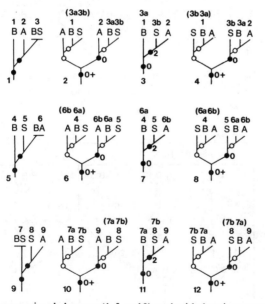

Figure 7.66. Three species-cladograms (1, 5, and 9), each with three interpretations under assumption 2 (2–4, 6–8, and 10–12). For each species-cladogram, one interpretation is economical (3, 7, and 11), and two interpretations are noneconomical (2 and 4, 6 and 8, 10 and 12).

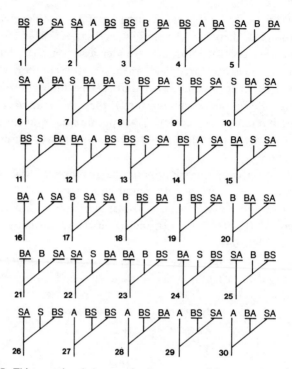

Figure 7.67. Thirty species-cladograms for three areas and three taxa, two of which occur in two areas.

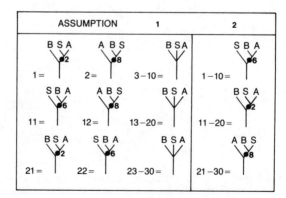

Figure 7.68. Summary of the component-analyses of the 30 species-cladograms of figure 7.67, under assumptions 1 and 2.

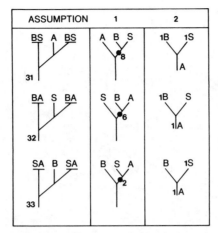

Figure 7.69. Three species-cladograms (for three areas and three taxa, two of which occur in two areas), and their component-analyses under assumptions 1 and 2 (continued from figure 7.67).

uninformative under assumption 1. Under assumption 2 each yields an informative component.

For 3-taxon 3-area species-cladograms having two widespread taxa (one in three areas and another in two areas), there are 27 possibilities (figures 7.72 and 7.73). Twenty-one are uninformative under assumption 1, but under assumption 2 each yields an informative component (figure 7.72). Three are informative under either assumption (figures 7.73.22–24), yielding either an informative component (assumption 1) or a half-component (assumption 2). Three are uninformative under assumption 1, but informative under assumption 2, yielding a half-component (7.73.25–27).

For 3-taxon 3-area species-cladograms having three widespread taxa (one in three areas and two in two areas), there are 15 possibilities (figures 7.74 and 7.75). All are uninformative under assumption 1. Under assumption 2, three yield an informative component (figure 7.74), and nine yield a half-component (figures 7.75.4–12). Three are uninformative under either assumption (figures 7.75.13–15).

For 3-taxon 3-area species-cladograms having three widespread taxa (two in three areas and one in two areas), there are six possibilities, all of which are uninformative under assumption 1 (figure 7.76). Under

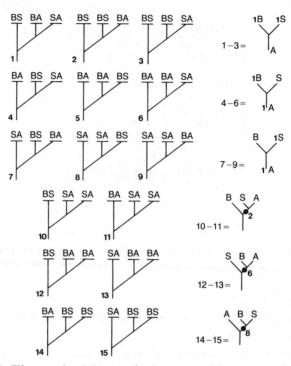

Figure 7.70. Fifteen species-cladograms (for three areas and three taxa, all of which occur in two areas), and their component-analyses under assumption 2. Under assumption 1, all 15 species-cladograms are uninformative.

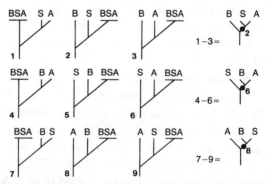

Figure 7.71. Nine species-cladograms (for three areas and three taxa, one of which occurs in three areas), and their component-analyses under assumption 2. Under assumption 1, all 9 species-cladograms are uninformative.

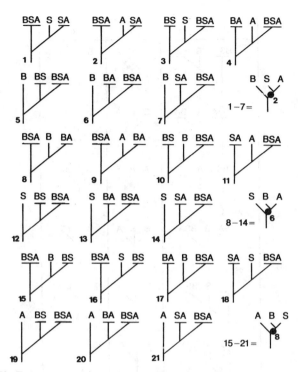

Figure 7.72. Twenty-one species-cladograms (for three areas and three taxa, one of which occurs in three areas and one of which occurs in two areas), and their component-analyses under assumption 2. Under assumption 1, all 21 species-cladograms are uninformative.

assumption 2, three yield a half-component (figures 7.76.1–3), and three are uninformative (figures 7.76.4–6).

A POSTSCRIPT ON REDUNDANCY

The examples of redundancy treated above (figures 7.59–7.76) all involve widespread taxa; it is possible, however, for species-clado-grams to include redundant endemic taxa. For 4-taxon 3-area species-cladograms involving no widespread taxa, there are 27 possibilities. Nine of these (figure 7.77, left) differ from the others in that the two taxa found in the same area constitute a single group. Because the information provided on the relationships of areas remains the same whether a

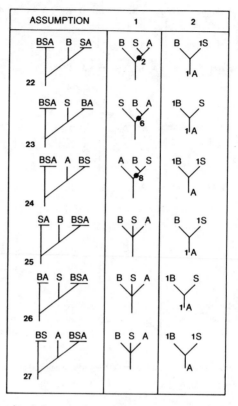

Figure 7.73. Six species-cladograms (for three areas and three taxa, one of which occurs in three areas and one of which occurs in two areas), and their component-analyses under assumptions 1 and 2 (continued from figure 7.72).

group is represented in a given area by a single species or by two or more species that are each other's closest relatives, these nine species-cladograms can be automatically reduced to area-cladograms (figure 7.77, right).

Assumptions 1 and 2 again emerge in new guises in the analysis of the remaining 18 species-cladograms (figure 7.78). Consider species-clado-gram 7.78.1. Under assumption 1, it has a single economical interpretation, yielding component 2 (figure 7.79.1). However, the component could conceivably result only from the assumption that both pieces of information regarding the area represented twice (Borneo) are true. We

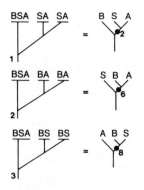

Figure 7.74. Three species-cladograms (for three areas and three taxa, one of which occurs in three areas and two of which occur in two areas), and their component-analyses under assumption 2. Under assumption 1 all three species-cladograms are uninformative.

might therefore assume instead that both pieces of information need not be true (assumption 2). If the area-relationships that are indicated by one occurrence, treated independently, are the same as those indicated by the second occurrence when treated independently, the species-cladogram might be interpreted as conveying the information common to both possibilities. For example, if the first occurrence of Borneo in

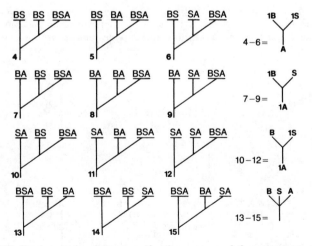

Figure 7.75. Twelve species-cladograms (for three areas and three taxa, one of which occurs in three areas and two of which occur in two areas), and their component-analyses under assumption 2. Under assumption 1, all 12 species-cladograms are uninformative.

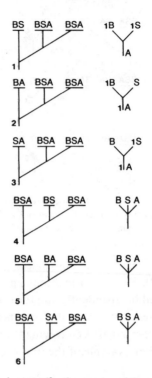

Figure 7.76. Six species-cladograms (for three areas and three taxa, two of which occur in three areas and one of which occurs in two areas), and their component-analyses under assumption 2. Under assumption 1, all six species-cladograms are uninformative.

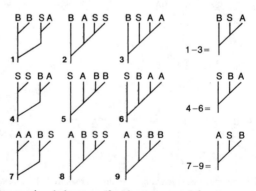

Figure 7.77. Nine species-cladograms (for three areas and four taxa, no one of which occurs in more than one area) and the area-cladograms they yield directly.

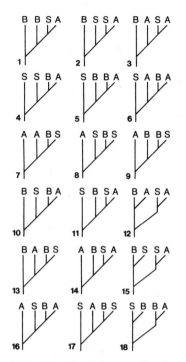

Figure 7.78. Eighteen additional species-cladograms (for three areas and four taxa, no one of which occurs in more than one area) which do not directly yield area-cladograms (continued from figure 7.77).

species-cladogram 7.78.1 is disregarded, component 2 is resolved, while if the second occurrence of Borneo is disregarded, component 2 is again resolved. In the case of species-cladogram 7.78.1, the same component is resolved under either assumption 1 or assumption 2.

Species-cladogram 7.78.2, however, is uninformative under assumption 1, because all three possible area-relationships yield equally economical interpretations. Under assumption 2, component 2 is resolved no matter which Sumatran occurrence is disregarded. Of the 18 species-cladograms under consideration, nine yield informative and nine yield uninformative component analyses under assumption 1, while nine yield fully informative and nine yield partially informative (one half-component) component analyses under assumption 2 (figure 7.79).

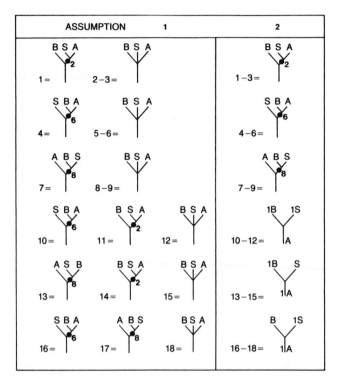

Figure 7.79. Summary of the component-analyses of the 18 species-cladograms of figure 7.78, under assumptions 1 and 2.

THE REAL WORLD

Let us suppose, then, that we want to test Wallace's concept of the interrelationships of Java, Borneo, Sumatra, and Asia (figure 7.1.1). We might investigate five groups of organisms, and we might find distributions and relationships like those shown in figure 7.80. Upon analysis, each species-cladogram would prove to be economically explainable (under each assumption) by some set of area-cladograms, which can be represented by a modified network (figure 7.80). By combining the networks, one or more roots and pathways may prove to provide economical explanations of all, or most, of the data; in the case of the data in figure 7.80, there is a single root and pathway that most economically explains all of the species-cladograms under either assumption. In this case, the generally economical explanation does correspond to Wallace's hypothesis of area-relationships. In other cases, no single explanation might emerge as most economical, and only a set of explanations might be indicated. Suppose that a single explanation is resolved, however; does it reflect the actual interrelationships of the areas?

In the case of the data in figure 7.80, one of the groups (7.80.3) allows only a single economical explanation under assumption 2. One might ask what the probability is of this area-cladogram also economically

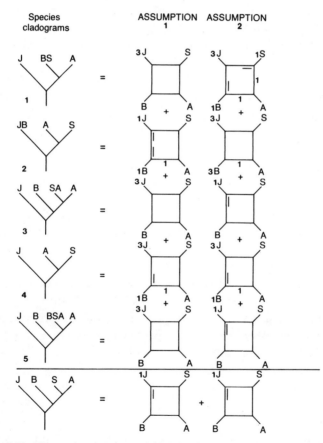

Figure 7.80. Five species-cladograms (left) and their component-analyses under assumptions 1 and 2 (right), together with the combination of their possibilities (bottom).

explaining the other four species-cladograms (under assumption 2) by chance alone. Group 7.80.1 allows for seven economical (out of 15 total) explanations; the probability of finding the unique economical explanation for group 7.80.3 indicated for group 7.80.1 also by chance alone is 7/15, or 0.47. Group 7.80.2 also allows for seven economical explanations; the probability of finding the unique economical explanation for group 7.80.3 in both groups 7.80.1 and 7.80.2 by chance alone is 7/15 × 7/15, or 0.22. Group 7.80.4 allows for five economical explanations; the probability of finding the same one again by chance alone is 0.07. Group 5 allows for only one economical explanation, bringing the probability of finding the same one again by chance alone down to 0.005. A similar analysis of the results under assumption 1 leads to a similar probability value; hence a nonrandom factor would seem to be operating in these data. The hypothesis that there is a nonrandom set of area-relationships can of course be checked by taking additional independent samples, i.e., investigating the distributions and interrelationships of other groups of organisms found in Java, Borneo, Sumatra, and Asia.

EVALUATION OF ASSUMPTIONS

We have seen that species-cladograms including widespread taxa are to some extent ambiguous, and that the information they convey depends at least in part on whether or not we make the assumption that whatever is true of the widespread taxon in one part of its range must also be true for its occurrence in other parts of its range. For species-cladograms including no redundant elements (figures 7.10–7.58), the results under assumption 1 (the species-cladogram is true for all occurrences of the widespread taxon) are a subset of those under assumption 2 (tables 7.4 and 7.6). For example, area-cladogram 1 (figure 7.21) is an economical explanation of species-cladograms 1, 10, 13, and 16 (figure 7.22) under assumption 1, and an economical explanation of species-cladograms 1, 4, 5, 7, 9, 10, 13, and 16 (figure 7.22) under assumption 2. Hence the difference between assumptions 1 and 2 might seem to be only one of degree. However, for species-cladograms including widespread redundant elements (figures 7.59–7.76), the results under assumption 1 are not a subset of those under assumption 2; in any case where a species-cladogram is informative under both assumptions, the results under assumption 1 contradict those under assumption 2. Obviously, then, the choice of assumption may substantially affect the results, and the question is therefore whether there might be reason to prefer one assumption over the other in those cases where they lead to different results. One indication might be that in those cases where the results of the two assumptions differ, the economical interpretations under assumption 1 do not include the informative component found in the result (figure 7.65); one might therefore suspect that the informative component is an artifact of the assumption, and not truly informative. We cannot test the validity of the assumptions against real data, because we would never know in advance what the true area-relationships are. We might, however, conduct a "thought experiment" by assuming that a certain set of area-relationships is true, determining what sorts of wide-spread taxa might then be expected to occur, and checking which (if either) of the assumptions would include the correct area-relationships as economical explanations of the data. Two different methods of generating widespread taxa will be examined.

Experiment 1:

In this experiment, let us assume that we have four areas (a–d) that were originally continuous (figure 7.81, top left), and that three subsequent events have occurred: the first, dividing area abcd by separating area a from area bcd; the second, dividing area bcd by separating area b from cd; and the third, separating areas c and d. If area abcd had originally been occupied by populations of an endemic species, and if the populations in each area had speciated after the isolation of the area, we might expect to find today a group of four species (A–D) endemic to the respective areas (a–d) and interrelated as shown by the species-cladogram (figure 7.81, top right). But it is possible that one or more of the populations might not have speciated, in which case we would expect to find a species widespread in two or more of the areas.

Suppose, for example, that no speciation occurred in response to the first event, because the new barrier between areas a and bcd was ineffective in preventing interchange of the organisms between those

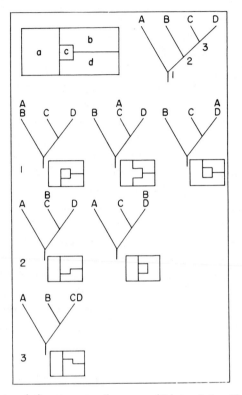

Figure 7.81. Four taxa in four corresponding areas, with interrelationships as in top right area-cladogram and distributions as in top left map. Row 1, possible species-cladograms and area configurations if a group failed to respond to event 1; row 2, possible species-cladograms and area configurations if a group failed to respond to event 2; row 3, possible species-cladogram and area configuration if a group failed to respond to event 3.

areas. If so, then at the time of the second event (separating area b from area cd) a single species would still have occurred in all four areas. If speciation did occur in area b as a result of the second event, there are two possibilities: either areas a and b retained effective interchange between themselves but not area cd, or areas a, c, and d retained effective interchange among themselves but not with area b. In the first case, we might obtain the first species-cladogram in row 1 (figure 7.81) if speciation occurred again after the third event; in the second case, we might obtain either the second or third species-cladogram in row 1 (figure 7.81), depending on whether the third event prevented effective interchange between areas c and d but not a and c, or a and d, respectively. Similarly, if speciation *had* occurred in response to the first event, but not in response to the second, we might obtain either species-cladogram of row 2 (figure 7.81), depending on whether the third event prevented effective interchange between areas c and d but not b and c, or b and d, respectively. Finally, if speciation had occurred in response to the first two events but not the third, we might obtain the species-cladogram shown in row 3 (figure 7.81).

A similar analysis for the structurally different primary 4-taxon cladogram is shown in figure 7.82. If speciation had not occurred in response to the separation of area a from area b, the cladogram in row 1 is expected; if it had not occurred in response to the separation of areas c and d, the cladogram in row 2 is expected. If speciation had not occurred in response to the separation of areas ab and cd, two alternatives appear. If the two later events were simultaneous, the species in areas a and b must be isolated from each other, and the species in areas c and d must be isolated from each other, but the species in the two pairs of areas need not be isolated from each other and could sort independently, producing either 2-taxon cladogram shown in row 3 (figure 7.82). If the two later events did not occur simultaneously, the species in each pair of areas (ab and cd) can only be associated with species in one member of the other pair of areas. Thus, if the separation of areas a and b preceded that of areas c and d, the four cladograms in row 4 could result, and if the separation of areas c and d occurred first, the four cladograms in row 5 could be found.

In the first case (figure 7.81), a total of six out of the 18 possible 3-taxon 4-area species-cladograms might arise from the failure of a group to speciate in response to one of the events; these six species-cladograms are numbers 1, 4, 7, 10, 13, and 16 of figure 7.22. In the second case (figure 7.82), a total of ten such species-cladograms (numbers 1, 4, 6, 7, 8, 10, 12, 13, 14, and 18) might arise. In the first case, comparison with tables 7.4 and 7.6 indicates that although the area-cladogram assumed to be true in our experiment (number 1 of figure 7.21) would be resolved as an economical explanation of all six resulting species-cladograms under assumption 2, it would be resolved as an economical explanation of only four of the resulting species-cladograms

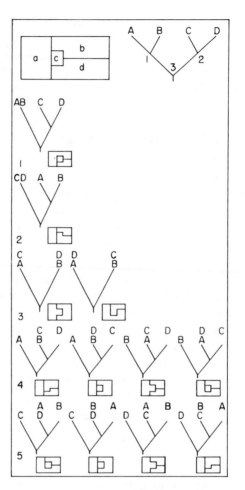

Figure 7.82. Taxa and distributions as in figure 7.81, with interrelationships as in top right area-cladogram. Possible species-cladograms and area configurations as in figure 7.81, with rows 3–5 possible if a group failed to respond to event 3.

(1, 10, 13, and 16) under assumption 1. Similarly, although the area-cladogram assumed to be true in the second case (number 13 of figure 7.21) would be resolved as an economical explanation of all ten resulting species-cladograms under assumption 2, it would be resolved as an economical explanation of only two of them (1 and 18) under assumption 1. We may therefore conclude that assumption 2 is the more realistic in those cases where widespread taxa are the result of the failure of a group to speciate in response to one or more events.

Experiment 2:

We might also expect to obtain a widespread taxon in cases where, for whatever reason, a taxon originally endemic to one area has been able to expand its range into one or more additional area. Let us assume, for the purposes of this experiment, that we have three species originally distributed in Borneo, Sumatra, and Asia, respectively, that the species in Sumatra and Asia are more closely related to each other than either is to the species in Borneo (as in Wallace's original concept, figure 7.1.1), and that one of the species has expanded its range to include one of the other areas as well. A total of six species-cladograms can be generated in this way; they are shown as numbers 1–6 in figure 7.63. A comparison with figure 7.64 shows that while under assumption 2, all six resulting species-cladograms can be economically explained only by the assumed area-relationships (i.e., they all resolve a Sumatra plus Asia component, component 2), under assumption 1 four of the six species-cladograms (numbers 3–6) are uninformative and the other two (numbers 1 and 2) resolve components not present in the assumed area-relationships. We may therefore conclude that assumption 2 is also the more realistic in those cases where widespread taxa are the result of a taxon having expanded its range.

8

BIOGEOGRAPHIC RESULTS:
REGIONS

BIOGEOGRAPHIC RELATIONSHIP: THE CLADISTIC ASPECT

A cladistic approach to biogeography presupposes that there are areas of endemism that are interrelated among themselves. Area-relationship is to be understood at a general level, in the sense that area-relationship includes, and can be broken down into, species-relationships among the members of particular groups of organisms that naturally occur in the areas. It is generally unknown even today whether area-relationship is either simple or complex: whether a single cladogram of areas includes all relevant species-relationships, or whether two or more cladograms are necessary to depict all of the species-relationships of taxa endemic to any one area.

Little attention here will be given to the concept of areas of endemism. The concept was defined long ago by A.-P. de Candolle; briefly, an area of endemism is delimited by the more or less coincident distributions of taxa that occur nowhere else.

Endemic taxa may be subspecies, species, or groups of species. An example of an area of endemism is the Hawaiian Islands. Taxa endemic to Hawaii include four species of the plant genus *Keysseria*—low perennial shrublets that in Hawaii occur in open bogs between 1000 and 1700 m; ten other species of the genus occur mainly to the west of Hawaii, with eight species known from New Guinea (figure 8.1). Endemic to Hawaii, also, is one species of the plant genus *Argemone*—spiny-leaved herbs growing to about one meter in upland areas; 26 other species of the genus occur to the east of Hawaii, mainly in North

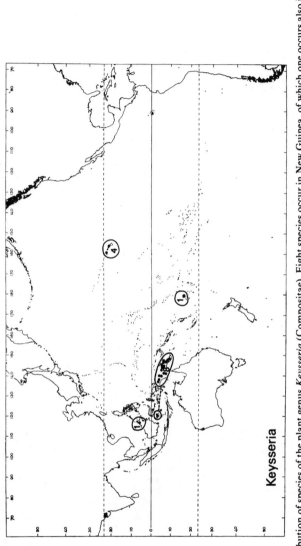

Figure 8.1. Distribution of species of the plant genus *Keysseria* (Compositae). Eight species occur in New Guinea, of which one occurs also in Celebes; one in Borneo; one in Fiji; and four in Hawaii. After M. M. J. Van Balgooy (1975), Pacific Plant Areas, vol. 3 (Leiden: Rijksherbarium), figure on p. 329.

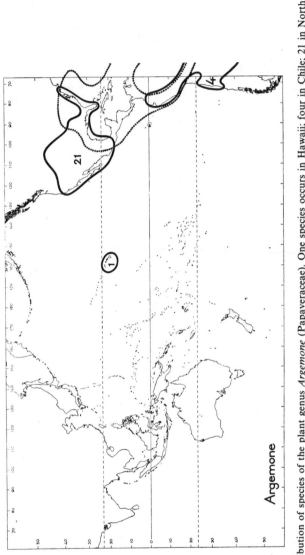

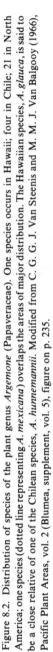

Figure 8.2. Distribution of species of the plant genus *Argemone* (Papaveraceae). One species occurs in Hawaii; four in Chile; 21 in North America; one species (dotted line representing *A. mexicana*) overlaps the areas of major distribution. The Hawaiian species, *A. glauca*, is said to be a close relative of one of the Chilean species, *A. hunnemannii*. Modified from C. G. G. J. Van Steenis and M. M. J. Van Balgooy (1966), Pacific Plant Areas, vol. 2 (Blumea, supplement, vol. 5), figure on p. 235.

America; the Hawaiian endemic, *A glauca*, is said to be closely related to the Chilean *A. hunnemannii* (figure 8.2). Endemic to Hawaii, also, is one species of the plant genus *Sophora* (section *Edwardsia*, series *Tetrapterae*)—shrubs or small trees to 15 m, growing in upland areas in forests, along streams, and on mountain slopes; 12 other species of this series (*Tetrapterae*) occur in scattered localities to the south of Hawaii (figure 8.3).

For any area of endemism, such as Hawaii, distributional information of this sort is abundant, as it has been for most areas since the mid-part of the nineteenth century. Rendered cladistically, however, the information is totally uninformative with respect to area-relationship—herein understood as the cladistic aspect of biogeographical relationship (figure 8.4).

At first glance, one might suppose that, as evidenced by *Keysseria*, the area-relationship of Hawaii is to some area of endemism to the west (figure 8.4.1); or, as evidenced by *Argemone*, to the east (figure 8.4.2); or, as evidenced by *Sophora*, to the south (figure 8.4.3). Given all three such types of evidence, one might conclude that the area-relationship of Hawaii is complex—that Hawaii is related to many places, indeed to all other areas of endemism the world over. And one might dismiss as absurd in the face of this complexity the possibility that Hawaii, and all other areas of endemism, are interrelated in a pattern specifiable in a single cladogram. It might seem, for example, that *Keysseria* implies one cladogram; *Argemone*, a second; and *Sophora*, a third. Yet the possibility is not in any sense contradicted by the three examples listed above, all of which would be interpretable, equally economically, by any one of the fifteen dichotomous cladograms possible for four areas, two of which are shown in figure 8.5. The examples, in short, contain no informative components of area-relationship as regards Hawaii and the other areas to the west, east, and south.

One problem with the three examples above is that each of them is only a 2-area statement (figures 8.4.1–3), and 2-area statements are uninformative, either singly or jointly, for they are, all of them, consistent with any interpretation of area-relationships. But even if each of these particular statements were to be made more detailed, for example in the form of a 3-area statement (figures 8.6.1–3), they would still be jointly uninformative, in the sense that they could not be

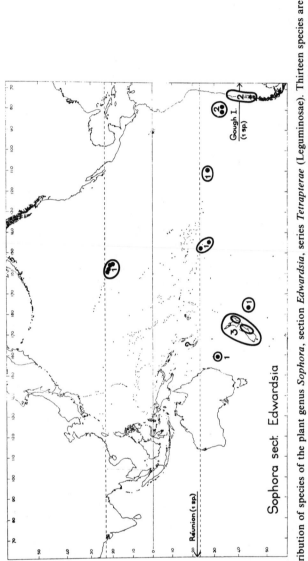

Figure 8.3. Distribution of species of the plant genus *Sophora*, section *Edwardsia*, series *Tetrapterae* (Leguminosae). Thirteen species are distributed on Réunion, Lord Howe, New Zealand, Chatham, Rapa, Raivave, Easter, Juan Fernandez, Chile, Gough, and Hawaii. After Van Steenis and Van Balgooy (1966; see caption for figure 8.2), figure on p. 285.

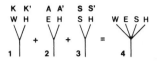

Figure 8.4. Three uninformative area-cladograms (1–3), wherein Hawaii (H) is indicated to be related to areas of endemism to the west (W, as represented by K and K', species of *Keysseria*), east (E, by A and A', species of *Argemone*), and south (S, by S and S', species of *Sophora*). The combination of the three results only in the uninformative tertiary area-cladogram (4).

combined in a single economical area-cladogram. Numerous clado-grams would be equally economical (figures 8.7.1–2). The problem with such 3-area statements is, of course, that they contain only one common element (H).

A comparison that might render these 3-area statements informative is, for example, with a species-cladogram of a group such as the plant genus *Xylosma*—about 100 species of shrubs and trees of varied habit and ecology (figure 8.8). *Xylosma* is represented by species endemic not only to Hawaii, but to areas to the west, east, and south as well. A detailed cladogram of the species of *Xylosma* could either agree or disagree, in its implications for area-relationships, with each of the 3-area statements of figures 8.6.1–3. In either case (agreement or disagreement) the 3-area statements would be informative.

Brundin (1966) attempted a cladistic approach to the area-relation-ships of the temperate portions of the southern continents (Australia, New Zealand, South America, and Africa), through study of species-relationships in three subfamilies of midges (Chironomidae). In one

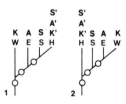

Figure 8.5. Two of the 15 possible dichotomous interpretations of the combined distributions of species of *Keysseria* (K, K'), *Argemone* (A, A'), and *Sophora* S, S'), with reference to the area-relationships of four areas of endemism: W (west), E (east), S (south), and H (Hawaii).

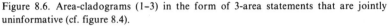

Figure 8.6. Area-cladograms (1–3) in the form of 3-area statements that are jointly uninformative (cf. figure 8.4).

subfamily (Diamesinae), he found a complex of intercontinental relationships for 20 species (figure 8.9). The species-cladogram may be simplified to show two items of information for area-relationships among the four areas: 1. New Zealand, Australia, and South America are more closely related among themselves than any one of them is to Africa (figure 8.10.1); 2. Australia and New Zealand are each more closely related to South America than to each other (figure 8.10.2), with the implication that the South American distributions are complex (that more than one area is involved).

In the second subfamily (Podonominae) he found intercontinental relationships for 126 austral species (figure 8.11). This species-cladogram may be simplified to show the same two items of information as in the previous case (figure 8.10). In the third subfamily (Aphroteniinae) he found intercontinental relationships for eight species (figure 8.11). This species-cladogram may be simplified to show one item of information: that Australia and South America are more closely related among themselves than either is to Africa (figure 8.12).

Brundin's results were of wide interest for many reasons that stemmed from his overall interpretation, which will be briefly reviewed here. Consider the area-cladogram of figure 8.12; it is represented, or implied, in the species-cladograms of all three subfamilies (figures 8.9, 8.11). That one and the same 3-area cladogram should be twice repeated in the

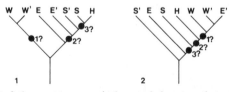

Figure 8.7. Two of the many economical area-cladograms that would result from combining jointly uninformative area-cladograms such as those in figure 8.6.

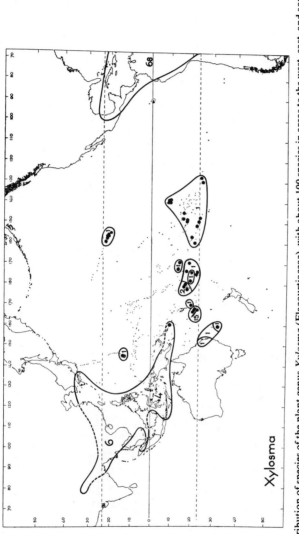

Figure 8.8. Distribution of species of the plant genus *Xylosma* (Flacourtiaceae), with about 100 species in areas to the west, east, and south of Hawaii. After Van Steenis and Van Balgooy (1966; see caption for figure 8.2), figure on p. 193.

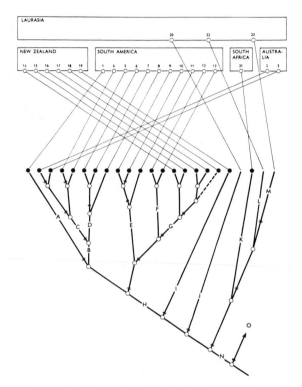

Figure 8.9. Cladogram for 20 austral species of the subfamily Diamesinae (Diptera, Chironomidae). After Brundin (1966), figure 635, p. 448.

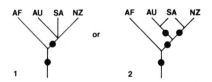

Figure 8.10. Simplified area-cladograms implied by the species-relationships of Diamesinae (cf. figure 8.9).

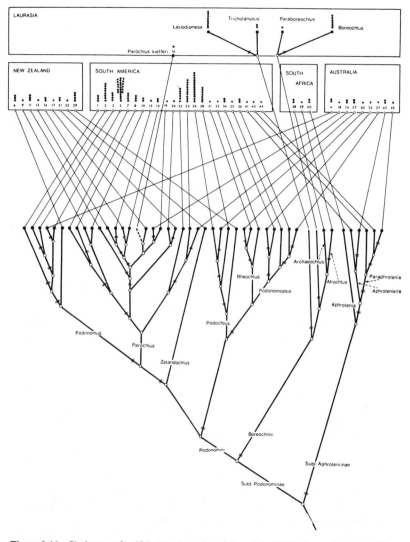

Figure 8.11. Cladogram for 126 austral species of the subfamily Podonominae and eight austral species of the subfamily Aphroteniinae (Diptera, Chironomidae). After Brundin (1966), figure 634, p. 442.

Figure 8.12. Simplified area-cladogram implied by the species-relationships of Aphroteniinae (cf. figure 8.11).

species-cladograms of three groups of dipterans might be dismissed as mere coincidence. But the probability of this concidence, due to chance alone, can be calculated, bearing in mind that for three areas (or three taxa), there are three possible dichotomous cladograms. Thus, the probability that the same cladogram should be repeated once, due to chance alone is 1 in 3, or 33 percent. The probability that it should be repeated twice, as in the above case, is 1 in 3^2, or 11 percent—a probability approaching the level (5 percent) at which scientists begin to doubt seriously that chance alone is at work.

Brundin pointed out, however, that the area-cladogram implied by his studies is the same as that held to be true by geologists studying the possibilities of continental drift in the southern hemisphere. At that time it was believed that all of the southern continents were assembled in a supercontinent (Gondwanaland), and that when this supercontinent began to fragment, the first split of relevance here was between Africa and a continental block that included South America, Antarctica, New Zealand, and Australia. Subsequent events, it was believed, led to the subsequent fragmentation of this continental block. Hence, the area-cladogram (figure 8.12) was seen by Brundin to agree also with continental-drift phenomena. The probability that the same cladogram should be repeated a third time is 1 in 3^3, or about 4 percent—convincing enough to Brundin to have permitted him to write as follows:

The theory of continental drift provides a background fitting all demands raised by the nature of the transantarctic relationships, as displayed by the chironomid midges. Indeed, the fit between the history told by the distribution patterns and reconstructed phylogenies on one hand, and the latest opinions concerning the geological nature and mutual connections of the Gondwana fragments, and the time-table affixed to the disruption of Gondwanaland on the other, is so close that there is agreement even in details. (1966:454)

The details mentioned by Brundin concern the dual relationships between Australia and New Zealand with South America, and the equally complex geological relationships among these areas—matters that will not be pursued further here.

The areas occupied by the midges studied by Brundin roughly comprise four major areas: southern Africa, southern and western South America with associated islands, New Zealand, and southeastern Australia-Tasmania (figure 8.13). Three of these major areas were

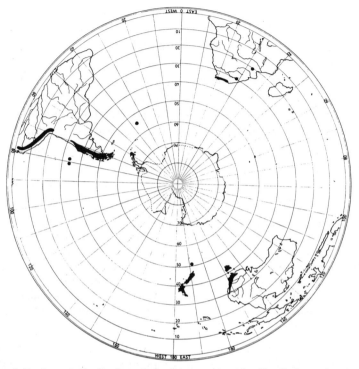

Figure 8.13. Austral distribution of the chironomid subfamilies Podonominae and Aphroteniinae. After Brundin (1966), figure 633, p. 441.

already recognized as areas of endemism by Candolle in 1820, and the fourth (New Zealand) was recognized by Candolle in 1838, when he expanded his list of regions from 20 to 40. Candolle also recognized the complexity in southern South America, which he subdivided into several regions (Chile, Peru, Tierra del Fuego, Juan Fernandez, Falklands). All of these areas have been merely confirmed by Brundin's studies of their midges, as well as by many other persons who have studied other groups of plants and animals during the many years between Candolle's time and Brundin's.

Relative to Candolle's vision of things, Brundin's work is novel in exploring the problem of area-relationships, a problem that Sclater posed over one hundred years ago, and a problem that Wallace tried to explore in the East Indies. Whereas Wallace approached the problem through a consideration of relative levels of endemism, Brundin attempted to deal with the cladistic aspect directly through study of the interrelationships of the endemic species. If the sense of Brundin's results is that there is a nonrandom pattern of area-relationships, then his remarks about continental drift fall into place: the evolution of the continental patterns becomes a causal explanation of the area-relationships as evidenced by the midges, and a causal explanation of the species-cladograms themselves insofar as they display common elements of pattern.

It is interesting in retrospect to note that Candolle considered why there should be areas of endemism (his "habitations"). He wrote:

This fact leads us to the idea, already stated above, that stations are determined uniquely by physical causes actually in operation, and that habitations are probably determined in part by geological causes that no longer exist today. According to this hypothesis one may easily conceive why plant species that are never found native in a certain area will nevertheless live there if they are introduced. But this theory is touched by the uncertainty, one must admit, of all of the ideas relating to the ancient state of our globe and to the primitive origin of living things. (1820:413, translated)

Candolle remarked also that "all of the theory of geographical botany rests on the particular idea one holds about the origin of living things and the permanence of species" (1820:417, translated). Candolle considered these matters very briefly, with particular reference to the then current opinion that species are permanent, and with reference to various theories about the origin of species through the action of external causes and hybridization. He did not consider any of these theories particularly convincing, and concluded that the causes for the differences between species are due to things yet unknown.

The idea that geological or geographic change might somehow be a causal factor in the emergence of new species, and the development of areas of endemism, is, thus, an old idea, germs of which are present in the writings of Buffon, Candolle, and others. The idea is an almost necessary adjunct to any theory of evolution, including the theory of natural selection that became popular following the publication of Darwin's *Origin of Species* in 1859. Yet what distinguishes the results of Brundin from the efforts of the nineteenth-century biologists is not evolutionary theory per se, even in its guise of causal explanation, but rather his focus on the cladistic aspect of area-relationships as displayed, on the one hand, by the species-cladograms of the three subfamilies of midges that he studied, and, on the other, by the area-cladogram that he abstracted from the geological literature. In short, he juxtaposed *biological* and *geological* area-cladograms (Rosen 1978). That these two sets of patterns, biological and geological, should agree among themselves, even in detailed elements of pattern, is, as mentioned, improbable due to chance alone. What is novel in Brundin's argument, therefore, is not his attempt at causal explanation with reference to the combined workings of biological and geological evolution, but rather the nature and structure of the phenomena that, he believed, demanded explanation. These phenomena, collectively and conceptually, constitute the cladistic aspect. They are the components of cladograms, which specify a hierarchical pattern of area-relationships.

The period between Sclater and Brundin spanned over one-hundred years. This was a period of considerable activity in systematics and biogeography, and the literature of this period is vast. Even so, the literature is virtually without focus on the cladistic aspect, especially as it applies to area-relationships. Instead there was a preoccupation with different concepts of area-relationship, each seemingly based on the old notion, revived by Darwin and Wallace, of dispersal (or migration) of organisms from one region to another. Three types of concepts were common:

(1) Levels of endemism, as exemplified by Wallace's discussion of the East Indies. The implication is that an area with a high degree of endemism has been isolated a longer time from other areas with lower degrees of endemism; or that strong barriers to dispersal are older and more permanent than weak ones.

(2) Biotic similarity, as exemplified by numerous authors who have counted the numbers of taxa in common between different areas (not necessarily areas of endemism). The implication is that areas with many taxa in common have been isolated from each other for a shorter time (or are not isolated from one another at all) than from areas with which they have fewer taxa in common.

(3) Dispersal relationships, as exemplified by numerous authors who

believed that they could discern routes of past migration of plants and animals between different areas. The implication is that areas that seem to have permitted much dispersal between themselves have been isolated from one another for a shorter time (or were not isolated at all) than from areas with which there has been less dispersal.

In certain respects these three concepts are similar and have been similarly applied. Each of them can lead to a similar result: a concept of area-relationship that can be expressed in a cladogram of areas. If they are areas of endemism, a prediction would be possible: namely, that the relationships specified in the cladogram will be true for all taxa endemic to the areas (or for more of them than might be expected by chance alone). To test any such prediction, one would need only to investigate the cladistic aspect of species-relationships of the endemic taxa—a test that seems rarely, if ever, to have been performed prior to the time of Brundin. One reason for this long delay in focusing upon the cladistic aspect is that the cladistic aspect was not clearly perceived, and, more often than not, in its place was one or another of the above three concepts. Another reason is that areas were commonly defined with reference only to geographical or political boundaries, or with reference to arbitrary divisions of latitude and longitude, but without reference to local areas of endemism. In some cases, areas of endemism and easily perceived geographical, or even political, boundaries might coincide, as in the case of islands, and the Hawaiian Islands in particular. But more often than not, areas—especially large ones—were defined and studied without reference to endemism at a local level. A large area defined in such a way, for example the northern hemisphere, North America, the United States, or the State of Utah, might harbor one or many local areas of endemism within its boundaries. If more than one area is therein contained, the several areas might have diverse area-relationships. And a mistake would have been made at the outset, in grouping the local areas together without reference to the area-relationships that they might have.

Thus the literature of this precladistic period demands critical appraisal in any particular case if one wishes to judge its relevance to the problem of area-relationships. A first step is to determine what, if any, connection there is between the areas an author might recognize and the local areas of endemism contained therein; a second step is to determine what concept of area-relationship an author is employing; a third step is to determine what concept of species-relationship an author is employing. The reader might then be in a position to judge the significance, if any, of the information, argument, and conclusions presented by the author of the piece in question.

Critical appraisal, of course, need not be reserved for papers published prior to 1966. A case in point has been furnished by Keast (1973), who published a commentary and review in response to Brundin's account of the austral midges. Keast concludes:

> The contemporary southern temperate biotas of South America and Australia are much more closely related than either is to the African one. Since this pattern is repeated in a wide variety of groups, of widely differing ecological requirements, there can be no doubt that this is a fundamental difference and is not an artifact resulting from secondary extinction in Africa. The biological data, hence, confirms the geological data that Africa separated off earlier from the Gondwana landmass. (1973:338)

Here Keast echoes the words of Brundin quoted above. Because Keast discusses and reviews the distributions of numerous groups of plants and animals, one might expect that some, at least, of these groups would be informative with respect to the relevant cladistic component (figure 8.12). Such, however, is not the case, as is revealed by inspection of his tables of data, for example his table summarizing the distributions of 39 groups of insects (figure 8.14). In no case is any cladistic component indicated. Most groups are shown to be distributed either in South America + Australasia (including Australia and New Zealand) or in South America + Australasia + Africa. For the Podonominae of Brundin, Keast states "African forms very distinct"—which says nothing explicit about their relationships. And for the Diamesinae of Brundin he states "very weak N. Z.–Aust. [relationships] and S. Amer.–African ones" when there is no cladistic component that corresponds to either of these "relationships" in Brundin's species-cladogram of the subfamily is question (figure 8.9). The concept of area-relationship employed by Keast evidently is number 2 above: biotic similarity, which in this case means that in his lists there are few taxa common to South America and Africa (or Australasia and Africa) that do not also occur in Australasia (or South America), whereas there are many taxa in common to South

TABLE 2   Some insect groups shared by South America, Australasia and /or Africa: disjuncts in southern cold temperate areas

XXX—diversified;   XX—moderately diversified;   X—relatively insignificant

Group	Distribution on southern landmasses			Authority and status
	South America	*Australasia*	*Africa*	
Plecoptera (stoneflies)				
Eustheniidae	XXX	XXX		Illies (1965, 1969). Last
Australoperlidae	XXX	XXX		sub-family apparently
Gripopterygidae	XXX	XXX		entered Africa from the
Notonemourinae	XXX	XXX	XXX	north (Illies, 1965).
(Capniidae)				
Megaloptera (scorpion-flies)				
Nannocho ristidae	XXX	XXX		Kuschel (1960)
Hemiptera (bugs)				
Peloridiidae	XXX	XXX		Evans (1958); Kuschel (1960)
Enicocephalidae	X	X		
(*Gamostolus*)				
Odonata (dragon-flies)				
Aeschnidae	XX	XX		Kimmins (in Holdgate, 1963)
Neopetalinae	X	XX		
Ephemeroptera (Mayflies)				
(Several genera)	XXX	XXX		Kuschel (1960)
Orthoptera (grasshoppers, crickets)				
Deinacridinae	XX	XX		Kuschel (1960)
Macropathinae	XX	XX	XX	
Cylindrachaetidae	XXX	XXX		
	Australian grasshopper fauna apparently derived from the north			Keay (1959)
Coleoptera (beetles)				
Camiaridae	X	X		Kuschel (1960)
Belidae	XXX	XXX		
Cylydrorhininae and Erirhininae	XXX	XX		
Carabidae	4 sub-tribes shared between S. Amer. and Australasia, none with African			Darlington (1965)
Diptera (flies)				
Various	9 families with one or more genera common to S. Amer. and Australia; only 2 that are also in Africa–Madagascar			Paramanov (1959)
Chironomidae				
Podonominae	Strong S. Amer.–Australian relationships, weak S. Amer.–N.Z. ones; African forms very distinct.			Brundin (1966)
Diamesinae	Close S. Amer.–N.Z. relationships, weak S. Amer.–Australian ones; very weak N.Z.–Aust., and S. Amer.–African ones.			
Trichoptera (caddis-flies)				
Triplectidinae	XX	XX		Ross (1967)
Hydropsychidae	X	X	X	
Hydrobiosinae	XXX	XX		Ross (1967)
Rhynchopsychidae	XX	XXX		
Philorheithridae	XXX	XXX		
Hymenoptera (wasps, ants)				
Megalyridae	XX	XX	XX	Kuschel (1960)
Thynnidae	XXX	XXX		
Plumariidae	XX		XX	

Figure 8.14. Austral distributions of 39 groups of insects. After Keast (1973), table 2, pp. 313–14. Reprinted with permission from D. H. Tarling and S. K. Runcorn, eds., *Implications of Continental Drift to the Earth Sciences*, 1:309–43. Copyright © 1973, Academic Press.

America and Australasia that do not occur in Africa. The problem with these "many taxa" is that they are uninformative with respect to area-relationship in the cladistic aspect *because* they are unrepresented in Africa. What would be informative are species-cladograms for the groups that occur in all three areas, such as the Notonemourinae, Macropathinae, Hydropsychidæ, and Megalyridae (figure 8.14). As regards Keast's concepts of species-relationships (which are, perhaps, not so important here inasmuch as Keast is merely summarizing the concepts of species-relationships as conceived by other workers), his comments about Brundin's midges are perplexing to say the least. And with respect to the distribution of a certain group of plants he writes:

Since the Proteaceae represent the "classic" example of a disjunct southern plant family they have often been quoted as showing the close phytogeographic relationships of Africa, Australasia and South America, and as supporting former direct land connections in the south (e.g. Beard, 1959). Johnson & Briggs (1963), however, in their detailed analysis of the history and relationships of the Proteaceae, confirm Burbidge (1960) that the relationships of the Australian and South American ones are decidedly closer than either is to the African members. Thus, within the closely-knit sub-family Grevilleoideae the genera *Lomatia, Oreocallis* and *Gevuina* are shared, and *Embothrium* of South America, a close relative of *Oreocallis*, is relatively similar to *Telopia* of Australia. The only member of this super-family in southern Africa is *Brabeium*, a specialized endemic genus related to *Macadamia* of Malaysia, New Caledonia and eastern Australia. (1973:319)

The reader might like to analyze this passage, perhaps with reference to Keast's publication, in order to discover exactly what concepts of area- and species-relationships its author is employing.

LEVELS OF ENDEMISM

Although endemism is generally acknowledged to reflect isolation of some area from other areas, there are various problems associated not only with the concept, but also with data exemplifying the concept. Briggs (1974), for example, compiled data pertaining to levels of endemism for the inshore marine fauna of various islands (table 8.1). There is variability between islands and variability between the various groups of organisms occurring on the same island. For fishes Briggs

Table 8.1. Percentage of Endemic Species of Various Groups of Inshore Marine Organisms (as compiled by Briggs 1974) and Land Plants (as compiled by Good 1974)

	E	F	C	M	O	P
Lord Howe	17	–	–	–	–	30
Norfolk	–	4	–	–	–	25
Hawaii	30	34	45	20	–	90
Easter	–	35	–	–	–	6
Marquesas	–	–	–	20	–	–
Cocos-Keeling	–	1	–	0	–	0
Palaus	–	–	–	–	28	–
Galapagos	–	23	15	16	38	42
Cocos	2	7	9	5	2	–
Clipperton	–	–	0	5	–	–
Revillagigedos	–	10	–	0	–	30
Guadalupe	–	–	–	10	–	–
Kermadecs	39	–	–	34	–	15
Juan Fernandez	80	56	27	52	32	70
Amsterdam/St. Paul	–	28	–	–	–	41
Tristan da Cunha/Gough	–	–	–	–	25	64
Chathams	–	–	0	15	7	20
Macquarie	83	0	–	64	60	10
Kerguelen/Heard/McDonald/Marion/Prince Edward/Crozets	37	66	–	–	30	–
South Georgia	12	34	–	60	9	6
Bouvet	8	50	–	40	–	–
Bermuda	0	5	0	–	–	8
Cape Verde	–	4	33	11	–	30
Azores	–	1	–	–	–	20
Madeira	–	3	–	–	–	20
St. Helena/Ascension	50	25	17	52	–	95

NOTE. Abbreviations: E, echinoderms; F, fishes; C, crustaceans; M, molluscs; O, other marine groups; P, land plants; –, no data reported.

found levels of endemism ranging between 0 percent (Macquarie) and 56 percent (Juan Fernandez); and for Macquarie he found levels of endemism ranging between 0 percent (fishes) and 83 percent (echinoderms). Much of the variability, he thought, was due to inadequate sampling.

It has long been realized that the absolute numbers of species on islands, or on continental areas, is a function of the size of the islands, or continental areas. In 1838, Candolle analyzed the distribution of 8,523 species of the plant family Compositae, in relation to 39 areas that he recognized as botanical regions, which include both islands and parts of continents. Apparent in his analysis is a gradual increase in number of species with area of the region (figure 8.15). Much the same relation holds between number of endemic species and area of the region (figure 8.16). It would seem that there is no difference between insular and continental areas of endemism in this respect.

Concern with islands has become fashionable as a result of the publication of MacArthur and Wilson (1967). One result is that the relation between species-number and island-area has been extensively explored, for example by Slud (1976), who compiled data on resident species of land birds on 163 islands in the warmer oceans (figure 8.17). His results seem much the same as Candolle's of 1838.

The concept of levels of endemism, however, is more problematical.

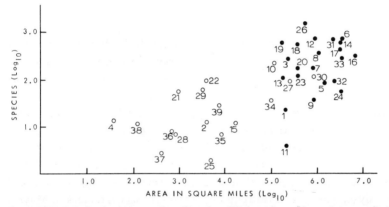

Figure 8.15. Relation between number of species of Compositae and area (square miles) of 39 regions of endemism, including insular (O) and continental (●) regions. The areas are in square miles, not in square leagues, as given by Candolle, by multiplying by a factor of nine. The regions are numbered as follows, in Candolle's terminology: 1. Terres-Magellaniques; 2. Iles Malouines; 3. Chili; 4. Ile Juan Fernandez; 5. Rio de la Plata; 6. Brésil; 7. Colombie; 8. Pérou; 9. Guiane; 10. Antilles; 11. Amérique centrale; 12. Mexique; 13. Californie; 14. Etats-Unis, Canada; 15. Iles Aleutiennes; 16. Sibérie; 17. Europe boréale et moyenne; 18. Europe méditerranéene; 19. Orient; 20. Barbarie; 21. Madère; 22. Canaries; 23. Egypte et Arabie; 24. Afrique équinoxiale; 25. Açores; 26. Cap de Bonne-Espérance; 27. Madagascar; 28. Ile Zanzibar; 29. Ile Maurice; 30. Iles de l'Inde; 31. Inde continentale; 32. Cochinchine, Chine, Japon; 33. Nouvelle-Hollande et Diémen; 34. Nouvelle-Zélande; 35. Nouvelle-Calédonie; 36. Iles de la Société; 37. Iles Tristan d'Acunha; 38. Ile Sainte-Hélène; 39. Iles Sandwich. Data from Candolle (1838), table IV.

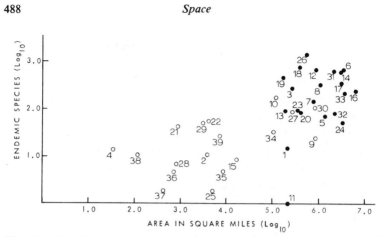

Figure 8.16. Relation between number of endemic species of Compositae and area (square miles) of 39 regions of endemism, including insular (O) and continental (●) regions (cf. figure 8.15). Data from Candolle (1838), table IV.

MacArthur and Wilson, for example, published a figure (from Mayr 1965b), indicating that there is a marked increase in the level of endemism with island-area (figure 8.18). This increase is a questionable matter, as may be appreciated by inspection of Candolle's (1838) data on plants (figure 8.19), for there is no discernible relation similar to that portrayed by MacArthur and Wilson. Their data pertain to species of birds, and a problem with their data might be one merely of taxonomic artifact, due to the ornithological practice of recognizing endemic forms at the level of subspecies. Hence, much real endemism might be masked if only endemic species are counted. In this respect, a remark of Mayr is significant: "These graphs must be considered as mere approximations, for the exact number of the resident, breeding, land-bird fauna is not certain for some of these islands, and in many cases an arbitrary decision must be made whether to call an endemic isolate a species or a subspecies" (1965b:587). Another problem might be that the sample of data points is biased in omitting small islands with a highly endemic fauna of resident birds, as was pointed out by Slud (1976), in reference to Cocos Island, with only four resident species, all endemic.

Whatever the case, it is apparent that levels of endemism vary considerably for reasons that are not well understood. Nor is it easy to judge how much of the variation is real and how much is spurious. For poorly collected and studied groups such as marine invertebrates, much of the variation may be due to inadequate sampling. For relatively well-collected groups such as birds, much variation may be due to taxonomic artifact.

With respect to the cladistic aspect, varied levels of endemism seem irrelevant; what is relevant are the interrelationships of the endemic taxa. With respect to the problem of their interrelationships, emphasis on relative levels of endemism, as exemplified by Wallace's consideration of the East Indies, would result only in associating islands such as Cocos-Keeling, Clipperton, Bermuda, and the Azores—an association that would probably prove meaningless in the cladistic aspect.

Concern with levels of endemism does, however, make certain islands stand out from the others. Hawaii and the Galapagos are cases in point. Of the four groups mentioned by Briggs, endemism varies between 20 percent and 45 percent for Hawaii, and between 15 percent and 38 percent for the Galapagos. Generally recognized as endemic, these biotas receive intensive study. It would not be surprising if continued study resulted in greater taxonomic precision, allowing the taxonomist to recognize a greater and greater percentage of organisms native to Hawaii or the Galapagos as distinct taxa, i.e., as endemic species or subspecies. Indeed, one might reasonably expect that most native Hawaiian marine organisms might ultimately be regarded as endemic, as is the case for the land plants. Whereas the above may seem a reasonable expectation for an area such as Hawaii, such would not be true for an island like Cocos or Bermuda, for which, relative to Hawaii, endemism seems less marked, or at least less easily perceived by the taxonomist. Still, one may ask, is the difference between these apparent levels of endemism one of degree or one of kind? If different levels of endemism (table 8.1) merely reflect relative ease of recognition of endemic taxa, then it would seem that future taxonomic investigation is destined to result in much higher, and perhaps more uniform, levels of endemism for all insular regions. Whatever the case, levels of endemism as such seem irrelevant to the matter of area-relationships, i.e., the cladistic aspect.

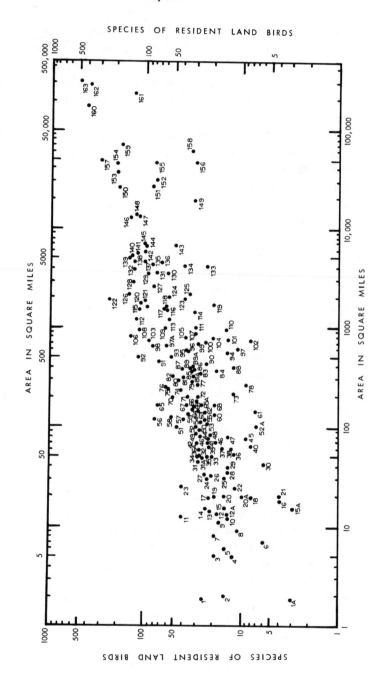

Figure 8.17. Relation between number of species of resident land birds and area (square miles) of 175 islands in the warmer oceans: 1, Santa Ana (Solomons); 1A, Laysan; 2, Jervis (Galapagos); 3, Saba; 4, Lord Howe; 5, Barbareta; 6, Annobon; 7, St. Eustatius; 8, Old Providence; 9, Anegada; 10, St. Andrew; 11, Gizo; 12, Virgin Gorda; 12A, Norfolk; 13, Culebra; 14, María Cleofas; 15, Mono; 15A, Henderson; 16, Cocos (E. Pacific); 17, St. John; 18, Bermuda; 19, Tortola; 20, Utila; 20A, Cayo Largo; 21, Clarion; 22, Mona; 23, San José (Pearl Is.); 24, Nissan; 25, Guanaja; 26, St. Thomas; 27, Montserrat; 28, Anguilla; 29, St. Martin; 30, Rodriguez; 31, Vieques; 31A, Car Nicobar; 32, Aneiteum; 33, Nevis; 34, María Magdalena; 35, Principe; 36, Tutuila; 36A, Camorta; 37, Marie Galante; 38, Aldabra; 39, Barbuda; 40, Christmas (Indian Ocean); 41, St. Kitts; 42, Aruba; 43, Charles; 44, Roatan; 45, Socorro; 46, Grand Cayman; 47, Guadalupe; 48, St. Croix; 49, Moheli; 50, Bonaire; 51, Rendova; 52, Maria Madre; 52A, Niue; 53, Antigua; 54, Hierro; 55, Aurora; 56, Tobago; 57, Grenada; 58, Cozumel; 59, St. Vincent; 60, Ponape; 61, Lanai; 62, Mayotte; 63, Anjouan; 64, Gomera; 65, Coiba; 65A, Enggano; 66, Kandavu; 67, Taveuni; 68, Barbados; 69, Curaçao; 70, Vella Lavella; 71, Efate; 72, Pentecost; 73, Guam; 74, St. Lucia; 75, Mafia; 76, Kolombangara; 77, Tanna; 78, Molokai; 79, Palma; 80, Gonave; 81, Dominica; 81A, São Tomé; 82, Pemba; 83, Lanzarote; 84, São Tiago; 85, Gran Comoro; 86, Ambrym; 87, Martinique; 88, Tahiti; 89, Rennell; 89A, Great Nicobar; 90, Upolu; 91, Margarita; 92, Basilan; 93, New Georgia; 94, Kauai; 95, Guadeloupe; 96, Gran Canaria; 97, Oahu; 97A, Simalur; 98, Zanzibar; 99, Fuerteventura; 100, Savaii; 101, Mauritius; 102, Maui; 103, Bunguran; 104, Indefatigable; 105, Tenerife; 106, Fernando Po; 107, Erromanga; 108, Japen; 109, Biak; 110, Reunion; 111, Malekula; 112, Waigeu; 113, Isle of Pines; 114, Socotra; 115, Bohol; 116, Choiseul; 117, Santa Isabel (Solomons); 117A, Nias; 118, San Cristobal (Solomons); 119, Albemarle; 120, Cebu; 121, Lombok; 122, Trinidad; 123, Espíritu Santo; 124, Malaita; 125, Vanua Levu; 126, Bali; 127, Guadalcanal; 128, Leyte; 129, Buru; 130, Puerto Rico; 131, Bougainville; 132, Mindoro; 133, Hawaii; 134, Vitu Levu; 135, Sumba; 136, Jamaica; 137, Panay; 138, Palawan; 139, Negros; 140, Samar; 141, Flores; 142, Sumbawa; 143, New Caledonia; 144, Ceram; 145, Halmahera; 146, Hainan; 147, Timor; 148, Taiwan; 149, Isla Grande (Tierra del Fuego); 150, Ceylon; 151, Tasmania; 152, Hispaniola; 153, Mindanao; 154, Luzon; 155, Cuba; 156, North Island (New Zealand); 157, Java; 158, South Island (New Zealand); 159, Celebes; 160, Sumatra; 161, Madagascar; 162, Borneo; 163, New Guinea. Modified from P. Slud (1976), Geographic and climatic relationships of avifaunas with special reference to comparative distribution in the Neotropics. Smithsonian Contrib. Zool. no. 212; figure 9, p. 23.

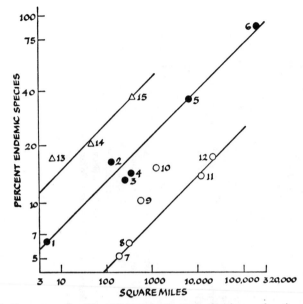

Figure 8.18. Percentage of resident bird species that are endemics as a function of island areas, in three kinds of islands. Dark circles (solitary, well isolated islands): 1, Lord Howe; 2, Ponape; 3, Rennel; 4, Chatham; 5, New Caledonia; 6, Madagascar. Open circles (single islands near mainlands or large archipelagos): 7, St. Matthias; 8, Pemba; 9, Manus; 10, Socotra; 11, Timor; 12, Tasmania. Triangles (islands in the Gulf of Guinea): 13, Annobon; 14, Principe; 15, San Tomé. Figure and legend after MacArthur and Wilson (1967), figure 60, p. 174. Copyright © 1967, Princeton University Press. Reprinted by permission of Princeton University Press; modified from Mayr (1965b), pp. 1587–88, figure 1.

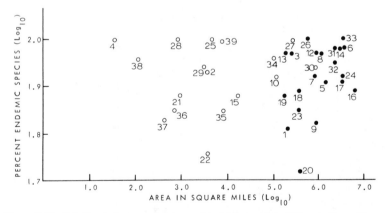

Figure 8.19. Relation between percentage of endemic species of Compositae and area (square miles) of 38 regions of endemism, including insular (○) and continental (●) regions (region 11 is omitted; cf. figures 8.15, 8.16). Data from Candolle (1838), table IV.

BIOTIC SIMILARITY

Similarity as a concept is ambiguous in the sense that all things are more or less similar, and their similarities depend upon the point of view of the observer. Biogeography is always a study of relative similarities, but the areas to be compared, and the measure of similarity to be employed, are open to diverse definition and interpretation. One possibility is to begin with areas of endemism and to approach the problem of area-relationships through study of the species-relationships of the endemic taxa. In this case, all areas would turn out to be more or less closely related, and their interrelationships would be specifiable in one (or more) area-cladogram. A second possibility is to begin with areas defined arbitrarily, for example in accordance with the division of the global landscape into 58 areas (figure 8.20). Given this scheme of division, one might still approach the problem of area-relationships through study of the species-relationships of the endemic taxa. In this case, too, all areas would turn out to be more or less closely related, and their interrelationships would be specifiable in one (or more) area-cladogram. A third possibility applicable to areas of endemism, or areas arbitrarily defined, is to count the number of taxa in common to different areas, and to consider this commonality a measure of area-relationship. This third possibility allows for diverse mathematical treatment, in accordance with one or another numerical index of similarity, of which there are some dozens. But in this case, too, all areas would be more or less closely related (or more or less similar), and their interrelationships would be specifiable in one (or more) area-cladogram. However, to achieve an area-cladogram through analysis of similarity values requires a further mathematical procedure (clustering), and there are many different clustering procedures.

It is not the purpose here to review the diverse mathematical indices and clustering procedures that might be employed in analyses of similarity, but to note only the nature of the results. Consider an example: an area-cladogram of 22 areas (figure 8.21), derived through a certain mathematical treatment of similarity values (figure 8.22), obtained through the use of a certain index of similarity for 820 species of butterflies. Certain of the areas are those considered by Wallace: Java, Borneo, Sumatra, and Malaya (Asia), and their relationships turn

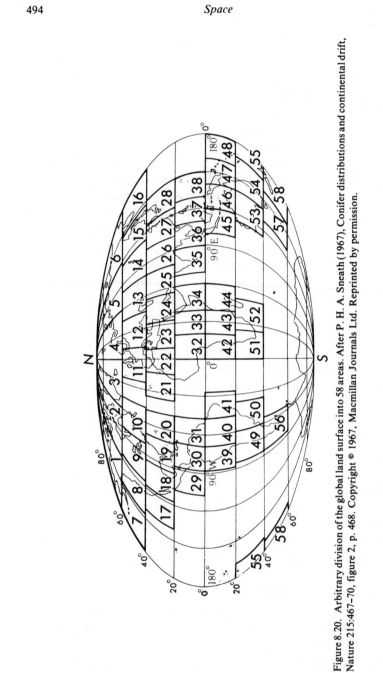

Figure 8.20. Arbitrary division of the global land surface into 58 areas. After P. H. A. Sneath (1967), Conifer distributions and continental drift, Nature 215:467–70, figure 2, p. 468. Copyright © 1967, Macmillan Journals Ltd. Reprinted by permission.

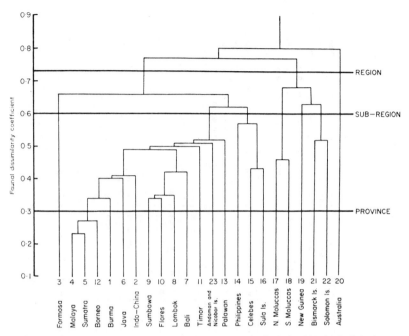

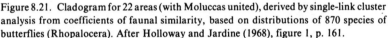

Figure 8.21. Cladogram for 22 areas (with Moluccas united), derived by single-link cluster analysis from coefficients of faunal similarity, based on distributions of 870 species of butterflies (Rhopalocera). After Holloway and Jardine (1968), figure 1, p. 161.

out to be the same as those initially estimated by Wallace, with reference to relative levels of endemism. Consider a second example: an area-cladogram of the same 22 areas (figure 8.23), similarly derived, for 876 species of birds. Consider a third example: an area-cladogram of 14 areas (figure 8.24), similarly derived, for 124 species of bats.

The three area-cladograms differ in several respects, which the reader might like to determine for himself. It may be noted here that the authors of the three cladograms (Holloway and Jardine) conclude that "all generalisations made on the basis of analysis of distributional data should be considered to be *strictly relative* to the particular area studied and to the particular taxonomic group studied" (1968:185). Despite this note of caution, the three area-cladograms may be subjected to cladistic analysis. An obstacle, however, arises from the fact that, although the same 22 areas appear in two of the cladograms (figures 8.21, 8.23), only

AVES

	1	2	3	4	5	6	7	8	9	10	11	12	13	14	15	16	17	18	19	20	21	22	23	
	_1_	0·838	0·482	0·655	0·599	0·711	0·825	0·788	0·832	0·897	0·662	0·788	0·786	0·859	0·918	0·968	0·968	0·970	1·000	0·969	1·000	1·000	0·689	Burma
	0·321	_2_	0·568	0·658	0·662	0·677	0·810	0·804	0·816	1·000	0·746	0·777	0·784	0·862	0·963	0·940	0·940	1·000	1·000	0·970	1·000	1·000	0·733	Indo-China
	0·395	0·790	_3_	0·926	0·922	0·934	0·971	0·943	0·915	0·916	0·943	0·929	0·943	0·871	0·910	0·941	1·000	1·000	1·000	1·000	1·000	1·000	0·887	Formosa
	0·696	0·672	0·926	_4_	0·212	0·524	0·777	0·747	0·839	0·848	0·943	0·888	0·415	0·782	0·880	0·904	0·965	0·926	0·963	0·971	0·965	0·963	0·725	Malaya
	0·338	0·813	0·922	0·212	_5_	0·391	0·424	0·594	0·639	0·672	0·763	0·695	0·282	0·529	0·780	0·878	0·904	0·936	0·936	0·965	0·969	0·963	0·734	Sumatra
	0·415	0·864	0·934	0·524	0·226	_6_	0·499	0·312	0·571	0·474	0·562	0·685	0·529	0·788	0·802	0·807	0·881	0·888	0·943	0·969	1·000	1·000	0·740	Java
	0·475	0·848	0·971	0·777	0·440	0·391	_7_	0·413	0·474	0·201	0·334	0·730	0·763	0·802	0·730	0·834	0·875	0·834	0·935	0·918	0·943	0·971	0·836	Bali
	0·689	0·887	0·943	0·747	0·391	0·499	0·413	_8_	0·201	0·334	0·512	0·763	0·722	0·836	0·889	0·859	0·859	0·834	0·918	0·922	0·942	0·943	0·835	Lombok
	0·730	0·862	0·915	0·839	0·654	0·312	0·474	0·201	_9_	0·204	0·512	0·763	0·722	0·836	0·889	0·861	0·861	0·834	0·913	0·888	0·913	0·942	0·834	Sumbawa
	0·696	0·862	0·916	0·848	0·730	0·571	0·201	0·334	0·204	_10_	0·494	0·817	0·804	0·836	0·764	0·884	0·888	0·836	0·882	0·862	0·915	0·971	0·888	Flores
	0·729	0·853	0·943	0·943	0·758	0·474	0·334	0·512	0·512	0·494	_11_	0·888	0·832	0·790	0·816	0·913	0·888	0·836	0·862	0·825	0·887	0·915	0·747	Timor
	0·862	0·878	0·929	0·888	0·812	0·562	0·592	0·667	0·667	0·467	0·512	_12_	0·541	0·897	0·853	0·864	0·913	0·836	0·825	0·971	0·967	0·964	0·836	Borneo
	0·473	0·709	0·943	0·415	0·266	0·685	0·756	0·768	0·768	0·329	0·512	0·541	_13_	0·644	0·816	0·887	0·864	0·965	0·862	1·000	1·000	0·971	0·777	Palawan
	0·682	0·864	0·871	0·782	0·304	0·654	0·830	0·812	0·812	0·513	0·897	0·618	0·644	_14_	0·577	0·794	0·853	0·843	0·904	0·908	0·936	0·971	0·804	Philippines
	0·734	0·884	0·910	0·880	0·616	0·499	0·689	0·816	0·816	0·795	0·862	0·752	0·780	0·577	_15_	0·461	0·816	0·856	0·774	0·886	0·941	0·936	0·862	Celebes
	0·816	0·901	0·941	0·904	0·672	0·730	0·816	0·850	0·850	0·882	0·897	0·812	0·804	0·794	0·461	_16_	0·777	0·824	0·769	0·895	0·957	0·912	0·942	Sula Is.
	0·882	0·936	0·893	0·965	0·711	0·750	0·786	0·788	0·788	0·874	0·904	0·888	0·882	0·882	0·400	0·777	_17_	0·434	0·688	0·689	0·901	0·910	0·942	N. Moluccas
	0·926	0·893	0·941	0·926	0·816	0·852	0·862	0·861	0·861	0·910	0·888	0·893	0·893	0·884	0·777	0·515	0·434	_18_	0·734	0·734	0·737	0·737	0·811	S. Moluccas
	0·929	0·942	0·941	0·963	0·904	0·878	0·906	0·939	0·939	0·965	0·965	0·918	0·882	0·780	0·769	0·682	0·734	0·734	_19_	0·807	0·639	0·734	0·734	New Guinea
	0·971	0·941	0·968	0·965	0·936	0·893	0·936	0·955	0·955	0·957	0·882	0·943	0·893	0·908	0·895	0·895	0·834	0·807	0·816	_20_	0·875	0·901	0·734	Australia
	0·967	0·968	0·969	0·963	0·971	0·904	0·966	0·934	0·934	0·936	0·922	0·967	0·971	0·971	0·963	0·889	0·726	0·723	0·816	0·847	_21_	0·554	1·000	Bismarck Is.
	0·957	0·969	0·971	0·965	0·957	0·918	0·942	0·971	0·971	0·940	0·918	1·000	1·000	0·934	0·904	0·884	0·834	0·636	0·847	0·855	0·502	_22_	1·000	Solomon Is.
	0·934	0·942	1·000	0·963	0·943	0·962	0·971	0·943	0·943	0·886	0·886	0·943	0·943	0·888	0·862	0·719	0·709	0·689	0·816	0·689	0·855	0·502	_23_	Andaman & Nicobar Is.
	0·524	0·862	0·622	0·695	0·836	0·761	0·777	0·853	0·730	0·734	0·825	0·832	0·887	0·940	0·939	0·912	0·967	0·943	0·915					

RHOPALOCERA

Figure 8.22. Similarity values. After Holloway and Jardine (1968), table 1, p. 160.

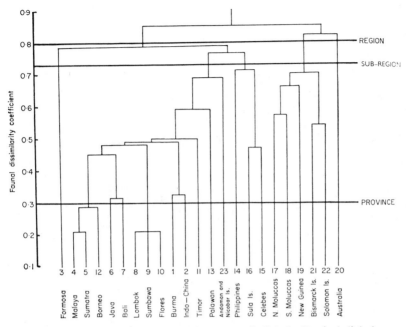

Figure 8.23. Cladogram for 22 areas (with Moluccas united) derived by single-link cluster analysis from coefficients of faunal similarity, based on distributions of 876 species of birds. After Holloway and Jardine (1968), figure 2, p. 162.

14 areas appear in the third cladogram (figure 8.24). Some of the 14 areas are the same as those in the other two cladograms (e.g., Malaya, Sumatra); some of the 14 areas are combinations of those in the other two cladograms (e.g., Borneo-Palawan, Java-Bali); one of the 14 areas does not appear in the other two cladograms (Sanghir Talaut). Areas common to all three cladograms include nine areas coded as follows: 1. Australia; 2. Moluccas; 3. Solomons; 4. New Guinea; 5. Celebes; 6. Philippines; 7. Indo-China; 8. Sumatra; 9. Malaya. The three cladograms may be simplified, to overcome the obstacle of noncomparabilities, in order to show what information they contain about the relationships of these nine areas. Thus, the butterflies are represented by figure 8.25; the birds, by figure 8.26; the bats, by figure 8.27. The components (numbered 1–12) of the simplified area-cladograms vary among the three, but some of them are replicated (table 8.2). The pattern of replication is improbable due to chance alone (P = 10^{-4}%). All of the

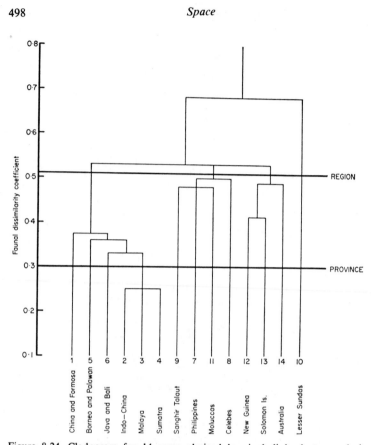

Figure 8.24. Cladogram for 14 areas, derived by single-link cluster analysis from coefficients of faunal similarity, based on distributions of 124 species of bats. After Holloway and Jardine (1968), figure 3, p. 163.

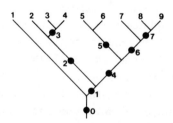

Figure 8.25. Simplified area-cladogram based on buttterflies (cf. figure 8.21).

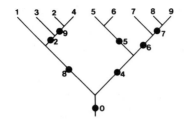

Figure 8.26. Simplified area-cladogram based on birds (cf. figure 8.23).

replicated components are combinable in a single general area-clado-gram (figure 8.28), which may be considered an indication of what the butterflies, birds, and bats jointly affirm about the relationships of the nine areas.

Except for area 1 (Australia) all nine areas occur in the general cladogram. Of the 12 different informative components of the simplified area-cladograms (figures 8.25–8.27), six do not appear in the general cladogram. Three of these (components 1, 8, 10) involve different relationships for Australia, and any two of the three conflict, one with the other; any one of the three could be added to the general cladogram without conflict, but no two of them could be added; these three components are, therefore, ambiguous components. Of the three remaining components (9, 11, 12), one (9) is derived from the birds (figure 8.26), and two (11, 12) are derived from the bats (figure 8.27). Component 9 conflicts with components 11 and 12; all three conflict with the general cladogram; these three components are, therefore, false components. These six components (ambiguous + false) represent vari-ability that seems random. Interestingly, all of the six informative compo-nents of the general cladogram appear also in the simplified area-cladogram for the butterflies.

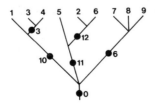

Figure 8.27. Simplified area-cladogram based on bats (cf. figure 8.24).

Table 8.2. Probabilities (P) of Replication of Components (C) in Area-Cladograms

Number of Terms	Number of Possible Components	8.25		8.26		8.27		P for Components of Same Number of Terms, %	P. Cumulative %
		C	P:%	C	P:%	C	P:%		
2	36	3	–	5	8.33	3	11.1	8.81×10^{-2}	8.81×10^{-2}
		5	–	7	9.52				
		7	–	9	–	12	–		
3	84	2	–	2	20.0	6	10.0	0.50	4.44×10^{-4}
		6	–	6	25.0	10	–		
						11	–		
4	126	4	–	8	–				10^{-4}
5	126	1	–	4	100.				10^{-4}
8	9								10^{-4}

NOTE. For a component that replicates a previously resolved component, the probability is that of replication due to chance alone, under the assumption that the components are resolved one at a time in the order listed for each cladogram. The probability fractions for replicates are: cladogram 8.26 (3/36, 2/21, 2/10, 1/4, 1/1), cladogram 8.27 (4/36, 2/20).

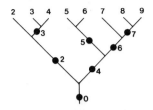

Figure 8.28. General area-cladogram (cf. figures 8.25–8.27).

Some of the remaining areas in the original three area-cladograms (figures 8.21, 8.23, and 8.24) may now be analyzed. They are coded as follows: 10. Formosa; 11. Palawan; 12. Timor; 13. Borneo; 14. Java; 15. Bali; 16. Lombok; 17. Flores; 18. Sumbawa. For purposes of comparison, the following steps are taken to render the troublesome items of the bat-cladogram comparable to the butterfly- and bird-cladograms: China-Formosa is considered equivalent to Formosa; the Lesser Sundas are considered equivalent to Lombok + Sumbawa + Flores + Timor; Borneo-Palwan and Java-Bali are considered as widespread areas. Three simplified area-cladograms result (figures 8.29, 8.30, and 8.31), with the component-analysis for the bat-cladogram performed under assumption 1. Four components are replicated (table 8.3), and the pattern of replication is improbable due to chance alone (P = 0.2%). All of the replicated components are combinable in a single general cladogram (figure 8.32), which may be considered an indication of what the butterflies, birds, and bats jointly affirm about the relationships of the nine areas.

Of the 11 different informative components of the simplified area-cladograms (figures 8.29, 8.30, and 8.31) six do not appear in the general cladogram (figure 8.32). Two of these (components 22 and 23), both derived from bats, conflict with the general cladogram; these two components are, therefore, false components. Of the four remaining

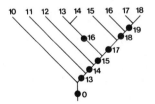

Figure 8.29. Simplified area-cladogram based on butterflies (cf. figure 8.21).

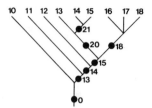

Figure 8.30. Simplified area-cladogram based on birds (cf. figure 8.23).

components (16, 17, 20, 21), each could be added singly to the general cladogram, but they conflict among themselves (16 with 21, 17 with 20 and 21, 20 with 17, 21 with 16 and 17); these four components are, therefore, ambiguous components. These six components (false + ambiguous) represent variability that seems random.

Component 19 is not replicated, but it has been added to the general cladogram so as to produce the most detailed estimate consistent with the pattern of replicated components. Component 19 does not conflict with any of the true or ambiguous components. But component 19 could represent random variation. Under the assumption that the four replicated components are jointly true in some sense, there is only a 1 in 3 probability that component 19 also is true in some sense. Interestingly, all five of the informative components of the general cladogram (figure 8.32) appear in the simplified area-cladogram for the butterflies.

These partial analyses of the original three area-cladograms (figures 8.21, 8.23, and 8.24) contain no common elements and cannot directly be combined in a single cladogram specifying the general relationships of all 18 areas. With the areas coded as above, and the same steps taken with respect to the bat cladogram, all 18 areas may be considered together. Three area-cladograms result, still somewhat simplified with respect to the three originals (figures 8.33–8.35). Eleven components are

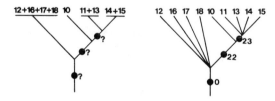

Figure 8.31. Simplified area-cladogram based on bats (cf. figure 8.24).

Table 8.3. Probabilities (P) of Replication of Components (C) in Area-Cladograms

Number of Terms	Number of Possible Components	Cladograms 8.29		Cladograms 8.30		Cladograms 8.31		P for Components of Same Number of Terms, %	P Cumulative %
		C	P:%	C	P:%	C	P:%		
2	36	16	–	21	–				
		19	–						
3	84	18	–	18	2.86			2.86	2.86
				20	–				
4	126	17	–			23	–		2.86
5	126					22	–		2.86
6	84	15	–	15	33.0			33.0	0.952
7	36	14	–	14	33.0			33.0	0.317
8	9	13	–	13	50.0			50.0	0.159

NOTE. For a component that replicates a previously resolved component, the probability is that of replication due to chance alone, under the assumption that the components are resolved one at a time in the order listed for each cladogram. The probability fractions for replicates are: cladogram 8.30 (1/35, 1/3, 1/3, 1/2).

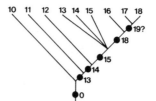

Figure 8.32. General area-cladogram (cf. figures 8.29–8.31).

replicated (table 8.4), and the pattern of replication is improbable due to chance alone ($P = 10^{-9}\%$). All of the replicated components are combinable in a single general cladogram (figure 8.36), which may be considered an indication of what the butterflies, birds, and bats jointly affirm about the relationships of the 18 areas.

Of the 29 different components of the simplified area-cladograms (figures 8.33–8.35), 16 do not appear in the general cladogram (figure 8.36). Nine of these (components 6, 9, 10, 11, 12, 35, 36, 37, 38), including one (9) derived from birds and eight derived from bats, conflict with the general cladogram; these nine components are, therefore, false components. Of the seven remaining components (8, 17, 21, 24, 30, 31, 34), each could be added to the general cladogram, but they conflict among themselves (8 with 24; 17 with 21 and 34; 21 with 17, 30, and 31; 24 with 8; 30 with 21 and 34; 31 with 21; 34 with 17 and 30); these seven components are, therefore, ambiguous components. These 16 components (false + ambiguous) represent variability that seems random.

Components 19 and 33 are not replicated, but they have been added to the general cladogram so as to produce the most detailed estimate

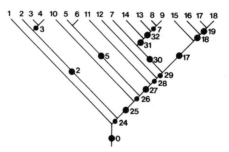

Figure 8.33. Area-cladogram based on butterflies (cf. figure 8.21).

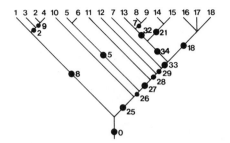

Figure 8.34. Area-cladogram based on birds (cf. figure 8.23).

consistent with the pattern of replicated components. Components 19 and 33 do not conflict with any of the true or ambiguous components, but they could represent random variation.

Under the assumption that the 11 replicated components are jointly true in some sense, there is only a 1 in 3 probability that component 19 also is true, and only a 1 in 5 probability that component 33 also is true; the probability that both are true is only 1 in 15. Interestingly, 12 of the 13 components of the general cladogram appear also in the simplified area-cladogram for butterflies.

The results of these analyses are open to various kinds of interpretation. Perhaps the most obvious conclusion, if not the most interesting, is that there is something fundamentally different between distributions of butterflies and birds, on the one hand, and those of bats, on the other:

One conclusion is immediate. The zoogeographic classifications derived from the study of distributions of taxa from the Rhopalocera [butterflies] and Aves differ widely from that derived for the Chiroptera. It is clear that zoogeographic classifications, at least in this area, should be considered relative to particular taxonomic groups. (Holloway and Jardine 1968:164)

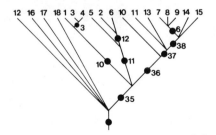

Figure 8.35. Area-cladogram based on bats (cf. figure 8.24).

Table 8.4. Probabilities (P) of Replication of Components (C) in Area-Cladograms

Number of Terms	Number of Possible Components	Cladograms						P for Components of Same Number of Terms, %	P. Cumulative %
		8.33		8.34		8.35			
		C	P:%	C	P:%	C	P:%		
2	153	3	–	5	2.61	3	3.92	2.56×10^{-3}	2.56×10^{-3}
		5	–	7	2.50	12	–		
		7	–	9	–				
		19	–	21	–				
3	816	2	–	2	1.88	6	–	2.23×10^{-3}	5.72×10^{-8}
		18	–	18	2.38	10	–		
		32	–	32	5.00	11	–		
4	3,060	17	–	8	–				10^{-7}
		31	–						
5	8,568	30	–	34	–	38	–		10^{-7}
7	31,824					37	–		10^{-7}
8	43,758			33	–	36	–		10^{-7}
9	48,620	29	–	29	25.0			25	1.43×10^{-8}
10	43,758	28	–	28	33.3			33	4.77×10^{-9}
11	31,824	27	–	27	50.0			50	2.38×10^{-9}
13	8,568	26	–	26	100.			100	10^{-9}
14	3,060	25	–	25	100.			100	10^{-9}
17	18	24	–						10^{-9}

NOTE. For a component that replicates a previously resolved component, the probability is that of replication due to chance alone, under the assumption that the components are resolved one at a time in the order listed for each cladogram. The probability fractions for replicates are: cladogram 8.34 (4/153, 3/120, 3/160, 2/84, 1/20, 1/4, 1/3, 1/2, 1/1, 1/1), cladogram 8.35 (6/153).

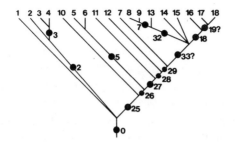

Figure 8.36. General area-cladogram (cf. figures 8.33–8.35).

These authors refer to widely different "zoogeographic classifications," which as can be seen from their area-cladograms (figures 8.21, 8.23, and 8.24), include various regions, subregions, and provinces. These designations refer to various levels of biotic similarity (represented by the horizontal lines across the branches of the cladograms). Thus the regions, subregions, and provinces correspond either to components of the cladograms or to terminal taxa. Because the terminal taxa are the same in the different cladograms, the only differences of note in the classifications are a function of the diverse components of the different area-cladograms. Hence, the concept of "zoogeographic classification" employed by these authors is directly relatable to the components of the area-cladograms. And it is true that the components derived from bats differ widely.

According to the above analysis, this peculiarity of the bats is "random" variation, i.e., variation that is unexplained within the limits of the analysis. Holloway and Jardine allude to the possibility that the bat data may indeed be random:

We discuss first the zoogeographic classification derived for the Chiroptera since this differs markedly from that derived for the Rhopalocera and Aves. The conclusion drawn by Tate (1946) that there is a distinct intercontinental or "Wallacean" region is confirmed, but the Lesser Sunda Islands are excluded from this, appearing as a distinct region. As Tate pointed out, the bat fauna of the Lesser Sundas is imperfectly known and small. Further information might well alter the position of the Lesser Sundas in the classification. (1968:164)

If the bat fauna of the Lesser Sundas is small, especially if it is highly endemic, the biotic similarity of the Lesser Sundas would be small relative to all other areas, which does seem to be the sense of the bat cladogram (figures 8.24, 8.31, and 8.35).

However, it is possible that in reality this apparent peculiarity of the bats is not a random effect, but rather a reflection of some real difference, not merely in distributional records, but perhaps in ecology and history as well. This possibility will not be pursued further here.

As noted above, biotic similarity refers to relative numbers of taxa occurring in different areas. Most of the areas in the above analyses are areas of endemism. Hence one may ask, whether the general cladogram (figure 8.36) might be true for the species-relationships of taxa endemic to the areas specified. If so, the general cladogram might be true for the relationships of all three groups (butterflies, birds, bats), or any two of them, or any one of them, or none at all. Without knowledge of the relationships of the endemic taxa of these groups, there is little more that can be said in response to this question. The question was not investigated by Holloway and Jardine. The general question, whether there is a relation between biotic similarity and relationships of taxa endemic to areas the world over, also remains uninvestigated. There seems no basis for any probabilistic statement applicable to the present case.

One may ask, further, whether the general cladogram might be true for the relationships of the areas in a historically geographic or geological sense. The areas involved in the above analysis have had a complex history such that a cladogram might be an appropriate and informative summary of their historical relationships. This question was not investigated by Holloway and Jardine, who did, however, make some attempt to summarize what was known at that time of the geological history of the Indo-Australian region. There is again a general question that remains uninvestigated: whether there is a relation between biotic similarity and the geological relationships of the areas occupied by the organisms. And again, there seems to be no basis for a probabilistic statement applicable to the present case, to the effect that one or another level of relation might be expected.

DISPERSAL RELATIONSHIPS

Presumed dispersal relationships have been the dominant, albeit controversial, mode of biogeographic analysis since the time of Wallace and Darwin. An example of presumed dispersal relationships is again furnished by the Hawaiian Islands, noted for their highly endemic biota, including many endemic insects (figure 8.37) and resident land birds (figure 8.38). It has been customary to imagine that the Hawaiian Islands were colonized, by means of long-distance migration, by ancestors of the endemic forms that live there now:

> Previous students of Hawaiian birds have concluded that all of the endemic genera, species, and subspecies evolved from 15 original ancestral species. Figure . . . [8.38] shows the postulated origins for these ancestral species. Such figures are misleading, however, in that they depict generalizations only, based on what ornithologists believe to be the closest relatives to Hawaiian birds on continental areas. (Berger 1972:17)

Routes of dispersal to Hawaii can easily be imagined, by inspection of Figures 8.1–8.3, showing the distributions of species of the plant genera *Keysseria, Argemone,* and *Sophora.* In each case most species of each genus occur elsewhere than Hawaii: 10 species of *Keysseria* to the west, 26 species of *Argemone* to the east, and 12 species of *Sophora* to the south. A little imagination easily leads to the idea that the few endemic

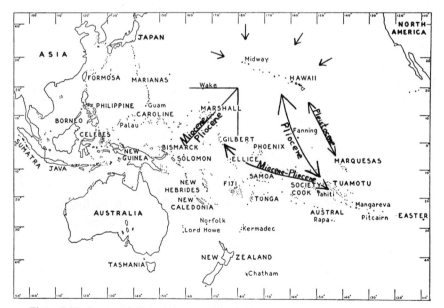

Figure 8.37. Presumed dispersal routes of the species ancestral to the endemic insects of the Hawaiian Islands. After E. C. Zimmerman (1948), Insects of Hawaii, vol. 1 (Honolulu: University of Hawaii Press), figure 26, p. 96.

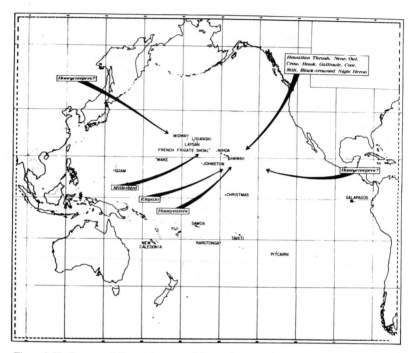

Figure 8.38. Presumed dispersal routes of the species ancestral to the endemic birds of the Hawaiian Islands. Modified from Berger (1972), figure 16, p. 17.

Hawaiian species of these plants resulted from colonization of Hawaii from these other areas, to the west, east, and south, at one or more times in the past. According to this idea, the occurrence of these genera in Hawaii is a secondary phenomenon, resulting from chance dispersal from various centers of origin—the truly native, or primary areas, of distribution of these genera.

Since the publication of MacArthur and Wilson (1967), it has become fashionable to visualize similar "islands" in continental areas, particularly with regard to areas of endemism. An example is furnished by Mayr and Phelps (1967), in an analysis of the birds occupying isolated upland areas (*tepuis*) in southern Venezuela (figure 8.39). They imagine the *tepuis* to have been colonized from four major areas: 1. the coastal ranges of northern Venezuela; 2. the Eastern Andes; 3. the Brazilian highlands; 4. the surrounding lowlands. Of the birds resident in the *tepuis*, Mayr and Phelps recognize several categories:

1. Endemic genera (2).
2. Endemic species without known relatives (4).
3. Endemic species, more or less closely related to an allopatric species (11).
4. Endemic species with a close relative (8).
5. Endemic species, members of a more widespread superspecies (4).
6. Nonendemic species with endemic subspecies (55).
7. Nonendemic species without endemic subspecies (12).

And they conclude:

From the data given above, it is evident that there is a completely even gradation from endemic genera to species that have not even begun to develop endemic subspecies. Fewer than one-third . . . of the characteristic upper zonal element of Pantepui . . . are endemic species. These facts are conclusive evidence for the continuity and long duration of the colonization of Pantepui. (p. 291)

The data presented by Mayr and Phelps are much like those for the

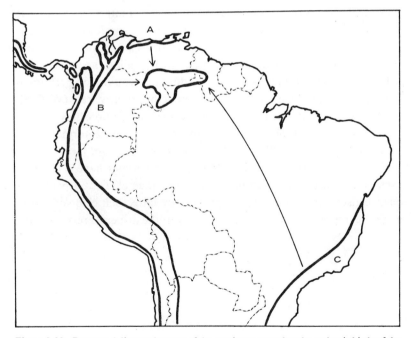

Figure 8.39. Presumed dispersal routes of the species ancestral to the endemic birds of the southern Venezuelan highlands. After Mayr and Phelps (1967), figure 1, p. 308.

Hawaiian plants and birds: some species are related to area A (north), some to area B (west), some to area C (south), and so on.

Such dispersal interpretations are problematical for several reasons, most of which concern the rules or clues used to recognize centers of origin, or source areas for presumed dispersals. But the main reason is that the interpretations might simply be false, in the sense that the endemic forms could represent the results, not of dispersal and colonization, but rather of *in situ* development. It is not the purpose here to evaluate these alternative possibilities for interpretation, but rather to focus on the cladistic aspect of dispersal interpretations. In the case of Hawaii, for example, the dispersal interpretation can be rendered as a tree-diagram (figure 8.40), which in itself is uninformative when viewed in the cladistic aspect. Being uninformative, it leaves open the possibility that Hawaii does have some definite relationship with other areas, such as could be expressed in a single area-cladogram. Nor does the tree diagram include information, even though such might exist or be obtainable, about the interrelationships of the primary areas—the presumed centers of origin (W, E, S). If these are interrelated in some particular way (as might be inferred by study of the species-relationships of their endemic taxa), then an informative area-cladogram could result from study of the species-relationships of the Hawaiian endemics (figure 8.41). The information in such a cladogram would not pertain to Hawaii, but rather to the interrelationships of the primary areas (W, E, S).

The point of concern here is that dispersal relationships can be rendered as cladograms, which at some level might be informative, or not, in any given case. "Informative," of course does not mean "true," but only that dispersal relationships may sometimes have a cladistic aspect. If the dispersal relationships are true, their cladistic aspect must be true also; if the dispersal relationships are false, their cladistic aspect might be, but is not necessarily, false also. If the dispersal relationships

Figure 8.40. 1: A tree-diagram reflecting dispersal of species into Hawaii (H) from source areas to the west (W), east (E), and south (S). 2: The cladistic aspect of the tree-diagram (1). 3: An uninformative area-cladogram for the four areas, derived from (2).

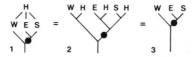

Figure 8.41. 1: A tree-diagram reflecting dispersal of species into Hawaii (H) from source areas to the west (W), east (E), and south (S), as well as the interrelationships of the colonizing species. 2: The cladistic aspect of the tree diagram (1). 3: An informative area-cladogram for three of the four areas, derived from (2).

are uninformative, as in the above case, this means only that their cladistic aspect is indeterminate.

The center-of-origin concept has long been associated with the idea that the major features of the earth's geography have been stable during the period of the development of the modern biota living upon it. But the concept has an older history going back to Linnaeus. Given a stable geography, and the observation that the species of a certain group occupy different parts of the world, it is easy to imagine that the history of the group has involved migration from one place to another, and evolution of new species along the way.

The rules, or clues, used to recognize centers of origin and dispersal relationships were reviewed by Cain (1944), who listed and discussed 13 such items:

1. Location of greatest differentiation of a type.
2. Location of dominance or greatest abundance of individuals.
3. Location of synthetic or closely related forms.
4. Location of maximum size of individuals.
5. Location of greatest productiveness and its relative stability in crops.
6. Continuity and convergence of lines of dispersal.
7. Location of least dependence upon a restricted habitat.
8. Continuity and directness of individual variations or modifications radiating from the center of origin along highways of dispersal.
9. Direction indicated by geographical affinities.
10. Direction indicated by the annual migration routes in birds.
11. Direction indicated by seasonal appearance.
12. Increase in the number of dominant genes toward the center of origin.
13. Center indicated by the concentricity of progressive equiformal areas.

Cain concluded:

Only one conclusion seems possible, and it carries implications far beyond the scope of the present discussion of criteria of center of origin. The sciences of geobotany (plant geography, plant ecology, plant sociology) and geozoology carry a heavy burden of hypothesis and assumption. . . . In many instances the assumptions . . . have so thoroughly permeated the science of geography and have so long been a part of its warp and woof that students of the field can distinguish fact from fiction only with difficulty. (pp. 210–11)

Cain's critique, thorough and well documented, gives some indication of the difficulties of the center-of-origin concept, considered also by Croizat et al. (1974). It is unnecessary to review this subject here, except to note one attempt to relate the concept to the cladistic aspect, under the heading of the "progression rule." This rule dates from Hennig (1950, 1966), who argued that a cladogram of species-relationships is sometimes sufficient in itself to indicate the center-of-origin and direction of dispersal (figures 8.42–8.43). The progression rule has been recently discussed by Ashlock (1974; figure 8.44):

Figure . . . [8.44] illustrates Hennig's hypothetical example, where taxon A is found in New Guinea, B in Queensland, C in Victoria, D in Tasmania, E in New Zealand, F in Tierra del Fuego, and G in Chile. If . . . [the species-cladogram is correct], then one would have to agree that the group A through G did indeed progress from New Guinea through Australia, Tasmania, New Zealand, and on directly to South America. (p. 85)

The argument in favor of the progression rule in this particular case is to the effect that one route of dispersal is more economical than any other:

To hypothesize an alternate route through the northern hemisphere, one would have to suppose a migration of the derived taxa (or their ancestors) through an area of more primitive species, without having left traces. (*Ibid.*)

But the progression rule, as applied by both Hennig and Ashlock, overlooks the possibility that dispersal need not necessarily have occurred. The rule seems another example of what Cain termed the "heavy burden" of assumption. By this Cain meant that because dispersal might be assumed to be a relevant factor in the history of life, it must therefore also be a relevant factor in some particular case, such as the one discussed by Ashlock, because the case exemplifies one or more criteria for determining a center-of-origin. In the cladistic aspect, of

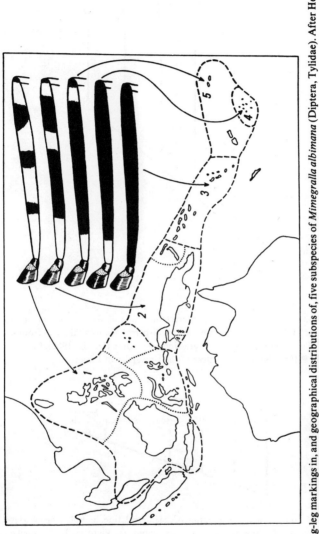

Figure 8.42. Hing-leg markings in, and geographical distributions of, five subspecies of *Mimegralla albimana* (Diptera, Tylidae). After Hennig (1966), figure 39, p. 135. Copyright © 1966, Board of Trustees of the University of Illinois.

516   *Space*

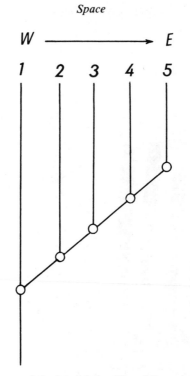

Figure 8.43. A cladogram of the interrelationships of five subspecies of *Mimegralla albimana* (cf. figure 8.42). The arrow implies dispersal of ancestral forms from west (W) to east (E), in accordance with the progression rule. After Hennig (1966), figure 40, p. 136. Copyright © 1966, Board of Trustees of the University of Illinois.

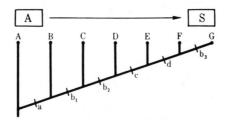

Figure 8.44. A hypothetical cladogram of the interrelationships of seven taxa (A–G) in the following areas: A, New Guinea; B, Queensland; C, Victoria; D, Tasmania; E, New Zealand; F, Tierra del Fuego; G, Chile. "A" and "S" refer to Australian and South American sectors, respectively; "a, b_1, b_2, c, d, b_3" designate defining characters of groups of taxa. After Ashlock (1974), figure 1, p. 85. Reproduced, with permission, from the Annual Review of Ecology and Systematics, vol. 5. Copyright © 1974, Annual Reviews Inc.

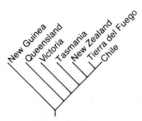

Figure 8.45. The area-cladogram implied by the species-cladogram of figure 8.44.

course, Ashlock's hypothetical species-cladogram would imply a certain area-cladogram, which might or might not be exemplified by one or more real groups (figure 8.45).

Dispersal interpretations were used in various ways by Linnaeus, Buffon, and Candolle. It was Darwin, however, who cast dispersal in a new role, as an all-pervasive factor explaining not merely the anomalous but the nonanomalous as well—not merely the widespread species of Candolle, but his regions, or areas of endemism, also:

The dissimilarity of the inhabitants of different regions may be attributed to modification through natural selection. . . . The degree of dissimilarity will depend on the migration of the more dominant forms of life from one region into another having been effected with more or less ease, at periods more or less remote. (1859:350)

I think all the grand leading facts of geographical distribution are explicable on the theory of migration (generally of the more dominant forms of life), together with subsequent modification and the multiplication of new forms. (1859:408)

For Darwin, dispersal was a way of tying together the sometimes far-flung distributions of the taxa of a natural group. But it was also something more—an argument on behalf of a geography stable in its major features.

The issue of stable versus unstable geography was raised in the British literature in a remarkable paper by Forbes (1846), who argued that the Old World once extended into the Atlantic far to the west, to include what are now islands such as the Azores, Canaries, Cape Verde, etc.:

My own belief is, that a great miocene land, bearing the peculiar flora and fauna of the type now known as Mediterranean, extended far into the Atlantic— past the Azores—and that, in all probability, the great semicircular belt of gulf-weed [*Sargassum*] ranging between the 15th and 45th degrees of north latitude, and constant in its place, marks the position of the coast-line of that ancient land, and had its parentage on its solid bounds. (pp. 348–49)

It is well known that Darwin came to reject all such views of change in the major geographical features of the earth; and because, in his view, the earth had not changed, the organisms *must* have migrated from their various areas of origin to the localities where they are now found. In doing so, he set the stage for a long-continuing tradition in biogeography, explicitly associated with his name, that culminated in books by Darlington (1957, 1965) and a revealing paper exposing his vision of the Darwinian tradition (Darlington 1959). What brought this tradition to an abrupt end was the development of continental-drift theory in its two modern forms: plate tectonics (e.g., Tarling and Runcorn 1973) and earth expansion (e.g., Carey 1976).

During the period of more than one hundred years when the Darwinian tradition was often asserted to be preeminent and preeminently reasonable, many biogeographers viewed the facts of distribution as tending to show that earthly geography must have changed, and that dispersal must have had far less importance than that attributed to it by Darwin and his many followers. In modern times the most notable exponent of this position, all too often labeled heterodox and dismissed in consequence, was Croizat (1958, 1964).

BIOGEOGRAPHIC REGIONS

During the history of biogeography there have been attempts at two sorts of classifications of areas: one example is Candolle's concern with areas of endemism, and he soon defined many such areas on the basis of plant distribution; another example is found in the writings of early zoologists, such as Prichard(1826), who was concerned with much the same object, but who defined far fewer areas. With respect to his objective Prichard wrote:

Hence by a reference to the geographical site of countries, we may divide the earth into a certain number of regions, fitted to become the abodes of particular groupes of animals; and we shall find on inquiry, that each of these provinces, thus conjecturally marked out, is actually inhabited by a distinct nation of quadrupeds [mammals], if we may use that term. (p. 54)

Prichard went on to define seven regions:

1. The Arctic region, much the same in the Old World and the New.
2. The temperate zone, divided into two districts, one in the Old World and one in the New.
3. The equatorial region, divided into three districts: Africa, America, and continental India.
4. The Indian isles, particularly the Sunda and Molucca Islands.
5. The Papuan region, including New Guinea, New Britain, New Ireland, the Solomons, New Hebrides, and the Pacific islands generally.
6. The Australian region.
7. The southern extremities of America and Africa.

Swainson (1835) commented favorably on Prichard's classification into regions:

The objections that may be stated against these divisions chiefly arise from the author not having kept in view the difference between affinity and analogy, as more particularly understood by modern naturalists. And we may illustrate this position by looking more attentively to the animals of two or three of these provinces. 1. The arctic regions of America, Europe, and Asia, indisputably possess the same genera, and in very many instances the same species; and if it should subsequently appear that these regions are sufficiently important in themselves to constitute a zoological province, then it is a perfectly natural one; for not only are the same groups, but even the same species, in several instances common to both. But can this be said of the second of these provinces, made to include the temperate regions of three continents? Certainly not. We find, indeed, *analogies* without end, between their respective groups of animals, but they have each a vast number of peculiar genera; and so few are the species common to all three, that the proportion is not perhaps greater than as 1 to 50. The genera, with but very few exceptions, are peculiar, but are represented by analogous genera; and each continent is distinctly separated in its animal productions by indications as certain and as indubitable as those which distinguish their respective inhabitants. Can we include temperate America in the same zoological province with the parallel regions of Europe, when there are not three land or rather perching birds common to both? and when more than two thirds of the genera found in America are totally unknown in Europe or in Asia? Look to the bears of the temperate regions of the three continents: those of America and Europe are similarly constructed, but the *species* are different; while those, again, of Asia, are formed upon a totally different model. We might fill pages with similar facts; all tending, as we conceive, to exemplify the necessity of preserving these relations as distinct in our view of animal geography, as we are compelled to do in threading the maze of natural arrangement. Dr. Prichard, however, has the great merit of having made the nearest approach to such a theory of animal distribution as is suggested by the natural geography of the earth; nor need we wonder that he has failed in the application, since others, who, from their peculiar studies, might be supposed more competent to the task, have erred from the very foundation. (pp. 13–14)

Swainson considered that Prichard's classification could be improved with reference to three propositions:

1. That the countries peopled by the five recorded varieties of the human species, are likewise inhabited by different races of animals, blending into each other at their confines.
2. That these regions are the true zoological divisions of the earth.
3. That this progression of animal forms is in unison with the first great law of natural arrangement, viz. the gradual amalgamation of the parts, and the circularity of the whole. (1835:14–15)

And Swainson considered that five regions would suffice:

1. The European or Caucasian.
2. The Asiatic or Mongolian.
3. The American.
4. The Ethiopian or African.
5. The Australian or Malay.

Although he felt that (p. 15) "the precise limits of the five zoological provinces here assumed, will not admit of accurate definition," he nevertheless described the regions as follows:

1. The European or Caucasian range includes the whole of Europe, properly so called, with part of Asia Minor, and the shores of the Mediterranean: in Northern Africa the zoological peculiarities of this region begin to disappear; they are lost to the eastward of the Caucasian mountains, and are blended with those of Asia and America to the north. 2. The Asiatic range: comprehending the whole of Asia east of the Ural mountains, a natural and well-defined barrier between the two continents. The chief seat of this zoological region is probably in central Asia; its western confines blend into the European towards Persia, and disappear on the west of the Caucasian chain; it is united to the African range among the provinces of Asia Minor; and is again connected with Europe, and also with America, by the arctic regions of the three continents; finally, its most southern limits are marked by the islands of Java and Sumatra, where the zoological character of the Australian region begins to be apparent. 3. The American range. United to Europe and Asia at its northern limits, this region or province comprehends the whole of the New World; but into which it blends at the other extremity is uncertain. 4. The next includes the whole of Africa south

of the Great Desert: a part, at least, of the countries bordering on the Mediterranean exhibit a decided affinity to the European range; while the absence of large animals in Madagascar, and the presence of genera peculiar to New Holland and the extreme point of Southern Africa, lead us to the fifth or the Australian range. 5. To this region nature has given peculiar characters, both in regard to its geographic situation and to its animal productions. New Guinea and the neighbouring islands mark its limit in that direction; Australia Proper is its chief seat, and it spreads over the whole of the numerous islands of the Pacific Ocean: whether this province blends with that of America or of Europe, remains for future discovery; but its connection with Africa and Asia has already been intimated. (1835:16–17)

In a paper that was to become a classic among zoologists, Sclater (1858) attempted a classification into two creations and six regions (figures 8.46–8.47):

Creatio Palaeogeana
  I. Regio Palaearctica (Europe and northern Asia)
 II. Regio Aethiopica (Africa south of the Sahara)
III. Regio Indica (India and southern Asia)
 IV. Regio Australiana (Australia and New Guinea)
Creatio Neogeana
  V. Regio Nearctica (North America)
 VI. Regio Neotropica (South America)

There followed a series of papers written on this subject by English-speaking zoologists who had ideas (and classifications) of their own, including noteworthies such as A. Günther, T. H. Huxley, A. Murray, W. T. Blanford, E. Blyth, and J. A. Allen. The results of this burst of enthusiasm for geographical classification were reviewed by Wallace (1876: 59–61), who was able to have the last word on the subject, in giving several "Reasons for adopting the six Regions first proposed by Mr. Sclater." Wallace's arrangement (figure 8.48) and nomenclature have been widely used ever since.

Candolle died in 1841, long before he would have had a chance to wonder what, if any, connection there is between the numerous areas of endemism in his list, and the involved discussion across the Channel about the primary divisions of the earth's surface. It was again left for Wallace to decide the matter, at least for the time being. And Wallace did so by recognizing a number of subregions, four per region (figure 8.49). With respect to his subregions Wallace noted:

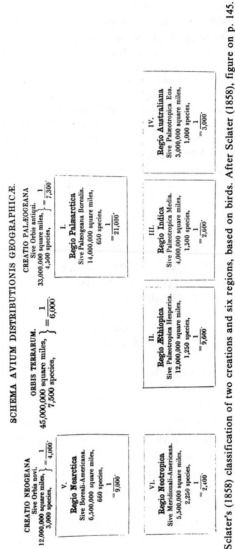

Figure 8.46. Sclater's (1858) classification of two creations and six regions, based on birds. After Sclater (1858), figure on p. 145.

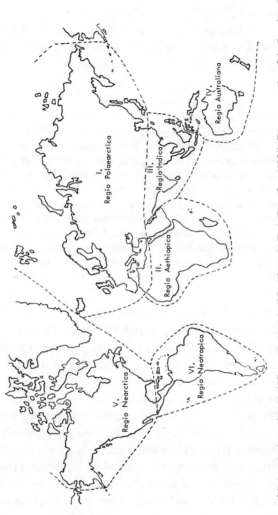
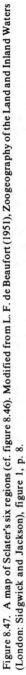

Figure 8.47. A map of Sclater's six regions (cf. figure 8.46). Modified from L. F. de Beaufort (1951), Zoogeography of the Land and Inland Waters (London: Sidgwick and Jackson), figure 1, p. 8.

	Regions.		
Neogæa	NEOTROPICAL ...	Austral zone.........	Notogæa.
Neogæa	NEARCTIC.........	Boreal zone	Arctogæa.
Palæogæa	PALÆARCTIC ...	Boreal zone	Arctogæa.
Palæogæa	ETHIOPIAN	Palæotropical zone	Arctogæa.
Palæogæa	ORIENTAL	Palæotropical zone	Arctogæa.
Palæogæa	AUSTRALIAN ...	Austral zone.........	Notogæa.

Figure 8.48. Wallace's (1876) modification of Sclater's two creations and six regions. After Wallace (1876), table on p. 1:66.

The twenty-four subregions here adopted were arrived at by a careful consideration of the distribution of the more important genera, and of the materials, both zoological and geographical, available for their determination; and it was not till they were almost finally decided on, that they were found to be equal in number throughout all the regions—four in each. As this uniformity is of great advantage in tabular and diagrammatic presentations of the distribution of several families, I decided not to disturb it unless very strong reasons should appear for adopting a greater or less number in any particular case. Such however have not arisen; and it is hoped that these divisions will prove as satisfactory and useful to naturalists in general as they have been to the author. Of course, in a detailed study of any region much more minute sub-division may be required; but even in that case it is believed that the sub-regions here adopted, will be found, with slight modifications, permanently available for exhibiting general results. (1876:80–81)

Most of Wallace's subregions correspond exactly to Candolle's areas of endemism based on plants (figure 8.49). Hence the various creations and regions of Prichard, Swainson, Sclater and others seem to be no more than attempts to group areas of endemism into a hierarchical classification in accordance with their relationships—attempts about which there was some disagreement, as noted above. In his 1876 account (figure 8.48) Wallace presented two different higher classifications of regions: one following Sclater's two creations (Neogaea, Palaeogaea) and one following Huxley's two major divisions (Notogaea, Arctogaea). These different schemes can be expressed by cladograms (figure 8.50). Wallace subsequently published a diagram (actually a network), about which he wrote: "The following arrangement of the regions will indicate their geographical position, and to a considerable extent their relation to each other" (1881:52). Viewed as a network (figure 8.51), the diagram allows for Sclater's classification (Neogaea, Palaeogaea), but not for Huxley's (Notogaea, Arctogaea).

One noteworthy feature of the period from Wallace to the present is

TABLE OF REGIONS AND SUB-REGIONS.

Regions.	Sub-regions.	Remarks.
I. Palæarctic...	1. North Europe. 2. Mediterranean (or S. Eu.) 3. Siberia. 4. Manchuria (or Japan)	 Transition to Ethiopian. Transition to Nearctic. Transition to Oriental.
II. Ethiopian ...	1. East Africa. 2. West Africa. 3. South Africa. 4. Madagascar.	Transition to Palæarctic.
III. Oriental......	1. Hindostan (or Central Ind.) 2. Ceylon. 3. Indo-China (or Himalayas) 4. Indo-Malaya.	Transition to Ethiopian. Transition to Palæarctic. Transition to Australian.
IV. Australian...	1. Austro-Malaya. 2. Australia. 3. Polynesia. 4. New Zealand.	Transition to Oriental. Transition to Neotropical.
V. Neotropical..	1. Chili (or S. Temp. Am.) 2. Brazil. 3. Mexico (or Trop. N. Am.) 4. Antilles.	Transition to Australian. Transition to Nearctic.
VI. Nearctic	1. California. 2. Rocky Mountains. 3. Alleghanies (or East U. S.) 4. Canada.	 Transition to Neotropical. Transition to Palæarctic.

Figure 8.49. Wallace's (1876) 24 subregions, four per region. Almost all (21 of 24) of the subregions correspond fairly exactly to areas of endemism previously recognized by Candolle (1838) on the basis of plant distribution: North Europe (Candolle's region 17); Mediterranean (18); Siberia (16); Manchuria (32, in part); East Africa (23); West Africa (24); South Africa (26); Madagascar (27); Hindostan (31); Ceylon (no counterpart); Indo-China (32, in part); Indo-Malaya (30); Austro-Malaya (no counterpart); Australia (33); Polynesia (36); New Zealand (34); Chili (3); Brazil (6); Mexico (12); Antilles (10); California (13); Rocky Mountains (no counterpart); Alleghanies (14, in part); Canada (14, in part). After Wallace (1876), table on pp. 1:81–82.

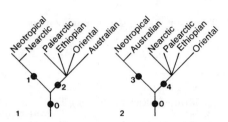

Figure 8.50. Area-cladograms illustrating Sclater's (1) and Huxley's (2) classifications of regions. 1, Neogaea; 2, Palaeogaea; 3, Notogaea; 4, Arctogaea.

the long series of contributions from zoologists who, following Wallace's lead, attempted to support the notion of six regions. Botanists are said to have gone their own way, but in a recent review Udvardy (1975) asserts that not much has changed in either camp since the time of Wallace (figure 8.52). This is certainly the case among zoologists, and there is no difficulty in finding the same concepts repeated over and over again (figure 8.53).

Concepts that persist in science do so for one of two reasons: either because they are found interesting enough to be repeatedly tested, and they subsequently withstand the tests; or because they cease to be interesting enough to test, and they endure through their own inertia. In the latter case, the test of time is no real test at all.

It is dubious if Wallace's mature classification of regions could be said to have been put seriously to the test—in the sense of having led to some predictions that future observations could either confirm or deny. One attempt in this direction is that of Darlington (1957), who summarized the relationships of the regions, based on compilations of the literature of vertebrates. He published diagrams, derived from fishes, frogs, and

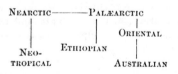

Figure 8.51. Wallace's (1881) diagram of the relations of the six regions. The diagram is a network, which if rooted between the "Nearctic" and the "Palaearctic" would result in Sclater's divisions of Neogaea and Palaeogaea (cf. figure 8.50.1). After Wallace (1881), figure on p. 52.

Floristic system (Kingdoms):	Faunistic system (Regions):
	P A L A E A R C T I C
B O R E A L	N E A R C T I C
---	---
	E T H I O P I A N
P A L A E O T R O P I C A L	
	O R I E N T A L
---	---
A U S T R A L I A N	A U S T R A L I A N
---	---
N E O T R O P I C A L	N E O T R O P I C A L
---	---

Figure 8.52. Udvardy's (1975) summary of the present status of biogeographic classification, based on plants and animals, respectively. After Udvardy (1975), table on p. 9.

mammals (figures 8.54–8.56), for each of which the cladistic aspect may be specified (figures 8.57–8.59). Whereas one of Sclater's components occurs in the cladistic analyses of the diagrams (component 1 = Neogaea), two other components occur that conflict with Sclater's other component (components 5! and 6! that conflict with component 2 = Palaeogaea), and with Huxley's components as well (3 = Notogaea, 4 = Arctogaea). Nevertheless, Darlington's components may be combined in a single area-cladogram (figure 8.60.1), which may be taken

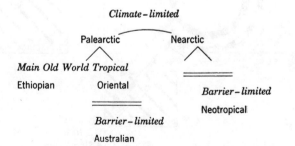

Figure 8.53. Darlington's (1957) diagram of the relationships of the six regions (cf. figure 8.51). With respect to this diagram Darlington states: "This is the main, average pattern of present distribution of animals, and something like it—radiation from the main Old World tropics—is also the apparent main pattern of animal dispersal." After Darlington (1957), figure 45, p. 426.

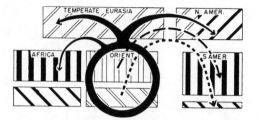

Figure 8.54. Darlington's (1957) summary of the "apparent history of dispersal of fresh-water fishes." After Darlington (1957), figure 18, p. 100.

as an indication of what Darlington's summaries have to say about the relationships of the regions. Curiously, Darlington advocated a classification that conflicts not only with this summary, but also with the classification of Sclater. The cladistic aspect of his classification is shown in figure 8.60.2; the classification is similar to that of Huxley (figure 8.50.2). But Darlington includes only one of Huxley's two

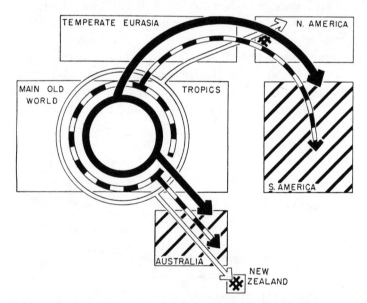

Figure 8.55. Darlington's (1957) summary of the "(hypothetical) successive dispersals of three families of frogs: Leiopelmidae (white circle and arrows) . . . ; Leptodactylidae (broken circle and arrows) . . . ; and Ranidae (black circle and arrows) . . . ." After Darlington (1957), figure 27, p. 157.

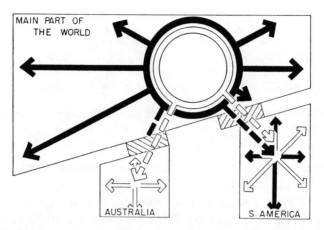

Figure 8.56. Darlington's (1957) summary of the "main pattern of radiation of mammals at the beginning of the Tertiary. Black circle and arrows represent placentals; white, marsupials; diagonally hatched bars, early Tertiary barriers." After Darlington (1957), figure 42, p. 362.

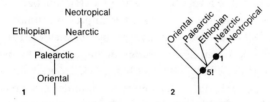

Figure 8.57. Darlington's (1957) summary of the fresh-water fishes in the form of a tree-diagram (1) and the area-cladogram it implies (2; cf. figure 8.54).

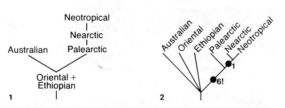

Figure 8.58. Darlington's (1957) summary of the frogs in the form of a tree-diagram (1) and the area-cladogram it implies (2; cf. figure 8.55).

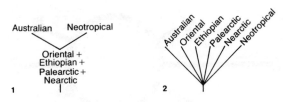

Figure 8.59. Darlington's (1957) summary of the mammals in the form of a tree-diagram (1) and the area-cladogram it implies (2; cf. figure 8.56).

components (4 = Arctogaea), and substitutes the term "Megagaea" for "Arctogaea" (figure 8.61).

In retrospect it is easy to see that Darlington's summaries are attempts to fit the facts of distribution to the preconceived notion of geographical stability. But his summaries were not without value, for his diagrams purported to show the actual relationships of organisms. To some persons (like Croizat 1964, and Brundin 1966) it was clear that the diagrams, and perhaps all that went with them, were simply false; and that something serious was wrong with the approach to biogeography that produced them. Yet these notions had the weight of tradition on their side—in this case a very long tradition, extending back to Darwin and Wallace, strengthened by what seemed impressive summaries of the mammalian fossil record put together first by Matthew (1915), and later added to by Simpson in a long series of papers that enjoyed some popularity (Simpson 1940a, 1940b, 1943, 1953, 1965, 1966).

Thus it was possible during this period to counter any objections to the notions of geographical stability and pervasive dispersal of organisms by citation of literature as if it had the force of scriptural authority. In the process, the objector often became an object of

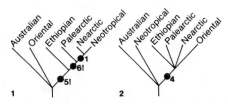

Figure 8.60. 1. The combination of Darlington's summaries of fresh-water fishes, frogs, and mammals in the form of an area-cladogram (cf. figures 8.54–8.59). 2. The area-cladogram corresponding to Darlington's classification of regions (cf. figure 8.61).

TABLE 10. THE ACCEPTED SYSTEM OF CONTINENTAL FAUNAL REGIONS

Realm Megagea (Arctogea): the main part of the world

1. Ethiopian Region: Africa (except the northern corner), with part of southern Arabia
2. Oriental Region: tropical Asia, with associated continental islands
3. Palearctic Region: Eurasia above the tropics, with the northern corner of Africa
4. Nearctic Region: North America, excepting the tropical part of Mexico

Realm Neogea

5. Neotropical Region: South and Central America with the tropical part of Mexico

Realm Notogea

6. Australian Region: Australia, with New Guinea, etc.

Figure 8.61. Darlington's classification of regions. After Darlington (1957), table 10, p. 425.

ridicule, and biogeography metamorphosed from science into an activity with strong sociological, political, and even religious overtones. Such seems to have been evident to Cain as early as 1944. But much to the discredit of biology as a whole, few persons ever had the interest or temerity to confront and publicly challenge this tradition and its power bases. Again, the most, and perhaps the only, significant challenge was that of Croizat, who came to believe that the problem was not endemic to biogeography but pandemic within the Darwinian tradition as a whole, extending beyond the concepts of geographical stability and dispersalism, to include the concepts of adaptation and natural selection—the stock-in-trade of evolutionary biologists even of today.

It would go too far here to delve in detail into these matters peripheral to science. These few notes serve merely as a reminder to the reader that scientists are human, and that human factors are apt to prevail in the public discussion of scientific matters. The history of biogeography, in short, is a case in point. It is a rich field for investigation, but it requires a critical approach.

BEYOND DISPERSALISM

The number of biogeographers who confidently drew dispersal routes on fixed continent maps ten or more years ago and now just as confidently draw dispersals of the same organisms on continental drift maps must cause us to seriously question the procedures of biogeographers. (Edmunds 1975:251)

Although many persons had doubts about the concepts of stable geography and pervasive dispersal, they lacked methods to produce within biogeography itself a refutation decisive enough to win much support within the community of biologists, who were generally busy with matters other than biogeography. What spelled the end of dispersalism was the revival of continental drift, and its rapid acceptance among geologists during the 1960s and 1970s. Whereas dispersalists for one hundred years were able to cite the majority of well-informed geologists in support of the concept of geographical stability, the situation soon changed so that the same majority could now be cited in support of the concept of geographical change on a massive scale. But elements of dispersalism still persist in those minds who believe that, whatever the reality of continental drift, all of it happened so long ago as to be irrelevant to the causal explanation of the distribution of any significant number of existing organisms. Besides, dispersal routes may be drawn on any map, even those that reflect geographical change in accordance with continental-drift theory. Hence, one may doubt that much has changed in biogeography as a result of recent developments in geology. True, biogeographers no longer assert that their data and interpretations prove that continental drift did not occur (as was the custom of many during the past century), but such seems no more than a slight shift in public posture.

What has remained generally unrecognized is that drift theory implies a different classification of biogeographical regions, and, indeed, poses a challenge to the very concept of regions, as they have been understood since early in the nineteenth century. Of course, few persons have ever been much interested in the classification of regions, and it might seem generally irrelevant whether the Notogaea concept or the Neogaea concept is to prevail in the classification of the future. But there is more at stake than these few concepts of groups of regions, namely the entire matter of areas of endemism and their interrelationships—in short, the real organization of the world biota.

It is well in this connection to remember that Wallace framed his regions as groups of areas of endemism. In this sense regions are higher taxa, the species of which are areas of endemism. There are a few questions that might be entertained in this connection: 1. How many areas of endemism are there? What endemic forms of life comprise them? And, where are they on the earth's surface? 2. Are the endemic forms interrelated among themselves in some common pattern or patterns? And if so, what is the pattern or patterns? And what are the causal factors that underlie it?

One stumbling block toward progress in this area is the belief that classification is no more than an artifice more or less useful for various purposes, but an artifice that conveys no precise information about the nature of the world. However Wallace might initially have conceived of his regions, he ultimately came to view them as artifice:

> The general principles on which the diversities of distribution among land animals depend, is already fairly well understood. What we require is to be able to work out the details in the different groups, and thus explain certain difficulties or anomalies. To detect anomalies it is essential to compare the distribution of the different groups by means of a common system of regions.
> . . .the only real interest of the study of geographical distribution lies in its giving us a clue to the causes which have brought about the very divergent and often conflicting distribution of the various species, genera and higher groups, and by thus being able to explain most of the anomalies of distribution. (1894: 612)

It seems never to have occurred to Wallace that, because certain anomalies emerge as by-products of his system of regions, the anomalies are possibly no more than methodological artifacts. More remarkable, perhaps, is that Wallace's arguments were supported sixty years later by Darlington, who likened the system of regions to a standard meter (the unit of linear measurement), and who advised that "the specialist should accept the standard, find how his animals differ from it, and try to find what the differences mean" (1957:422).

A point worth noting here is that continental-drift theory immediately offered a causal explanation for the many interrelationships of the widely displaced austral temperate biota, with the implication that at least parts of Australia, New Zealand, South America, and Africa are more closely related among themselves than to any other continental areas—even those areas, farther to the north, within the same continents. What is the implication for a classification of areas, of regions? Should the whole of Australia be grouped with the Neotropical and Ethiopian regions into some super-region with a new and exotic name?

Or should the continents be broken up into pieces, each corresponding to an area of endemism, that could be classified differently, in accordance with their respective relationships? Temperate Australia, for example, might be classed with temperate South America and temperate Africa in some region that does not also include the tropical parts of these continental masses.

Yet it might seem that, with the austral temperate biota aside, continental-drift theory does not offer much of a challenge to Wallace's classification of regions. According to Wegener's concept, for example, in which all of the continents were initially massed together, we seem to observe the gradual differentiation of Wallace's regions as we follow the development through time (figure 8.62). Modern studies have changed Wegener's classic conception in some details, but the substance remains. Nor would the situation change much with an expanding earth model, according to which the continental masses became more and more isolated as the ocean widened (figure 8.63).

There remains one great imponderable in continental-drift theory, one which allows for a complex origin of continents and continental biotas generally: namely the origin and development of the Pacific Ocean. According to Nur and Ben Avraham (1977), the present Pacific Ocean developed between the rifted margins of a Pacifica continent, whose fragments eventually became embedded, after collision, within the continental margins now bordering the Pacific (figure 8.64). The implication is, for example, that North America is a geologic composite, with the western part belonging to a Pacific sector, and the eastern part to an Atlantic. So also with South America. So also, with the directions reversed, Eurasia.

Drift theory has long permitted continental complexity through collision of continental fragments, with the classic case being India's northward drift and collision with Asia. Nur and Ben Avraham suggest that continental collision is the rule rather than the exception.

If it be allowed that present continents are of complex origin tectonically, what of their biotas? The idea of a Pacific continent is not a new one, having been conceived already by biogeographers confronting biotic relationships spanning the Pacific at tropical latitudes. Croizat (1958:5), for example, went so far as to assert that classic continental-drift theory is better left aside by the biogeographer, because it utterly fails to account for trans-Pacific distributions (figures 8.65–8.66).

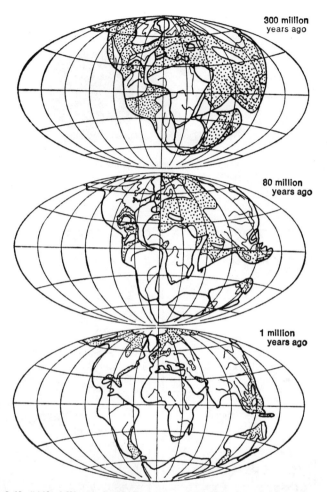

300 million
years ago

80 million
years ago

1 million
years ago

Figure 8.62. "Alfred Wegener's (1915) concept of the evolution of the continents. This reconstruction was based on evidence from many scientific fields and is remarkably similar to our present understanding of the evolution of the southern continents. . . . The shaded areas indicate where shallow seas covered the continents." Figure and legend after D. and M. Tarling (1971), Continental Drift: A Study of the Earth's Moving Surface (New York: Anchor), figure 4, p. 6. Copyright © 1971, G. Bell. Reprinted by permission of Doubleday.

Figure 8.63. Carey's (1976) analogy between an expanding earth and an inflating rubber ball. After Carey (1976), figure 25, p. 48.

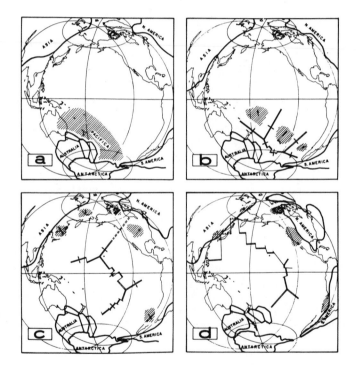

Figure 8.64. "Schematic model of the breakup of Pacifica and the resulting collision events. Possible ages of the stages: a, 225 Myr; b, 180 Myr; c, 135 Myr; d, 65 Myr. Fine lines mark the present day continental outline. Heavy lines mark the location of the various continental areas through the geological evolution." Figure and legend after Nur and Ben-Avraham (1977), fig. 1, p. 42.

What are the prospects for progress? It would seem that biogeography is at the beginning created for it by the work of Candolle and others concerned with areas of endemism on the most local level. The system of regions formalized by Wallace was one attempt to go beyond the beginning. Now, one hundred years later, the attempt seems to have ended in failure. The lesson to be learned? Well. . ., perhaps it is time to try again, in the hope that something, after all, has been learned in the effort, and that the cladistic aspect of area relationship might be approached directly, by component analysis of data on the distribution and interrelationships of the world's biota.

EPILOGUE

This book has explored the question of order in nature on two levels. Even the most elementary observations of the attributes of living organisms reveal a bewildering array of characters (similarities and differences). These elementary observations might seem to reveal only chaotic or random associations of organisms (taxa), each indicated, and supported, by different characters. One might believe, therefore, that the world, in that sense, is complex. But the questions arise: might the world really be simple, rather than complex? Might there be a single pattern of relationships (a general cladogram of taxa) for all organisms, such that any character whatsoever would define a component of that pattern? Those biologists of the eighteenth century who believed in the existence of a natural system, and those of the nineteenth and twentieth centuries who also believed in the existence of a natural system (produced by evolution), have answered that question affirmatively. And the result of their labors (our current classifications of organisms) disclose enough nonrandom character distributions, and have enough predictive value, to make their answer convincing.

Similarly, even the most elementary observations of the ranges of living organisms reveal a bewildering array of distributions (showing similarities and differences). These elementary observations might seem to reveal only chaotic or random associations of areas (regions), each indicated, and supported, by the distributions of different groups of organisms. One might believe, therefore, that the world, in that sense, is complex. But the questions again arise: might the world really be simple, rather than complex? Might there be a single pattern of relationships (a general cladogram of areas) for all groups of organisms, such that the distribution of any group whatsoever would define a component of that pattern? Biologists of the eighteenth, nineteenth, and twentieth centuries are alike in being unable to answer that question, affirmatively or not. But it *is* possible that the world is as orderly with respect to the distribution of organisms as it is with respect to their characters, and that nature, as Croizat maintained, "forever repeats," in the distribution of group after group, the same pattern of area relationships, caused by the same events.

Let us suppose, for the moment, that the second question, like the first, could be answered affirmatively, and that an understanding of nature in that sense (a general cladogram of areas of endemism) could be achieved. One of the characters of humankind, a desire to understand the world around us, might thus be evidenced and even satisfied. But humans desire to understand not only the world around us, but also our own part in it. What might the answers to these questions tell us about ourselves, and our history?

One might ask, for example, whether prior to the commerce and traffic of recorded historical times, there existed natural populations of humans which were effectively restricted, by geography or ecology, to certain areas. There is some evidence that this might be the case. Biogeographers interested in the neotropics, for example, have identified numerous areas of relatively small extent that are each occupied by endemic species and subspecies of birds (Haffer 1974, 1978), "reptiles" and amphibians (Vanzolini 1970; Müller 1973), butterflies (Brown 1976), and plants (Prance 1973). These workers argue that correlations exist between those areas of endemism and the relative expansions and contractions of tropical rain forest and savannah, respectively, during the wet and dry periods associated with the advance and retreat of Pleistocene glaciations. The areas of endemism, in this interpretation, represent small patches of rain forest which were isolated from each other at the height of the dry periods. Workers studying the languages of native South American peoples have found distributions of the languages (as well as archaeological and ethnographic features) that correlate with those of the endemic birds, herptiles, butterflies, and plants, and have drawn a causal inference:

The recognition that Amazonia has suffered successive episodes of fragmentation of the forest during and since the Pleistocene provides biologists with a mechanism for explaining the extensive speciation, disjunct distributions of taxa, and other biogeographical features that could not be accounted for in the absence of natural barriers to interbreeding and dispersal. The same kinds of patterns are discernible in linguistic, archaeological and ethnographic data, implying that human beings were similarly affected by the ebb and flow of the forest. Although geological, palynological, biogeographical and cultural types of evidence are all limited and often tenuous, the correspondences in patterning and chronology are too close to be coincidental. (Meggers 1977:300)

One might wonder, of course, whether the similar distributions of organisms and languages could have the same cause (Whitten 1979; Meggers 1979). If the distributions of the various groups of organisms are all the result of the same events, one would expect to find repeated in each group the same pattern of relationships among the taxa restricted to the various areas of endemism. Unfortunately, no one knows, yet, whether there is a general cladogram for the inhabitants of the forest refugia or not. But what about the languages, and the peoples who use them? It is possible to construct cladograms both for spoken languages and for series of written versions (manuscripts) of the same texts (Hoenigswald 1960; Maas 1958); indeed cladistic methods were used in such studies long before their explicit use in biology. Unfortunately, no one knows, yet, whether the South American languages are interrelated in the same way as the inhabitants of the refugia.

But suppose that it were possible to draw a general cladogram for the areas of endemism of the world, and to draw a cladogram for the various natural populations of humans, whether it be based on languages, or on morphology, or on biochemistry, or indeed be a general cladogram encompassing all the available characters differentiating human populations (figure 8.67). The cladogram of humans might, or might not, correspond to some segment of the general cladogram of areas (figure 8.68). If it does, the implication would be that the same events might have caused the nonrandom pattern of relationships in all groups of organisms, including humans, and that each of the groups of organisms involved, including humans, might be at least of the same minimal age (if not older).

Suppose further that the interrelationships of natural human populations in widely separated areas of endemism, say in Australia, New Zealand, Amazonia, and South Africa, correspond to the interrelation-

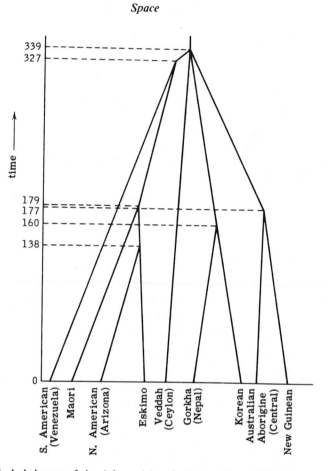

Figure 8.67. A cladogram of nine Asian and American populations of humans, based on data concerning the frequency of blood-group alleles at five loci. After E. A. Thompson (1975), Human Evolutionary Trees (Cambridge: Cambridge University Press), figure 5.1(b), p. 99.

ships of the birds, herptiles, butterflies, and plants of those areas. And suppose that the relevant portions of the general area-cladogram can be correlated with sea-floor spreading events separating the areas, say, 80 million years ago. Would we not have to consider the possibility that humans, also, are that old, and have been affected by the same events? Estimates of the age of the lineage leading to *Homo* already vary tremendously, from around 3.5 to 35 million years (various authors in

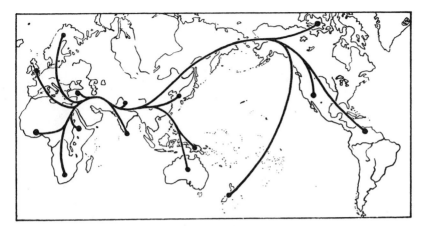

Figure 8.68. An unrooted cladogram of 15 populations of humans superimposed on their geographic distribution, based on a larger sample of data on the same loci as in figure 8.67. After A. W. F. Edwards and L. L. Cavalli-Sforza (1964), Reconstruction of evolutionary trees. In V. H. Heywood and J. McNeill, eds., Phenetic and Phylogenetic Classification (London: The Systematics Association, pub. no. 6), pp. 67–76; figure 1, p. 75.

Goodman and Tashian 1976). How accurate are those estimates? No one knows.

What we have, then, are unanswered questions: about organisms in general, about their interrelationships, about areas of endemism, about *their* interrelationships, about human populations, and about *their* interrelationships. We hope that this volume may at least focus attention on the questions, and perhaps lead to some answers as well.

SELECTED REFERENCES

Adams, E. N. 1973. Consensus techniques and the comparison of taxonomic trees. Syst. Zool. 21:390–397.

Adanson, M. 1763–1764. Familles des plantes. 2 vols. Paris: Vincent.

Ashlock, P. D. 1974. The uses of cladistics. In R. F. Johnston, P. W. Frank, and C. D. Michener, eds., Annual Review of Ecology and Systematics (Palo Alto, Calif.: Annual Reviews), 5:81–99.

Baldwin, J. M., ed. 1911. Dictionary of Philosophy and Psychology, vol. 2. New York: Macmillan.

Balme, D. M. 1962. ΓΕΝΟΣ and ΕΙΔΟΣ in Aristotle's biology. Classical Quart. 56 (new ser., vol. 12): 81–98.

Berger, A. J. 1972. Hawaiian Birdlife. Honolulu: University of Hawaii Press.

Bernard, F. 1895. The principles of palaeontology. Ann. Rep. State Geol. New York 1894:129–215.

Blackwelder, R. E. 1967. Taxonomy: A Text and Reference Book. New York: Wiley.

Bowler, P. J. 1975. The changing meaning of "evolution." J. Hist. Ideas 36:95–114.

Boyden, A. 1973. Perspectives in Zoology. Oxford: Pergamon Press.

Bremekamp, C. E. B. 1953. Linnés views on the hierarchy of the taxonomic groups. Acta Botanica Neerlandica 2:242–53.

Briggs, J. C. 1974. Marine Zoogeography. New York: McGraw-Hill.

Brown, K. S., Jr. 1976. Geographical patterns of evolution in Neotropcal Lepidoptera. J. Ent. 44B:201–42.

Brundin, L. 1966. Transantarctic Relationships and Their Significance, as Evidenced by Chironomid Midges. Kungl. Svenska Vetenskapsakademiens Handlingar. Fjärde series, 11:1–472.

Buffon, C. de. 1761. Histoire naturelle, générale et particulière, vol. 9. Paris.

—— 1766. Histoire naturelle, générale et particulière, vol. 14. Paris.

Caesius, F. 1651. Phytosophicarum tabularum. In F. Hernandez, ed., Rerum medicarum novae hispaniae thesaurus seu plantarum animalium mineralium mexicanorum historia, pp. 901–52. Rome: Mascardi.

Cain, S. A. 1944. Foundations of Plant Geography. New York: Harper.

Candolle, A.-P. de. 1820. Géographie botanique. In Dictionnaire des sciences naturelles, 18:359–422. Strasbourg.

—— 1828. Mémoire sur la famille des Crassulacées. Paris: Treuttel and Würtz.

—— 1838. Statistique de la famille des Composées. Paris and Strasbourg: Treuttel and Würtz.

—— 1844. Théorie élémentaire de la botanique. 3d ed. Paris: Roret.

Carey, S. W. 1976. The Expanding Earth. Amsterdam: Elsevier.

Carpenter, W. 1841. Principles of General and Comparative Physiology. London: John Churchill.

Chambers, R. 1844. Vestiges of the Natural History of Creation. London: John Churchill.

Clifford, H. T. and W. Stephenson. 1975. An Introduction to Numerical Classification. New York: Academic Press.

Croizat, L. 1945. History and nomenclature of the higher units of classification. Bull. Torrey Bot. Club 72:52–75.

—— 1958. Panbiogeography. Caracas: Published by the author.

—— 1964. Space, Time, Form: The Biological Synthesis. Caracas: Published by the author.

—— 1976. Biogeografía Analítica y Sintética ("Panbiogeografía") de las Américas. Caracas: Biblioteca de la Academia de Ciencias Físicas, Matemáticas y Naturales, vols. 15–16.

Croizat, L., et al. 1974. Centers of origin and related concepts. Syst. Zool. 23:265–87.

Crowson, R. A. 1970. Classification and Biology. New York: Atherton Press.

Danser, B. H. 1950. A theory of systematics. Bibliogr. Biotheor. 4: 117–80.

Darlington, P. J., Jr. 1957. Zoogeography: The Geographical Distribution of Animals. New York: Wiley.

—— 1959. Darwin and zoogeography. Proc. Amer. Phil. Soc. 103:307–319.

—— 1965. Biogeography of the Southern End of the World. Cambridge, Mass.: Harvard University Press.

Darwin, C. 1859. On the Origin of Species (London: John Murray; reprint, New York: Atheneum, 1967).

De Beer, G. R. 1951. Embryos and Ancestors. London: Oxford University Press.

Dunal, M.-F. 1817. Monographie de la Famille des Anonacées. Paris: Treuttel and Würtz.

Dupuis, C. 1979. La "Systématique Phylogénétique" de W. Hennig (historique, discussion, choix de références). Cahiers Nat. 34:1–69.

Edmunds, G. F., Jr. 1975. Phylogenetic biogeography of mayflies. Ann. Missouri Bot. Garden 62:251–63.

Farris, J. S. 1977. On the phenetic approach to vertebrate classification. In M. K. Hecht, P. C. Goody, and B. M. Hecht, eds., Major Patterns in Vertebrate Evolution, pp. 823–50. New York: Plenum Press.

—— 1979. On the naturalness of phylogenetic classification. Syst. Zool. 28:200–214.

—— 1980. The information content of the phylogenetic system. Syst. Zool. 28:483–520.

Fleming, C. A. 1958. Darwinism in New Zealand: some examples, influences, and developments. Proc. Roy. Soc. New Zealand 86:65–86.

Fleming, D. 1829. [Review of] On systems and methods in natural history. Quart. Review 41:302–27.

Forbes, E. 1846. On the connexion between the distribution of the existing fauna and flora of the British Isles. Mem. Geol. Surv. Great Britain 1:336–432.

Garstang, E. 1922. The theory of recapitulation: a critical re-statement of the biogenetic law. J. Linnean Soc., Zool. 35:81–101.

Ghiselin, M. T. 1969. The Triumph of the Darwinian Method. Berkeley: University of California Press.

Gingerich, P. D. 1976. Cranial anatomy and evolution of early Tertiary Plesiadapidae (Mammalia, Primates). Univ. Michigan Pap. Paleont. 15:1–140.

Giseke, P. D. 1792. Praelectiones in ordines naturales plantarum. Hamburg: Benj. Gottl. Hoffmann.

Good, R. 1974. The Geography of the Flowering Plants. London: Longman.

Goodman, M. 1975. Protein sequence and immunological specificity. In W. P. Luckett and F. S. Szalay, eds., Phylogeny of the Primates, pp. 219–48. New York: Plenum Press.

Goodman, M. and G. W. Moore. 1971. Immunodiffusion systematics of the Primates. I. The Catarrhini. Syst. Zool. 20:19–62.

Goodman, M. and R. C. Tashian, eds. 1976. Molecular Anthropology. New York: Plenum Press.

Gould, S. J. 1969. An evolutionary microcosm: Pleistocene and recent history of the land snail P. (*Poecilozonites*) in Bermuda. Bull. Mus. Comp. Zool. 138:407–531.

—— 1973. Systematic pluralism and the uses of history. Syst. Zool. 22:322–24.

—— 1977. Ontogeny and Phylogeny. Cambridge, Mass.: Harvard University Press.

Griffiths, G. C. D. 1974. On the foundations of biological systematics. Acta Biotheoretica 23:85–131.

Haffer, J. 1974. Avian speciation in tropical South America. Publ. Nuttall Ornithol. Club 14:1–390.

—— 1978. Distribution of Amazon forest birds. Bonn. Zool. Beitr. 29:38–78.

Hennig, W. 1950. Grundzüge einer Theorie der phylogenetischen Systematik. Berlin: Deutscher Zentralverlag.

—— 1966. Phylogenetic Systematics. D. Dwight Davis and Rainer Zangerl, tr. Urbana: University of Illinois Press.

—— 1969. Die Stammesgeschichte der Insekten. Frankfurt: Waldemar Kramer.

Hoenigswald, H. M. 1960. Language Change and Linguistic Reconstruction. Chicago: University of Chicago Press.

Holloway, J. D. and N. Jardine. 1968. Two approaches to zoogeography: a study based on the distributions of butterflies, birds, and bats in the Indo-Australian area. Proc. Linnean Soc. London 179:153–88.

Hort, A. 1916. Theophrastus. Enquiry into Plants and Minor Works on Odours and Weather Signs, vol. 1. Cambridge, Mass.: Harvard University Press.

Hort, A., and M. L. Green. 1938. The "Critica Botanica" of Linnaeus. London: Ray Society.

Hull, D. L. 1965. The effect of essentialism on taxonomy. British J. Philos. Sci. 15:314–26, 16:1–18.

—— 1980. The limits of cladism. Syst. Zool. 28:416–40.

Humboldt, A. de. 1816. Sur les lois que l'on observe dans la distribution des formes végétales. Ann. Chimie Physique, ser. 2, 1:225–39.

Huxley, J. 1957. The three types of evolutionary process. Nature 180:454–55.

—— 1958. Evolutionary processes and taxonomy with special reference to grades. In O. Hedberg, ed., Systematics of To-day. Uppsala Universitets Årsskrift 1958:(6)21–39 (Uppsala: A.-B. Lundquistska Bokhandeln).

Huxley, L. 1900. Life and Letters of Thomas Henry Huxley, vol. 1. New York: Appleton.

ICN. 1964. International Code of Zoological Nomenclature. London: International Trust for Zoological Nomenclature.

—— 1975. International Code of Nomenclature of Bacteria. Washington: International Association of Microbiological Societies.

—— 1978. International Code of Botanical Nomenclature. Utrecht: Bohn, Scheltema, and Holkema.

Jardine, N. and R. Sibson. 1971. Mathematical Taxonomy. London: Wiley.

Keast, A. 1973. Contemporary biotas and the separation sequence of the southern continents. In D. H. Tarling and S. K. Runcorn, eds.,

Implications of Continental Drift to the Earth Sciences, 1:309–43. London: Academic Press.

Lankester, E. R. 1870. On the use of the term homology in modern zoology, and the distinction between homogenetic and homoplastic agreements. Ann. Mag. Nat. Hist., ser. 4, 6:34–43.

Larson, J. L. 1967. Linnaeus and the natural method. Isis 58:304–20.

Latreille, P. 1817. Introduction à la géographie générale des arachnides et des insectes. Mem. Mus. Hist. Nat. 3:37–67.

Linnaeus, C. 1735. Systema naturae sive regna tria naturae. Leiden: Haak.

—— 1736. Fundamenta botanica. Amsterdam: Schouten.

—— 1738. Classes plantarum. Leiden: Wishoff.

—— 1751. Philosophia botanica. Stockholm: Kiesewetter.

—— 1754. Genera plantarum. 5th ed. Stockholm: Salvii.

—— 1781. Selected Dissertations from the Amoenitates Academicae. London.

—— 1783. A system of vegetables, vol. 1. Lichfield: Botanical Society.

Lovejoy, A. O. 1936. The Great Chain of Being: A Study of the History of an Idea. Cambridge, Mass.: Harvard University Press.

Lyell, C. 1832. Principles of Geology, vol. 2. London.

Maas, P. 1958. Textual Criticism. Oxford: Oxford University Press.

MacArthur, R. H. and E. O. Wilson. 1967. The Theory of Island Biogeography. Princeton, N. J.: Princeton University Press.

McKenna, M. C. 1975. Toward a phylogenetic classification of the Mammalia. In W. P. Luckett and F. S. Szalay, eds., Phylogeny of the Primates, pp. 21–46. New York: Plenum Press.

Macleay, W. S. 1819. Horae entomologicae, vol. 1. London: S. Bagster.

—— 1829–30. On the dying struggle of the dichotomous system. Philos. Mag., n. ser., 7:431–45; 8:53–57, 134–40, 200–207.

—— 1842. On the natural arrangement of fishes. Ann. Mag. Nat. Hist. 9:197–207.

Matthew, W. D. 1915. Climate and evolution. Ann. New York Acad. Sci. 24:171–318.

Mayr, E. 1957. Species concepts and definitions. Amer. Assoc. Adv. Sci. Publ. 50:1–22.

—— 1961. Cause and effect in biology. Science 134:1501–6.

—— 1965a. What is a fauna? Zool. Jb. Syst. 92:473–86.

—— 1965b. Avifauna: turnover on islands. Science 150:1587–88.

—— 1969. Principles of Systematic Zoology. New York: McGraw-Hill.

—— 1971. The nature of the Darwinian revolution. Science 176:981–89.

—— 1974. Cladistic analysis or cladistic classification? Z. Zool. Syst. Evolut.-forsch. 12:94–128.

Mayr, E., E. G. Linsley, and R. L. Usinger. 1953. Methods and Principles of Systematic Zoology. New York: McGraw-Hill.

Mayr, E., and W. H. Phelps, Jr. 1967. The origin of the bird fauna of the south Venezuelan highlands. Bull. Amer. Mus. Nat. Hist. 136:269–328.

Meggers, B. J. 1977. Vegetational fluctuation and prehistoric cultural adaptation in Amazonia: some tentative correlations. World Archaeol. 8:287–303.

—— 1979. Climatic oscillation as a factor in the prehistory of Amazonia. Amer. Antiquity 44:252–66.

Michener, C. D. 1977. Discordant evolution and the classification of allodapine bees. Syst. Zool. 26:32–56.

Mickevich, M. F. 1978. Taxonomic congruence. Syst. Zool. 27:143–158.

Mitchell, P. C. 1901. On the intestinal tract of birds; with remarks on the valuation and nomenclature of zoological characters. Trans. Linnean Soc. London, Zool., ser. 2, 8:173–275.

Moravcsik, J. M. E. 1973. The anatomy of Plato's divisions. In E. N. Lee, A. P. D. Mourelatos, and R. M. Rorty, eds., Exegesis and Argument: Studies in Greek Philosophy Presented to Gregory Vlastos, pp. 324–48. Assen: Van Gorcum.

Morison, R. 1672. Plantarum umbelliferarum distributio nova, per tabulas cognationis et affinitatis ex libro naturae observata et detecta. Oxford: Sheldon Theatre.

Müller, P. 1973. The Dispersal Centres of Terrestrial Vertebrates in the Neotropical Realm. The Hague: Junk.

Nordenskiöld, E. 1928. The History of Biology: A Survey. New York: Knopf.

Nur, A. and Z. Ben-Avraham. 1977. Lost Pacifica continent. Nature 270:41–43.

Ogle, W. 1882. Aristotle on the Parts of Animals: Translated with introduction and notes. London: Kegan Paul, Trench.

—— 1912. De partibus animalium. In J. A. Smith and W. D. Ross, eds., The Works of Aristotle Translated into English, Oxford: Oxford University Press.

Parker, F. H. 1967. The Story of Western Philosophy. Bloomington: Indiana University Press.

Patterson, C. 1977. The contribution of paleontology to teleostean phylogeny. In M. K. Hecht, P. C. Goody, and B. M. Hecht, eds., Major Patterns in Vertebrate Evolution, pp. 579–643. New York: Plenum Press.

Patterson, C. and D. E. Rosen. 1977. Review of ichthyodectiform and other Mesozoic teleost fishes and the theory and practice of classifying fossils. Bull. Amer. Mus. Nat. Hist. 158:81–172.

Popper, K. R. 1959. The Logic of Scientific Discovery. New York: Harper and Row.

Prance, G. T. 1973. Phytogeographic support for the theory of Pleistocene forest refugia in the Amazon basin. Acta Amazonica 3(3):5–28.

Prichard, J. C. 1826. Researches into the Physical History of Mankind, 2nd ed. London.

—— 1836. Researches into the Physical History of Mankind, vol. 1, 3d ed. London.

Quine, W. V. 1969. Ontological Relativity and Other Essays. New York: Columbia University Press.

Raven, C. E. 1942. John Ray Naturalist: His Life and Works. Cambridge: Cambridge University Press.

Ray, J. 1682. Methodus plantarum nova. London: Faitborne and Kersey.

—— 1703. Methodus plantarum emendata et aucta. London: Royal Society.

Rosa, D. 1918. Ologenesi. Florence: R. Bemporad.

—— 1931. L'Ologénèse. Paris: Alcan.

—— 1933. Le due strade della biologia pura. Riv. Biol. 15:437–44.

Rosen, D. 1978. Vicariant patterns and historical explanation in biogeography. Syst. Zool. 27:159–88.

Ross, H. H. 1974. Biological Systematics. Reading, Pa.: Addison-Wesley.

Sachs, J. V. 1890. History of Botany (1530–1860). Oxford: Oxford University Press.

Schaefer, C. W. 1976. The reality of the higher taxonomic categories. Z. Zool. Syst. Evolut.-forsch. 14:1–10.

Scharff, R. F. 1899. The History of the European Fauna. London.

—— 1912. Distribution and Origin of Life in America. New York: Macmillan.

Sclater, P. L. 1858. On the general geographical distribution of the members of the class Aves. J. Linnean Soc., Zool. 2:130–45.

Simpson, G. G. 1940a. Mammals and land bridges. J. Washington Acad. Sci. 30:137–63.

—— 1940b. Antarctica as a faunal migration route. Proc. Sixth Pacific Sci. Congr. 2:755–68.

—— 1943. Mammals and the nature of continents. Amer. J. Sci. 241:1–31.

—— 1945. The principles of classification and a classification of mammals. Bull. Amer. Mus. Nat. Hist. 85:i–xvi, 1–350.

—— 1953. Evolution and Geography. Eugene: Oregon State System of Higher Education.

—— 1961. Principles of Animal Taxonomy. New York: Columbia University Press.

—— 1963. The meaning of taxonomic statements. In S. L. Washburn, ed., Classification and Human Evolution, pp. 1–31. Chicago: Aldine.

—— 1965. The Geography of Evolution. Philadelphia: Chilton.

—— 1966. Mammalian evolution on the southern continents. N. Jb. Geol. Paläont. Abh. 125:1–18.

—— 1971. Remarks on immunology and catarrhine classification. Syst. Zool. 20:369–70.

Skemp, J. B. 1952. Plato's Statesman. New Haven, Conn.: Yale University Press.

Slud, P. 1976. Geographic and climatic relationships of avifaunas with special reference to comparative distribution in the neotropics. Smithsonian Contrib. Zool. 212:1–149.

Smith, J. E. 1814. An Introduction to Physiological and Systematical Botany. 3d ed. London: Longman, Hurst, Reese, Orme, Brown.

Sneath, P. H. A. and R. R. Sokal. 1973. Numerical Taxonomy. San Francisco: Freeman.

Sokal, R. R. and P. H. A. Sneath. 1963. Principles of Numerical Taxonomy. San Francisco: Freeman.

Stafleu, F. A. 1963. Adanson and his "Familles des plantes." In Anon., ed., Adanson, vol. 1, pp. 123–264. Pittsburg: Hunt Botanical Library.

—— 1971. Linnaeus and the Linnaeans. Utrecht: International Association for Plant Taxonomy.

Stearn, W. T. 1957. An introduction to the *Species Plantarum* and cognate works of Carl Linnaeus. In Carl Linnaeus Species Plantarum: A Facsimile of the First Edition 1753, pp. i–xiv, 1–176. London: Ray Society.

Strickland, H. E. 1841. On the true method of discovering the natural system in zoology and botany. Ann. Mag. Nat. Hist. 6:184–94.

Swainson, W. 1834. A Preliminary Discourse on the Study of Natural History. London: Longman.

—— 1835. A Treatise on the Geography and Classification of Animals. London: Longman.

Tarling, D. H. and S. K. Runcorn, eds. 1973. Implications of Continental Drift to the Earth Sciences, 2 vols. London: Academic Press.

Udvardy, M. D. F. 1975. A Classification of the Biogeographical Provinces of the World. Occasional Paper no. 18, International Union for Conservation of Nature and Natural Resources, Morges, Switzerland.

Vanzolini, P. E. 1970. Zoologia sístemática, geografia e a origem das espécies. Inst. Geogr. Univ. São Paulo, Sér. Teses Monogr. 3:1–56.

Virey, J.-J. 1825. Remarques sur l'identité de certaines lois générales. Bull. Sci. Nat. Geol. 4:275–78.

Wallace, A. R. 1855. On the law which has regulated the introduction of new species. Ann. Mag. Nat. Hist., ser. 2, 16:184–96.

—— 1856. Attempts at a natural arrangement of birds. Ann. Mag. Nat. Hist., ser. 2, 18:193–216.

—— 1863. On the physical geography of the Malay Archipelago. J. Roy. Geograph. Soc. London 33:217–34.

—— 1876. The Geographical Distribution of Animals, 2 vols. London: Macmillan.

—— 1881. Island Life. New York: Harper.

—— 1894. What are zoological regions? Nature 49:610–13.

Whitten, R. G. 1979. Comments on the theory of Holocene refugia in the culture history of Amazonia. Amer. Antiquity 44:238–51.

Wiley, E. O. 1979. An annotated Linnaean hierarchy, with comments on natural taxa and competing systems. Syst. Zool. 28:308–37.

Wilkins, J. 1668. An Essay towards a Real Character and a Philosophical Language. London: Royal Society.

Zangerl, R. 1948. The methods of comparative anatomy and its contribution to the study of evolution. Evolution 2:351–74.

AUTHOR INDEX

SUBJECT INDEX

Adaptation, 531
Analogy, 122
Archaeopteryx, 154
Area–cladogram, 52, 57, 221, 398, 410, 468
Area–relationship: biotic similarity, 57,
481, 493; cladistic aspect, 52, 410, 468;
dispersal, 481, 509, 528; level of
endemism, 401, 481, 485
Artifact, 145, 204, 231, 320, 462, 533

Binomial nomenclature, 13, 35, 81
Biogenetic law, 37, 160, 331, 344
Biogeography: as core subdiscipline, 6; as
test of phyletic tree, 221; dispersal
interpretation, 44, 359, 481, 508;
ecological vs. historical, 42, 364;
evidence of evolution, 222; historical
origin, 357; island, 487, 510; period of
stasis, 393, 481; relation to earth history,
43, 55, 480; relation to systematics, 42,
221, 343, 392, 468, 511; superabundant
literature, 392, 481; vicariance
interpretation, 45

Center of origin: as explanation, 359;
Cain's rules, 513; Garden of Eden, 358;
indicated by phylogram, 44, 514; one vs.
many, 370; progression rule, 514; small
vs. large, 398; stable geography, 44, 513,
518, 530
Character: absolute number, 235;
analogous, 122; as classification, 16; as
represented in branching diagram, 18,
183, 199, 208, 227; as set of character
states, 301; as symplesiomorphy, 138; as
synapomorphy, 137, 154, 164; as theory,
301; defining, 69, 90, 154, 164, 195, 304,
324; homologous, 122, 223; loss, 30;

misperceived identity, 28, 195, 304;
misperceived level of generality, 28, 196,
304; negative, 29, 67, 71, 118, 184, 193,
225; parallel, 340; primitive vs.
advanced, 37, 137, 209, 331; problem of
incongruence, 23, 189, 216, 304;
reinterpretation, 197; relative generality,
15, 154, 160, 183, 196, 300, 331
Character transformation: implied by
homology, 157, 209, 301; implied by tree,
206; inferred by comparison, 27, 37;
numerous possibilities, 39; ontogenetic,
25, 37, 120, 159, 331; paleontological
evidence, 41, 333; positive and negative
occurrences, 227, 275; relation to
classification, 154
Character–type: absences, 204; as general
form of data, 183; asymmetry as basis for
inference, 223; conflict in positive
occurrences, 189, 194; generalization of
distribution, 215, 223; implied by
transformation, 206; in 3-taxon
problem, 238; points of view: gradistic,
234, phenetic, 233, phyletic, 234; positive
and negative occurrences, 185, 225;
represented in cladogram, 183;
represented in tree, 200; true and false
positive occurrences, 194
Circular arrangement, 110
Circularity, 224, 227, 237, 353
Cladism, 139
Cladogram: actual information content,
183; agreement in geographic dimension,
221; analysis for two taxa, 170, 183;
analysis for three taxa, 172, 184, 238;
analysis for four taxa, 174; as basis for
classification, 139, 148, 158, 172, 177,
271; as basis for inference, 189; as basis